Dieter Bauer (Ed.)
Computational Strong-Field Quantum Dynamics
De Gruyter Graduate

Also of Interest

Advances in Optical Physics. Volume 3: Advances in Nonlinear Optics
X. Chen, G. Zhang, H. Zeng, Q. Guo, W. She, 2014
ISBN 978-3-11-030430-5, e-ISBN (PDF) 978-3-11-030449-7,
e-ISBN (EPUB) 978-3-11-038282-2

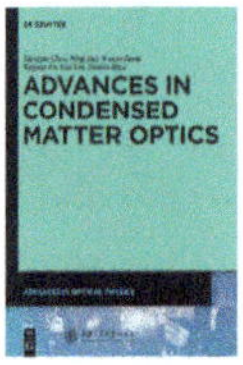

Advances in Optical Physics. Advances in Condensed Matter Optics
L. Chen, N. Dai, X. Jiang, K. Jin, H. Liu, H. Zhao, 2014
ISBN 978-3-11-030693-4, e-ISBN (PDF) 978-3-11-030702-3,
e-ISBN (EPUB) 978-3-11-038818-3

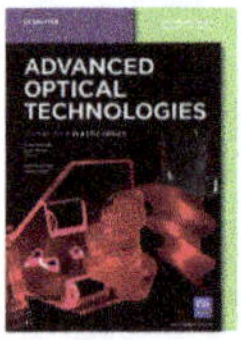

Advanced Optical Technologies
Michael Pfeffer (Editor-in-Chief)
ISSN 2192-8576, e-ISSN 2192-8584

Open Physics. formerly Central European Journal of Physics
DE GRUYTER OPEN
e-ISSN 2391-5471

Computational Strong-Field Quantum Dynamics

Intense Light-Matter Interactions

Edited by
Dieter Bauer

DE GRUYTER

Primary: 02.70.-c, 03.65.-w; Secondary: 33.60.+q, 32.30.-r, 33.20.Xx

Editor
Prof. Dieter Bauer
Universität Rostock
Institut für Physik
18051 Rostock
dieter.bauer@uni-rostock.de

ISBN 978-3-11-041725-8
e-ISBN (PDF) 978-3-11-041726-5
e-ISBN (EPUB) 978-3-11-041934-4

Library of Congress Cataloging-in-Publication Data
A CIP catalog record for this book has been applied for at the Library of Congress.

Bibliographic information published by the Deutsche Nationalbibliothek
The Deutsche Nationalbibliothek lists this publication in the Deutsche Nationalbibliografie; detailed bibliographic data are available on the Internet at http://dnb.dnb.de.

Cover image: from V. Mosert, D. Bauer. Photoelectron spectra with Qprop and t-SURFF. Comput. Phys. Commun, 207, 452, 2016.
Typesetting: Konvertus, Haarlem
Printing and binding: CPI books GmbH, Leck
♾ Printed on acid-free paper
Printed in Germany

www.degruyter.com

Contents

Preface

This book is about computational methods to simulate the interaction of strong laser fields with matter. By definition, "strong" in this context means that the electric field of the laser brings the state of the target system far away from the initial, typically the ground-state, configuration. As a result, "standard" textbook quantum mechanical perturbation theory is not applicable, and interesting nonperturbative phenomena such as high-order above-threshold ionization or high-harmonic generation are observed. The shortness of the laser pulses allows to study ultrafast atomic and molecular processes directly in the time domain.

Experimentally, strong-field laser physics started booming in the late 1980s, after the invention of the chirped-pulse amplification of laser pulses. On the theoretical side, strong-field laser physics was founded soon after the invention of the laser in the 1960s by Reiss, Keldysh, Popov, and others, introducing and developing variants of what nowadays is subsumed under "strong-field approximation." The laser intensities anticipated in these works must have appeared as science fiction at the time. However, those were the times when theory was ahead of experiment. Nowadays, theory is almost always lagging behind: there are several examples where experiments revealed "surprising" features, not expected from the strong-field approximation or tunneling theories. Moreover, at the latest with the development of short-wavelength sources like free-electron lasers or high harmonics, the scope of strong-field physics widens beyond the single-active valence electron. In fact, laser-generated or laser-driven, correlated many-body physics comprises a multitude of research areas, ranging from core-hole motion in atoms, giant resonances, imaging of strongly correlated plasma, warm dense matter, and matter under extreme conditions.

The targeted readership comprises graduate students commencing a Master or a PhD project in the field of computational laser-matter interaction. The book provides an overview of common methods to numerically propagate single-particle wavefunctions according to the time-dependent Schrödinger (Chapter I) or relativistic (Chapter III) wave equations and to calculate the typical observables of interest (Chapter II). Further, three time-dependent quantum many-body approaches are discussed, which all have their pros and cons with regard to certain aspects of intense laser-matter interaction: time-dependent density functional theory (Chapter IV), multiconfigurational time-dependent Hartree–Fock (Chapter V), and time-dependent configuration interaction singles (Chapter VI). The semianalytical strong-field approximation is the workhorse of strong-field laser physics and introduced in Chapter VII, including the saddle-point method and the resulting intuitive interpretation in terms of interfering quantum orbits. Intense lasers turn mesoscopic or macroscopic targets into plasmas that influence the further laser pulse propagation. Simulations on a quantum level with a reasonable incorporation of correlation are not feasible for such target sizes.

Hence, a microscopic particle-in-cell approach that allows to treat nonideal plasmas and solves Maxwell's equations self-consistently is covered in Chapter VIII.

Portions of Chapters I, II, and IV originate from lecture notes for a course on Computational Physics Thomas Fennel and myself teach at Rostock University since 2011. Part of the final examination is a programming project. The book may thus also serve as a source for lecturers giving similar courses, even if they do not primarily focus on strong-field laser topics.

I would like to thank all contributing authors; they are the *real* experts in the topic of their chapter, and it has been a pleasure to collaborate so smoothly. I am grateful to Martins Brics and Yaroslav Lutsyshyn for proofreading.

Rostock, October 2016 — Dieter Bauer

List of abbreviations

1D	one-dimensional (analogously 2D, 3D)
ABC	absorbing boundary condition
ADI	alternating direction implicit
ATAS	attosecond transient absorption spectroscopy
ATI	above-threshold ionization
au	atomic units
CA	cell address
CAP	complex absorbing potential
CASSCF	complete active space self-consistent field
CI	configuration interaction
CIC	cloud-in-cell
CID	configuration interaction doubles
CIS	configuration interaction singles
CISD	configuration interaction singles doubles
CMF	constant mean-field
CN	Crank–Nicolson
DFT	density functional theory
DVR	discrete variable representation
FCI	full configuration interaction
FD	finite difference
FDTD	finite difference time domain
FEL	free electron laser
FEM	finite element method
FFT	fast Fourier transform
FFTed	fast-Fourier transformed
FFTW	fastest Fourier transform in the west
HATI	high-order above-threshold ionization
HF	Hartree–Fock
HHG	high-harmonic generation
HOMO	highest occupied molecular orbital
IDM	ion density matrix
IER	ionic eigenstate-resolved
ISFA	improved strong-field approximation
KS	Kohn–Sham
LDA	local density approximation
LSD	local spin density approximation
MCHF	multi-configurational Hartree–Fock
MCTDHF	multi-configurational time-dependent Hartree–Fock
MD	molecular dynamics
MicPIC	microscopic particle-in-cell
NIR	near infrared
NSDI	nonsequential double ionization
OEP	optimized effective potential
PBC	periodic boundary condition
PC	particle count
PEC	perfect electric conductor
PIC	particle-in-cell
PIL	particle index list
PML	perfectly matched layer
pt	point
PZ	Perdew–Zunger

SAE	single active electron approximation
SCF	self-consistent field
SFA	strong-field approximation
SIC	self-interaction correction
SPM	saddle-point method
TDCI	time-dependent configuration interaction
TDCIS	time-dependent configuration interaction singles
TDDFT	time-dependent density functional theory
TDHF	time-dependent Hartree–Fock
TDKS	time-dependent Kohn–Sham
TDSE	time-dependent Schrödinger equation
t-SURFF	time-dependent surface flux method
UPML	uniaxial perfectly matched layer
VMF	variable mean-field
WKB	Wentzel–Kramers–Brillouin
xc	exchange-correlation
XFEL	x-ray free electron laser
x-LDA	exchange-only local density approximation
x-only	exchange-only
XUV	extreme ultraviolet

Dieter Bauer and Thomas Fennel

I How to propagate a wavefunction?

The correct answer to the question in the title of this chapter is clearly “it depends.” There are many algorithms for solving the time-dependent Schrödinger equation (TDSE) or equations that can be converted to TDSE-like form (diffusion, fluid, or Maxwell). We do not aim at presenting an exhaustive overview of wavefunction propagation, which would fill $>10^3$ pages and be extremely dull to read (and to write). Instead, we focus on approaches that are adequate for strong-field problems where electrons “explore” the continuum while undergoing large excursions due to the oscillatory nature of the strong laser field. We therefore do not discuss methods that rely on expansions in unperturbed eigenstates of the target system under study. Moreover, we are interested in methods that allow for (almost) arbitrary vector potentials, in particular ultrashort pulses. For that reason, the Floquet approach [13] is not covered. We also restrict ourselves to nonrelativistic problems although the methods can be applied to relativistic wave equations as well, as the reader will recognize in Chapter III. From a conceptual viewpoint, the extension to more degrees of freedom or particles is simple once one knows how to propagate a wavefunction for a few degrees of freedom—as outlined in this chapter. However, the exponential scaling of the numerical effort with the number of particles is the buzzkiller here. How to solve the many-electron TDSE approximately is the topic of Chapters V and VI. The residual selectivity applied in this chapter is based on a mixture of experience and ignorance. We therefore apologize in advance if we forgot to mention *your* favorite TDSE propagation method. We will be grateful if you notify us about it, in particular if you are sure that your method outperforms everything we present in this chapter.

Authors writing on computational aspects have the choice of presenting pedagogical toy examples the reader can readily implement within one afternoon, or the “real stuff”—highly optimized, with all the technical details. We believe that the details should not be swept under the rug. For instance, the algorithms described in Section 1.5 are rather involved and cumbersome to implement but extremely efficient and, in fact, frequently used in the strong-field community. Hence, we give all the details but provide the code, as we cannot expect that the majority of readers has the time, desire, or energy to implement them.

Dieter Bauer: Institute of Physics, University of Rostock, 18051 Rostock, Germany; email: dieter.bauer@uni-rostock.de
Thomas Fennel: Institute of Physics, University of Rostock, 18051 Rostock, Germany; email: thomas.fennel@uni-rostock.de

De Gruyter Graduate – Computational Strong-Field Quantum Dynamics, Volume 5, 2017, pp. 1–44.
DOI 10.1515/9783110417265-001

Finally, we want to stress that this chapter is about the *propagation* of wavefunctions only. Typically, the propagation is from time $t = 0$, at which the system is in an unperturbed eigenstate (which needs to be calculated in the first place), up to time $t = T_{\text{final}}$ after the laser pulse, when the wavefunction might be spread over hundreds or thousands of Bohr radii due to ionization and laser-assisted scattering. In principle, the full (quantum mechanically allowed) knowledge about the system is contained in this wavefunction. Numerically, some of the information is necessarily sacrificed. How to calculate observables (or not-so-observable entities that are nevertheless of interest) from the numerically represented wavefunction during or after the laser pulse is the topic of chapter II.

1 Time-dependent Schrödinger equation

The TDSE in position-space representation reads as

$$i\hbar\frac{\partial}{\partial t}\Psi(x,t) = \hat{H}(t)\Psi(x,t) = \left[-\frac{\hbar^2}{2m}\frac{\partial^2}{\partial x^2} + \hat{V}(x,t)\right]\Psi(x,t). \tag{1}$$

Here, we allow for an explicitly time-dependent Hamiltonian, and we restrict ourselves to one spatial dimension (1D) for the moment. In general, the potential operator $\hat{V}(x,t)$ may contain spatial derivatives[1] and thus is not necessarily diagonal in position-space representation. Given $\Psi(x,t)$, the formal solution is

$$\Psi(x,t') = \hat{U}(t',t)\Psi(x,t), \tag{2}$$

where $\hat{U}(t',t)$ is the time-evolution operator from time t to time t'. For small time steps Δt, we can assume that the Hamiltonian is piecewise constant so that

$$\Psi(x,t+\Delta t) = \hat{U}(t+\Delta t,t)\Psi(x,t), \qquad \hat{U}(t+\Delta t,t) = \exp\left[-\frac{i}{\hbar}\hat{H}(t+\Delta t/2)\Delta t\right]. \tag{3}$$

Let the integer upper index n denotes the time step such that $t = n\Delta t$, and

$$\Psi^{n+1} = \hat{U}(\Delta t)\Psi^n, \qquad \hat{U}(\Delta t) = \exp\left(-i\hat{H}\Delta t\right) \tag{4}$$

with all spatial arguments suppressed. Moreover, it is understood that always the current Hamiltonian $\hat{H} = \hat{H}[(n+1/2)\Delta t]$ is employed in $\hat{U}(\Delta t)$, and units in which $\hbar = m = 1$ are used.

1 For instance, in velocity-gauge coupling to an external field, see Section 1.5.2 below.

1.1 Time propagation and stability

A straightforward discretization of the TDSE in time is

$$i\frac{\Psi^{n+1}-\Psi^{n}}{\Delta t}=\hat{H}\Psi^{n} \tag{5}$$

and thus

$$\Psi^{n+1}=(1-i\Delta t\hat{H})\Psi^{n}, \tag{6}$$

which is equivalent to expanding $\hat{U}(\Delta t)$ up to first order in Δt. It is easily seen that this so-called explicit Euler forward scheme is nonunitary and unconditionally unstable.

Nonunitary propagation affects the norm, i.e., if $\langle\Psi^{n}|\Psi^{n}\rangle = 1$, in general $\langle\Psi^{n+1}|\Psi^{n+1}\rangle \neq 1$. It is very desirable to have an unconditionally stable propagation algorithm that avoids explosions of the norm, independent of how physically insane time step or spatial grid spacing might be chosen. Moreover, numerical tricks to reduce the spatial grid size (e.g., by introducing absorbing boundaries [17], see also Section 3.6 in Chapter VI) affect the norm on purpose and even employ the reduction in norm to "measure" the ionization degree. Hence, one should ensure that the norm does only change because of the boundary conditions but not because of the propagation algorithm itself.

We will now introduce the so-called Crank–Nicolson propagation algorithm [2, 14], which is unconditionally stable. It is easily derived from the identities

$$\Psi^{n+1/2}=\hat{U}(-\Delta t/2)\Psi^{n+1}=\hat{U}^{\dagger}(\Delta t/2)\Psi^{n+1}=\hat{U}(\Delta t/2)\Psi^{n}=\Psi^{n+1/2}. \tag{7}$$

Expanding $\hat{U}(\pm\Delta t/2)=(1\mp i\hat{H}\,\Delta t/2)+\mathcal{O}(\Delta t^2)$, one obtains

$$\left(1+\frac{i\Delta t}{2}\hat{H}\right)\Psi^{n+1}=\left(1-\frac{i\Delta t}{2}\hat{H}\right)\Psi^{n} \tag{8}$$

and formally

$$\Psi^{n+1}=\hat{U}_{\mathrm{CN}}(\Delta t)\Psi^{n},\qquad \hat{U}_{\mathrm{CN}}(\Delta t)=\frac{1-\frac{i\Delta t}{2}\hat{H}}{1+\frac{i\Delta t}{2}\hat{H}}, \tag{9}$$

which is unitary. The reader may check that $\hat{U}(\Delta t)=U_{\mathrm{CN}}(\Delta t)+\mathcal{O}(\Delta t^3)$.

1.2 Spatial discretization

There are various ways to discretize the space on which the wavefunction Ψ is defined. The perhaps most straightforward method is to directly discretize position space. This leads to the so-called finite difference (FD) approach, which we will follow in this section. But before we do so, we note that one could also expand $\Psi(x,t)$ in some complete[2] basis $\{\varphi_m(x)\}$,

$$\Psi(x,t)=\sum_m c_m(t)\varphi_m(x). \tag{10}$$

The problem of propagating Ψ according to the TDSE (1) then translates to the propagation of the set of coefficients $\{c_m(t)\}$ according to a set of first-order ordinary differential equations. In practice, the basis functions $\varphi_m(x)$ should be chosen such that scalar products, the relevant matrix elements of the operators in the Hamiltonian, and the observables of interest can be calculated accurately and efficiently as quadratures. Examples for $\{\varphi_m(x)\}$ used in such discrete variable representations (DVRs), also known as pseudo-spectral methods, are polynomials [20], splines [9], or, in spectral methods, plane waves (see Section 1.4.3 and Section 3.2 of Chapter III). DVR may be combined with finite element methods (FEMs) so that Ψ is expanded piecewise on spatial intervals (see [22] for a discussion of FEM, FEM-DVR, and FD).

In either of these approaches, one ends up with some kind of discrete numerical grid. The simplest FD scheme in 1D employs a discretized spatial grid with N_x grid points and uniform spacing Δx,

$$x_s=x_0+s\Delta x,\qquad s=0,1,2,\dots N_x-1, \tag{11}$$

where x_0 is an offset. The wavefunction at a given time step n can be organized in a vector

$$\Psi^n=(\Psi_0^n,\Psi_1^n,\Psi_2^n,\dots,\Psi_{N_x-1}^n)^\top,\qquad \Psi_s^n=\Psi^n(x_s). \tag{12}$$

Hence, the grid index s directly labels the spatial grid point x_s. In DVR, the grid may be more abstract, as the index m labels, e.g., the polynomial or the spline. In FEM-DVR, there might be more indices, labeling the domains. However, mathematically, there are no fundamental differences between all these approaches although they may differ substantially in computational efficiency.

Continuing the FD way, the 3-pt stencil for the second derivative reads as

$$\left.\frac{\partial^2}{\partial x^2}\Psi^n(x)\right|_{x_s}+\mathcal{O}(\Delta x^2)=\frac{\Psi_{s+1}^n-2\Psi_s^n+\Psi_{s-1}^n}{\Delta x^2}, \tag{13}$$

leading to the discretized Hamiltonian

$$\mathbf{H}=-\frac{1}{2\Delta x^2}\begin{pmatrix}-2 & 1 & & & 0\\ 1 & -2 & 1 & & \\ & \ddots & \ddots & \ddots & \\ & & 1 & -2 & 1\\ 0 & & & 1 & -2\end{pmatrix}+\begin{pmatrix}V_0 & & & & \\ & V_1 & & & \\ & & \ddots & & \\ & & & V_{N_x-2} & \\ & & & & V_{N_x-1}\end{pmatrix}. \tag{14}$$

2 The basis does not need to be orthonormal but may be chosen so for convenience.

Here, we tentatively assume that $\hat{V}(x,t)$ is diagonal in position space. The upper-right and lower-left corner elements in the kinetic-energy matrix are set to zero, corresponding to reflecting boundary conditions (i.e., as if the potential jumps from V_0 at x_0 to ∞ at $x_0 - \Delta x$ and analogously at $x_0 + N_x \Delta x$ from V_{N_x-1} to ∞). For reflecting boundary conditions, the discretized Hamiltonian is tridiagonal. For periodic boundary conditions, the upper-right and lower-left corner elements are both 1, and the tridiagonality of $\mathbf{H}$ is broken.

Defining

$$\mathbf{A}_{\pm} = 1 \mp \frac{i\Delta t}{2}\mathbf{H}, \tag{15}$$

the discretized version of (8) becomes

$$\mathbf{A}_{-}\Psi^{n+1} = \mathbf{A}_{+}\Psi^{n} = \Psi^{n+1/2}. \tag{16}$$

The right-hand side can be evaluated numerically by simple matrix-vector multiplication. Assuming reflecting boundary conditions, also the matrices $\mathbf{A}_{\pm}$ are tridiagonal. Hence,

$$\mathbf{A}_{-}\Psi^{n+1} = \Psi^{n+1/2} \tag{17}$$

can be solved for Ψ^{n+1} by forward-backward substitution.

1.2.1 Forward-backward substitution

The problem of solving (17) for Ψ^{n+1} is of a kind that is ubiquitous in computational physics: a known matrix times an unknown vector $\mathbf{f}$ is a known right-hand side $\mathbf{d}$:

$$\begin{pmatrix} b_0 & c_0 & & & & \\ a_1 & b_1 & c_1 & & & \\ & a_2 & b_2 & c_2 & & \\ & & \ddots & \ddots & \ddots & \\ & & & \ddots & \ddots & c_{N-2} \\ & & & & a_{N-1} & b_{N-1} \end{pmatrix} \begin{pmatrix} f_0 \\ f_1 \\ \vdots \\ \vdots \\ \vdots \\ f_{N-1} \end{pmatrix} = \begin{pmatrix} d_0 \\ d_1 \\ \vdots \\ \vdots \\ \vdots \\ d_{N-1} \end{pmatrix}. \tag{18}$$

The fact that the matrix is not densely populated but tridiagonal simplifies the problem significantly. To solve the equation for $\mathbf{f} = (f_0, f_1, \ldots, f_{N-1})^\top$, we begin with a forward sweep to eliminate the subdiagonal. This leads to a modified diagonal and right-hand side via

$$b'_0 = b_0, \qquad b'_n = b_n - \frac{a_n}{b'_{n-1}} c_{n-1}, \qquad d'_0 = d_0, \qquad d'_n = d_n - \frac{a_n}{b'_{n-1}} d'_{n-1}. \tag{19}$$

The remaining problem now has the form

$$\begin{pmatrix} b'_0 & c_0 & & & & \\ & b'_1 & c_1 & & & \\ & & b'_2 & c_2 & & \\ & & & \ddots & \ddots & \\ & & & & \ddots & c_{N-2} \\ & & & & & b'_{N-1} \end{pmatrix} \begin{pmatrix} f_0 \\ f_1 \\ \vdots \\ \vdots \\ \vdots \\ f_{N-1} \end{pmatrix} = \begin{pmatrix} d'_0 \\ d'_1 \\ \vdots \\ \vdots \\ \vdots \\ d'_{N-1} \end{pmatrix}, \tag{20}$$

which can be solved by backward substitution,

$$f_{N-1} = \frac{d'_{N-1}}{b'_{N-1}}, \qquad f_n = \frac{d'_n - c_n f_{n+1}}{b'_n}. \tag{21}$$

For periodic boundary conditions, the matrix in (18) has nonvanishing upper-right and lower-left elements, rendering direct forward-backward substitution inapplicable. However, using the Sherman–Morrison formula [14, 18], the problem can be reduced to applying forward-backward substitution to two auxiliary equations involving tridiagonal matrices only.

1.2.2 Numerical dispersion analysis of the Crank–Nicolson approximant

To estimate the accuracy of the Crank–Nicolson approach, we inspect its numerical dispersion. To that end, we consider the free TDSE (i.e., just with the kinetic energy $\hat{T}$ but without potential in the Hamiltonian) in 1D

$$i\partial_t \Psi(x,t) = \hat{T}\Psi(x,t) = -\frac{1}{2}\frac{\partial^2}{\partial x^2}\Psi(x,t). \tag{22}$$

The analytical solution, dispersion relation, and group velocity read as

$$\Psi(x,t) = e^{i(kx-\omega t)}, \qquad \omega(k) = \frac{k^2}{2}, \qquad v_g = \frac{\partial \omega}{\partial k} = k, \tag{23}$$

respectively. Hence, analytically, $\Psi(x,0) = \Psi_0(x) = e^{ikx}$, and

$$\Psi(x,\Delta t) = e^{i(kx-\omega\Delta t)} = e^{-i\omega\Delta t}\Psi_0(x), \tag{24}$$

and with Crank–Nicolson,

$$\Psi(x,\Delta t) = \frac{1 - \frac{i\Delta t}{2}\hat{T}(k)}{1 + \frac{i\Delta t}{2}\hat{T}(k)}\Psi_0(x) =: e^{-i\omega_{\mathrm{CN}}\Delta t}\Psi_0(x). \tag{25}$$

The next step is to solve this equation for the angular velocity ω_{CN}. Expanding the exponential function and

$$\omega_{\mathrm{CN}}(\Delta t) = \omega_{\mathrm{CN}}^{(0)} + \omega_{\mathrm{CN}}^{(1)}\Delta t + \omega_{\mathrm{CN}}^{(2)}\frac{\Delta t^2}{2} + \cdots, \tag{26}$$

$$\frac{1}{1+x} = 1 - x + x^2 - x^3 + \cdots, \tag{27}$$

one obtains up to orders of Δt^3

$$\begin{aligned} 1 - i&\left(\omega_{\mathrm{CN}}^{(0)} + \omega_{\mathrm{CN}}^{(1)}\Delta t + \omega_{\mathrm{CN}}^{(2)}\frac{\Delta t^2}{2}\right)\Delta t \\ -\frac{1}{2}&\left({\omega_{\mathrm{CN}}^{(0)}}^2 + 2\omega_{\mathrm{CN}}^{(0)}\omega_{\mathrm{CN}}^{(1)}\Delta t\right)\Delta t^2 + \frac{i}{6}{\omega_{\mathrm{CN}}^{(0)}}^3\Delta t^3 \\ = &\left(1 - \frac{i\Delta t}{2}T(k)\right)\left(1 - \frac{i\Delta t}{2}T(k) - \frac{\Delta t^2}{4}T^2(k) + \frac{i\Delta t^3}{8}T^3(k)\right). \end{aligned}$$

Sorting in powers of Δt yields $1 = 1$, $\omega_{\mathrm{CN}}^{(0)} = T(k)$, $\omega_{\mathrm{CN}}^{(1)} = 0$, and $\omega_{\mathrm{CN}}^{(2)} = -\frac{1}{6}T^3$, so that

$$\omega_{\mathrm{CN}}(k, \Delta t) = T(k)\left(1 - \frac{1}{12}T^2(k)\Delta t^2 + \cdots\right). \tag{28}$$

What is still missing is the spatial discretization of the kinetic-energy operator $\hat{T} = -\frac{1}{2}\frac{\partial^2}{\partial x^2}$. Considering the 3-pt stencil and a plane wave, we have

$$\hat{T}e^{ikx} = -\frac{1}{2}e^{ikx}\frac{[e^{-ik\Delta x} - 2 + e^{ik\Delta x}]}{\Delta x^2} =: T(k, \Delta x)e^{ikx} \tag{29}$$

and thus

$$T(k, \Delta x) = \frac{1 - \cos(k\Delta x)}{\Delta x^2} = \frac{k^2}{2}\left(1 - \frac{1}{12}k^2\Delta x^2 + \cdots\right). \tag{30}$$

From (28), then follows the numerical dispersion relation for discretized time and space

$$\omega_{\mathrm{CN}}(k, \Delta t, \Delta x) = \frac{k^2}{2} - \frac{k^4}{24}\Delta x^2 - \frac{k^6}{96}\Delta t^2 + \cdots \tag{31}$$

and hence the numerical group velocity

$$v_{g,\mathrm{CN}}(k, \Delta t, \Delta x) = k - \frac{k^3}{6}\Delta x^2 - \frac{k^5}{16}\Delta t^2 + \cdots. \tag{32}$$

These results show that both temporal and spatial steps have to be chosen properly to ensure an accurate[3] propagation and spreading of the wavefunction. Numerical

---

3 Mind the difference between stability and accuracy! Crank–Nicolson is unconditionally stable also for choices Δx and Δt for which the error terms in (31) are *not* small. Convergence checks with respect to Δx and Δt are indispensable.

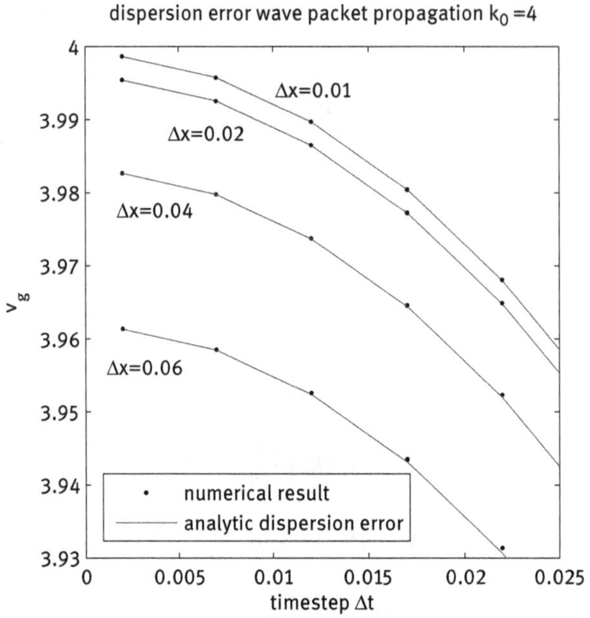

Fig. 1. Comparison of numerically and analytically obtained Crank–Nicolson group velocity for $k = k_0 = 4$. Deviation from $v_g = k$ is due to the numerical dispersion relation different from $\omega(k) = k^2/2$ and depending on both Δx and Δt, see (31).

dispersion relation (31) reminds of those found for the band structure of crystal lattices. Real physical effects there, such as Bloch oscillations, have their artificial, numerical counterparts on computational grids. A comparison of analytical Crank–Nicolson group velocity (32) with the numerically determined is shown in Figure 1, proving excellent agreement. Spectral analysis for a variety of FD schemes can be found in [8].

1.2.3 Numerov boost in accuracy

It turns out that one can boost the accuracy with respect to the spatial discretization without decreasing Δx and with almost no additional computational overhead. This is one of the very rare "free lunches" in (computational) physics that one should gratefully exploit.

Let us first write (13) more specifically as

$$\frac{\partial^2}{\partial x^2} + \frac{\Delta x^2}{12}\frac{\partial^4}{\partial x^4} + \frac{\Delta x^4}{360}\frac{\partial^6}{\partial x^6} + \cdots = \mathbf{D}, \tag{33}$$

where the matrix

$$\mathbf{D} = \frac{1}{\Delta x^2} \begin{pmatrix} -2 & 1 & & & \\ 1 & -2 & 1 & & \\ & \ddots & \ddots & \ddots & \\ & & 1 & -2 & 1 \\ & & & 1 & -2 \end{pmatrix} \tag{34}$$

acts on a discretized wavefunction as in the right-hand side of (13). Rewriting (33) as

$$\mathbf{D} = \left(1 + \frac{\Delta x^2}{12}\frac{\partial^2}{\partial x^2} + \frac{\Delta x^4}{360}\frac{\partial^4}{\partial x^4} + \dots\right)\frac{\partial^2}{\partial x^2}, \tag{35}$$

we find

$$\mathbf{D} = \left(1 + \frac{\Delta x^2}{12}\mathbf{D} + \mathcal{O}(\Delta x^4)\right)\frac{\partial^2}{\partial x^2}$$

so that

$$\frac{\partial^2}{\partial x^2} = \mathbf{M}^{-1}\mathbf{D} + \mathcal{O}(\Delta x^4) \tag{36}$$

with

$$\mathbf{M} = 1 + \frac{\Delta x^2}{12}\mathbf{D}. \tag{37}$$

We have found a more accurate expression for the second spatial derivative than the usual 3-pt stencil. However, although **M** is tridiagonal, its inverse will be dense. Therefore, the scheme would not be very useful if $\mathbf{M}^{-1}$ appears too often in the Crank–Nicolson scheme, making intermediate forward-backward sweeps necessary. Fortunately, $\mathbf{M}^{-1}$ cancels completely. With

$$\mathbf{H} = -\frac{1}{2}\mathbf{M}^{-1}\mathbf{D} + \mathbf{V}, \tag{38}$$

we have

$$\left(1 + \frac{i\Delta t}{2}\left[-\frac{1}{2}\mathbf{M}^{-1}\mathbf{D} + \mathbf{V}\right]\right)\Psi^{n+1} = \left(1 - \frac{i\Delta t}{2}\left[-\frac{1}{2}\mathbf{M}^{-1}\mathbf{D} + \mathbf{V}\right]\right)\Psi^{n}$$

and thus

$$\left(\mathbf{M} + \frac{i\Delta t}{2}\left[-\frac{1}{2}\mathbf{D} + \mathbf{MV}\right]\right)\Psi^{n+1} = \left(\mathbf{M} - \frac{i\Delta t}{2}\left[-\frac{1}{2}\mathbf{D} + \mathbf{MV}\right]\right)\Psi^{n}. \tag{39}$$

The new propagation matrices

$$\mathbf{A}_{\pm} \rightarrow \mathbf{A}'_{\pm} = \mathbf{M} \mp \frac{i\Delta t}{2}\left(-\frac{1}{2}\mathbf{D} + \mathbf{MV}\right) \tag{40}$$

for (16) are still tridiagonal (if **V** is diagonal), forward-backward substitution can be applied, and we get the higher accuracy practically “for free”. The improved, spatially discretized Crank–Nicolson propagator reads as

$$\mathbf{U}_{\mathrm{CN}}(\Delta t) = \frac{\mathbf{M} - \frac{i\Delta t}{2}\left[-\frac{1}{2}\mathbf{D} + \mathbf{MV}\right]}{\mathbf{M} + \frac{i\Delta t}{2}\left[-\frac{1}{2}\mathbf{D} + \mathbf{MV}\right]}. \tag{41}$$

1.3 Imaginary-time propagation

The propagation of a wavefunction according to the TDSE is an initial-value problem. Often, the ground state of the unperturbed system (described by $\hat{H}_0$, which is time independent) serves as the initial state at $t = 0$, and the driving field acts only during times $t > 0$. Once a code for propagating a wavefunction is implemented, it can be used to find the ground state of the unperturbed system as well. By switching to imaginary time,

$$t \to -it, \qquad t \in \mathbb{R}^+, \tag{42}$$

the free, real-time evolution of a superposition of eigenstates $|\Psi_k\rangle$ of $\hat{H}_0$ with eigenenergies ϵ_k, $k = 0, 1, 2, \ldots$,

$$|\Psi(t)\rangle = e^{-i\epsilon_0 t}|\Psi_0\rangle + e^{-i\epsilon_1 t}|\Psi_1\rangle + \cdots, \tag{43}$$

becomes

$$|\Psi(t)\rangle = e^{-\epsilon_0 t}|\Psi_0\rangle + e^{-\epsilon_1 t}|\Psi_1\rangle + \cdots, \tag{44}$$

i.e., the excited states will be damped stronger than the ground state by a factor $e^{-t(\epsilon_k - \epsilon_0)}$. Another way to interpret imaginary time in this context is by analogy with statistical physics: imaginary time corresponds to an inverse temperature $\beta = 1/k_{\mathrm{B}}T$, and the enhancement factor $e^{-\beta\epsilon}$ is the statistical weight. Thus, imaginary time $\beta \to \infty$ corresponds to zero temperature, at which the system is in the ground state.

In terms of a time step-based evolution with Δt, we have to make the replacement $\Delta t \to -i\Delta t$ in the numerical propagator. In particular, the imaginary-time Crank–Nicolson propagator for the unperturbed system described by $\hat{H}_0$ reads as

$$\hat{U}_{\mathrm{itCN}}(\Delta t) = \frac{1 - \frac{\Delta t}{2}\hat{H}_0}{1 + \frac{\Delta t}{2}\hat{H}_0}. \tag{45}$$

As imaginary-time evolution is not unitary but we want a ground state that is normalized to unity, we renormalize,

$$\left|\Psi^{n+1}\right\rangle \to \frac{\left|\Psi^{n+1}\right\rangle}{\sqrt{\left\langle\Psi^{n+1}\middle|\Psi^{n+1}\right\rangle}}, \tag{46}$$

after each time step

$$\left|\Psi^{n+1}\right\rangle = U_{\text{itCN}}(\Delta t)\left|\Psi^{n}\right\rangle. \tag{47}$$

After a sufficient number of iterations, only the ground state will "survive." Of course, in the interest of fast convergence, one should avoid that the initial guess $\Psi(x, t = 0)$ is far away from or, in the worst case, orthogonal to the ground state.[4]

The only parameter that determines convergence for a given discretized $\hat{H}_0$ is the imaginary time step Δt. The enhancement factor of eigenstate k for a single Crank–Nicolson step in imaginary time is

$$\alpha_k = \frac{1 - \frac{\Delta t}{2}\epsilon_k}{1 + \frac{\Delta t}{2}\epsilon_k} = \frac{1-\delta}{1+\delta} = \alpha(\delta), \qquad \delta = \frac{\Delta t}{2}\epsilon_k. \tag{48}$$

The absolute value of the enhancement factor should decrease with increasing δ (i.e., energy). However, this is ensured only in the range $-1 < \delta < 1$, see Figure 2. Optimally fast convergence occurs for Δt and ϵ_k such that $\delta = -1 + \varepsilon$ because the enhancement is maximal (a small ε avoids the singularity of $\alpha(\delta)$). Note that for $\delta > 1$, the enhancement

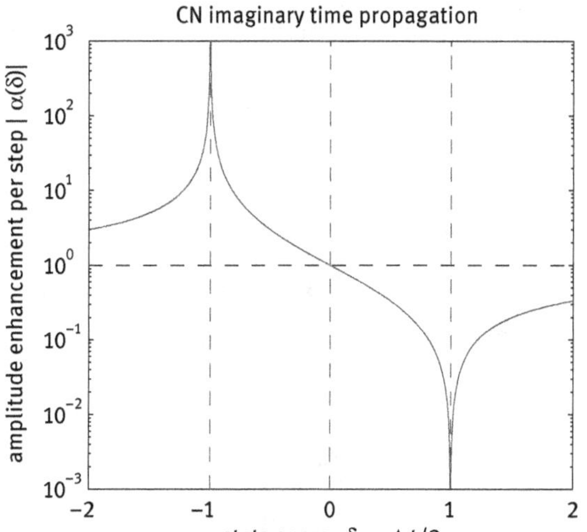

Fig. 2. Enhancement factor for the imaginary-time propagation using the Crank–Nicolson propagator. Only for states with eigenenergies within the interval $-1 < \delta < 1$, the imaginary-time propagation will converge to the lowest energy. Hamiltonian and time step must be chosen such that no states with $\delta < 0$ exist. States with $\delta > 1$ can be accepted, if the interested state lies within $-1 < \delta < 0$.

4 Interestingly, even for an orthogonal guess (e.g., an odd function for an even binding potential), often the numerical errors ultimately cause convergence toward the ground state.

increases again. In any case, good convergence can be realized by shifting (i.e., adding a constant to $\hat{H}_0$) and scaling (via Δt) such that $-1 < \delta < 0$ for the state of interest.

1.3.1 Convergence check

We would like to terminate the imaginary-time propagation once the error drops below a desired value. To that end, we introduce as a measure for the error the residual

$$R = \langle \Psi^n_{\text{res}} | \Psi^n_{\text{res}} \rangle \,, \tag{49}$$

where

$$|\Psi^n_{\text{res}}\rangle = (\hat{H}_0 - E^n)|\Psi^n\rangle, \qquad E^n = \langle \Psi^n | \hat{H}_0 | \Psi^n \rangle \tag{50}$$

with $|\Psi^n\rangle$ normalized. Clearly $R = 0$ if $\hat{H}_0|\Psi^n\rangle = E^n|\Psi^n\rangle$, i.e., an exact eigenstate was found.

Let us consider a $|\Psi^n\rangle$ that still contains two eigenstates, $|\Psi^n\rangle = c_0|\Psi_0\rangle + c_1|\Psi_1\rangle$, so that $E^n = |c_0|^2\epsilon_0 + |c_1|^2\epsilon_1$ and $R = \left[\epsilon_0 - E^n\right]^2 |c_0|^2 + \left[\epsilon_1 - E^n\right]^2 |c_1|^2$. The residual can only vanish if the square-bracket terms for all populated states vanish. For nondegenerate states, this can only be true for one term in the sum, which therefore must be fully occupied ($|c| = 1$). In this case, the energy expectation value E^n is automatically equal to the eigenenergy, and the variance vanishes.

1.3.2 Excited states

Without additional constraint, imaginary-time propagation yields the ground state. The mth excited state can be obtained by projecting out all $m-1$ previously calculated eigenstates after each imaginary-time propagation step,

$$|\Psi_m^{n+1}\rangle - \sum_{k=0}^{m-1} \langle \Psi_k | \Psi_m^{n+1} \rangle |\Psi_k\rangle \to |\Psi_m^{n+1}\rangle. \tag{51}$$

Here, the $|\Psi_k\rangle$, $k = 0, 1, \ldots, m-1$ are the previously found states, and we assume that all states have been normalized before (51) is performed. Degeneracy does not pose a problem here, as a complete basis of the respective degenerate subspace is found before the next higher eigenenergy is addressed.

Although not strictly guaranteed, in practice, it usually works to determine the lowest m eigenstates Ψ_k^0, $k = 0, 1, \ldots, m-1$, by initializing them differently (possibly with the correct parity and other known features) and propagating them in parallel in imaginary time, performing each imaginary-time step a Gram-Schmidt orthogonalization (51). All states involved should be immediately (re-) normalized to unity according to (46) after any operation that changes their norm (i.e., an imaginary-time propagation step or outprojection of other states).

Although imaginary-time propagation in combination with Gram-Schmidt orthonormalization is a very efficient method to determine eigenstates without the need to diagonalize huge matrices, it is very cumbersome if the eigenenergies lie closer and closer with increasing m, as, for instance, in the case of high Rydberg states or in a continuum. In Section 2.2.5 of Chapter II, we will introduce a complementary method that is better suited for such cases.

1.4 More dimensions: Operator splitting

So far, we restricted ourselves to 1D. In two dimensions (2D), the TDSE (again in units where $\hbar = m = 1$) reads as

$$i\frac{\partial}{\partial t}\Psi(x,y,t) = \hat{H}(t)\Psi(x,y,t) = \left[-\frac{1}{2}\left(\frac{\partial^2}{\partial x^2} + \frac{\partial^2}{\partial y^2}\right) + \hat{V}(x,y,t)\right]\Psi(x,y,t). \quad (52)$$

Here x, y could be the coordinates of one 2D-particle or the two positions for two particles, each in 1D. In general, $\hat{V}(x,y,t)$ can be written neither as a sum nor as a product of potentials depending only on x or y. Nevertheless, one may split

$$\hat{H}(t) = \hat{H}_x(t) + \hat{H}_y(t) \quad (53)$$

with

$$\hat{H}_x(t) = -\frac{1}{2}\left(\frac{\partial^2}{\partial x^2} - \hat{V}(x,y,t)\right), \qquad H_y(t) = -\frac{1}{2}\left(\frac{\partial^2}{\partial y^2} - \hat{V}(x,y,t)\right), \quad (54)$$

keeping in mind that $\hat{H}_x$ depends also on y, and $\hat{H}_y$ depends also on x.

The extension of discretization (12) to 2D is obvious, leading to a discretized wavefunction with elements $\Psi^n_{sq} = \Psi^n(x_s, y_q)$ with $s = 0, 1, \ldots, N_x - 1$ as before in 1D and an additional index $q = 0, 1, \ldots, N_y - 1$. Organizing the spatially discretized $\Psi(x,y,t)$ in the form (fast-running x-index s)

$$\Psi^n = (\Psi^n_{0,0}, \Psi^n_{1,0}, \ldots, \Psi^n_{N_x-1,0}; \Psi^n_{0,1}, \ldots, \Psi^n_{N_x-1,1}; \ldots; \Psi^n_{0,N_y-1}, \ldots \Psi^n_{N_x-1,N_y-1})^\top \quad (55)$$

or (fast-running y-index q)

$$\Psi^n = (\Psi^n_{0,0}, \Psi^n_{0,1}, \ldots, \Psi^n_{0,N_y-1}; \Psi^n_{1,0}, \ldots, \Psi^n_{1,N_y-1}; \ldots; \Psi^n_{N_x-1,0}, \ldots, \Psi^n_{N_x-1,N_y-1})^\top,$$

$\hat{H}_x$ or $\hat{H}_y$ alone are tridiagonal matrices, respectively. The sum of both, however, gives a band matrix that has elements far off the diagonal. If we apply the usual real-time Crank–Nicolson scheme

$$\left(1 + \frac{i\Delta t}{2}\hat{H}\right)\Psi^{n+1} = \left(1 - \frac{i\Delta t}{2}\hat{H}\right)\Psi^n = \Psi^{n+\frac{1}{2}}, \quad (56)$$

the evaluation of the explicit part on the right-hand side is unproblematic in the sense that the numerical effort is of the smallest expected order $\mathcal{O}(N_x N_y)$. However, the matrix $\hat{A}_- = 1 + \frac{i\Delta t}{2}\hat{H}$ is not tridiagonal.[5] As a consequence, the implicit step in the Crank–Nicolson scheme can no longer be accomplished in terms of efficient forward-backward substitution. Subsections 1.4.1–1.4.3 aim at "factorizing" this problem of propagation in higher dimensions into consecutive propagation steps in 1D each.

1.4.1 Peaceman–Rachford alternating direction implicit method

A scheme that still allows to benefit from tridiagonal matrices is the alternating direction implicit (ADI) method proposed by Peaceman and Rachford [12]. The idea is to interleave propagation in x and y directions in the Crank–Nicolson style:

$$\left(1 + \frac{i\Delta t}{2}\hat{H}_y\right)\Psi^{n+\frac{1}{2}} = \left(1 - \frac{i\Delta t}{2}\hat{H}_x\right)\Psi^n, \tag{57}$$

$$\left(1 + \frac{i\Delta t}{2}\hat{H}_x\right)\Psi^{n+1} = \left(1 - \frac{i\Delta t}{2}\hat{H}_y\right)\Psi^{n+\frac{1}{2}}. \tag{58}$$

In order to determine how accurate this scheme is, we multiply the last equation by $(1 + \frac{i\Delta t}{2}\hat{H}_y)$ and combine both to

$$\left(1 + \frac{i\Delta t}{2}\hat{H}_y\right)\left(1 + \frac{i\Delta t}{2}\hat{H}_x\right)\Psi^{n+1} = \left(1 - \frac{i\Delta t}{2}\hat{H}_y\right)\left(1 - \frac{i\Delta t}{2}\hat{H}_x\right)\Psi^n. \tag{59}$$

Factoring out gives

$$\left(1 + \frac{i\Delta t}{2}\hat{H}\right)\Psi^{n+1} = \left(1 - \frac{i\Delta t}{2}\hat{H}\right)\Psi^n + \frac{\Delta t^2}{4}\hat{H}_y\hat{H}_x\left(\Psi^{n+1} - \Psi^n\right).$$

With $\Psi^{n+1} - \Psi^n = \Delta t \frac{\partial}{\partial t}\Psi^{n+\frac{1}{2}}$, we find

$$\left(1 + \frac{i\Delta t}{2}\hat{H}\right)\Psi^{n+1} = \left(1 - \frac{i\Delta t}{2}\hat{H}\right)\Psi^n + \mathcal{O}(\Delta t^3). \tag{60}$$

Hence, the ADI scheme has a single step error of the same order $\mathcal{O}(\Delta t^3)$ as a Crank–Nicolson step with the full (i.e., unsplit) Hamiltonian would have anyway. Unfortunately, the ADI scheme spoils unitarity (unless $\hat{H}_x$ and $\hat{H}_y$ commute).

5 At least if a straightforward grid representation of the wavefunction such as (55) is used.

1.4.2 Operator splitting

We can preserve unitary time propagation using “operator splitting.” Let us consider the operator

$$\hat{U}(\alpha) = e^{\alpha(\hat{A}+\hat{B})}, \tag{61}$$

where $\hat{A}$ and $\hat{B}$ do not commute. In cases of interest to us, $\alpha = -i\Delta t$ and $\hat{H} = \hat{T} + \hat{V}$ or $\hat{H} = \hat{H}_x + \hat{H}_y$. Using adequate discretization, the operators $\hat{A}$ and $\hat{B}$ become matrices $\mathbf{A}$ and $\mathbf{B}$ (say, diagonal or tridiagonal). We are able to efficiently solve $\mathbf{Ax} = \mathbf{b}$ and $\mathbf{By} = \mathbf{c}$ for $\mathbf{x}$ and $\mathbf{y}$, respectively. Therefore, we like to split the time evolution in products of exponentials containing only one of the operators in the exponent. Let us inspect the expansion of the full operator

$$\hat{U}(\alpha) = e^{\alpha(\hat{A}+\hat{B})} = 1 + \alpha(\hat{A}+\hat{B}) + \frac{\alpha^2}{2}(\hat{A}^2 + \hat{B}^2 + \hat{A}\hat{B} + \hat{B}\hat{A}) + \mathcal{O}(\alpha^3). \tag{62}$$

The most simple approximation is

$$\begin{aligned} e^{\alpha\hat{A}}e^{\alpha\hat{B}} &= \left(1 + \alpha\hat{A} + \frac{\alpha^2}{2}\hat{A}^2 + \dots\right)\left(1 + \alpha\hat{B} + \frac{\alpha^2}{2}\hat{B}^2 + \dots\right) \\ &= 1 + \alpha(\hat{A}+\hat{B}) + \frac{\alpha^2}{2}(\hat{A}^2 + \hat{B}^2 + 2\hat{A}\hat{B}) + \mathcal{O}(\alpha^3), \end{aligned}$$

which deviates already in order $\mathcal{O}(\alpha^2)$ from (62) since $\hat{A}\hat{B} + \hat{B}\hat{A} \neq 2\hat{A}\hat{B}$ unless $\hat{A}$ and $\hat{B}$ commute. A better approximation is

$$e^{\frac{\alpha}{2}\hat{A}}e^{\alpha\hat{B}}e^{\frac{\alpha}{2}\hat{A}} = 1 + \alpha(\hat{A}+\hat{B}) + \frac{\alpha^2}{2}(\hat{A}^2 + \hat{B}^2 + \hat{A}\hat{B} + \hat{B}\hat{A}) + \mathcal{O}(\alpha^3),$$

which also covers the α^2 terms correctly and deviates (not shown) from order $\mathcal{O}(\alpha^3)$ on from (62). Higher-order expressions can be derived using Suzuki–Trotter decomposition [19, 21]. Here, we are content with

$$\hat{U}(\alpha) = e^{\alpha(\hat{A}+\hat{B})} = e^{\frac{\alpha}{2}\hat{A}}e^{\alpha\hat{B}}e^{\frac{\alpha}{2}\hat{A}} + \mathcal{O}(\alpha^3), \tag{63}$$

which, for our problem at hand, turns into

$$\hat{U}(\Delta t) = e^{-i\frac{\Delta t}{2}\hat{H}_x}e^{-i\Delta t\hat{H}_y}e^{-i\frac{\Delta t}{2}\hat{H}_x} + \mathcal{O}(\Delta t^3). \tag{64}$$

Such a factorization in exponentials yields a unitary time propagation if $\hat{H}_x$ and $\hat{H}_y$ are Hermitian and if each exponential is approximated by the corresponding Crank–Nicolson propagator (or some other exactly unitary approximant).

1.4.3 Avoiding discretization of derivatives: The Feit–Fleck–Steiger approach

Applying the splitting (63) directly to the TDSE (52), we can write, for instance,

$$\begin{aligned}\Psi(t+\Delta t) &= e^{-i\frac{\Delta t}{2}[-(\partial^2_{xx}+\partial^2_{yy})/2]}\, e^{-i\Delta t\, V(t+\Delta t/2)}\, e^{-i\frac{\Delta t}{2}[-(\partial^2_{xx}+\partial^2_{yy})/2]}\, \Psi(t) + \mathcal{O}(\Delta t^3)\\ &= e^{-i\frac{\Delta t}{4}[\hat{p}_x^2+\hat{p}_y^2]}\, e^{-i\Delta t\, V(t+\Delta t/2)}\, e^{-i\frac{\Delta t}{4}[\hat{p}_x^2+\hat{p}_y^2]}\, \Psi(t) + \mathcal{O}(\Delta t^3),\end{aligned} \tag{65}$$

where the momentum operators $\hat{p}_x = -i\partial_x$ and $\hat{p}_y = -i\partial_y$ have been used in the second line. We can split further the kinetic-energy parts without introducing additional errors because $\hat{p}_x$ and $\hat{p}_y$ commute, i.e.,

$$\Psi(t+\Delta t) = e^{i\frac{\Delta t}{4}\partial^2_{xx}}\, e^{i\frac{\Delta t}{4}\partial^2_{yy}}\, e^{-i\Delta t\, V(t+\Delta t/2)}\, e^{i\frac{\Delta t}{4}\partial^2_{xx}}\, e^{i\frac{\Delta t}{4}\partial^2_{yy}}\, \Psi(t) + \mathcal{O}(\Delta t^3). \tag{66}$$

Discretization of the spatial derivatives in combination with Crank–Nicolson approximants for the exponentials can be avoided by representing the wavefunction in a band-limited, discrete Fourier series,

$$\Psi(x,y,t) = \sum_{m=-N_x/2+1}^{N_x/2} \sum_{n=-N_y/2+1}^{N_y/2} \bar{\Psi}_{mn}(t)\, e^{2\pi i(mx/L_x+ny/L_y)}, \tag{67}$$

where the even integers $N_{x,y}$ are the number of grid points in x and y directions, respectively, and $L_{x,y}$ are the corresponding spatial grid sizes. Note that this is an expansion of the wavefunction in form (10). The two first exponentials applied to $\Psi(x,y,t)$ in (66) act

$$e^{i\frac{\Delta t}{4}\partial^2_{xx}}\, e^{i\frac{\Delta t}{4}\partial^2_{yy}}\, \Psi(x,y,t) = \sum_{m=-N_x/2+1}^{N_x/2} \sum_{n=-N_y/2+1}^{N_y/2} \bar{\Psi}_{mn}(t+\Delta t/2)\, e^{2\pi i(mx/L_x+ny/L_y)}, \tag{68}$$

with

$$\bar{\Psi}_{mn}(t+\Delta t/2) = e^{-i\frac{\Delta t}{4}(\frac{2\pi}{L_x})^2 m^2}\, e^{-i\frac{\Delta t}{4}(\frac{2\pi}{L_y})^2 n^2}\, \bar{\Psi}_{mn}(t). \tag{69}$$

Equation (67) reads with x and y discretized on a spatial grid

$$x_s = s\Delta x, \qquad s = -\frac{N_x}{2}+1, -\frac{N_x}{2}+2, \dots, \frac{N_x}{2}, \qquad \Delta x = \frac{L_x}{N_x}, \tag{70}$$

$$y_l = l\Delta y, \qquad l = -\frac{N_y}{2}+1, -\frac{N_y}{2}+2, \dots, \frac{N_y}{2}, \qquad \Delta y = \frac{L_y}{N_y}, \tag{71}$$

$$\Psi_{sl}(t) = \sum_m \sum_n \bar{\Psi}_{mn}(t)\, e^{2\pi i(ms/N_x+nl/N_y)}. \tag{72}$$

This is the discretized version of $\langle xy|\Psi(t)\rangle = \iint \mathrm{d}p_x\,\mathrm{d}p_y\,\langle xy|p_x p_y\rangle\,\langle p_x p_y|\Psi(t)\rangle$. The inverse of (72) is

$$\bar{\Psi}_{mn}(t) = \frac{1}{N_x N_y}\sum_s \sum_l \Psi_{sl}(t)\, e^{-2\pi i(ms/N_x + nl/N_y)}, \tag{73}$$

corresponding to $\langle p_x p_y|\Psi(t)\rangle = \iint \mathrm{d}x\,\mathrm{d}y\,\langle p_x p_y|xy\rangle\,\langle xy|\Psi(t)\rangle$.

The propagation of the wavefunction proceeds via application of the exponentials in (66) in the space in which they are diagonal [5]:

1. Transform $\Psi_{sl}(t)$ according (73) to momentum space $\bar{\Psi}_{mn}(t)$.
2. Apply $e^{-i\frac{\Delta t}{2}\hat{p}_x^2}\, e^{-i\frac{\Delta t}{2}\hat{p}_y^2}$, i.e., multiply with appropriate phase factors according (69) to obtain $\bar{\Psi}_{mn}(t+\Delta t/2)$.
3. Transform $\bar{\Psi}_{mn}(t+\Delta t/2)$ according (72) to position space $\Psi_{sl}(t+\Delta t/2)$.
4. Apply $e^{-i\Delta t\, V(x_s, y_l, t+\Delta t/2)}$, which is multiplicative in position space, giving $\Psi'_{sl}(t+\Delta t)$.
5. Transform $\Psi'_{sl}(t+\Delta t)$ according (73) to momentum space $\bar{\Psi}'_{mn}(t+\Delta t)$.
6. Apply $e^{-i\frac{\Delta t}{2}\hat{p}_x^2}\, e^{-i\frac{\Delta t}{2}\hat{p}_y^2}$ once more to obtain $\bar{\Psi}_{mn}(t+\Delta t)$.
7. Transforming $\bar{\Psi}_{mn}(t+\Delta t) \longrightarrow \Psi_{sl}(t+\Delta t)$ according (72) gives the desired propagated wavefunction $\Psi_{sl}(t+\Delta t)$ in position space.

If a subsequent propagation step is to follow, one does not need to do the last step but can directly continue with step 2.

The discrete Fourier transforms (73), (72) can be performed using the "Fast Fourier Transform" (FFT) [6, 14], which is of order $\mathcal{O}(N \log N)$ with N being the total number of grid points, i.e., in the 2D case considered here $N = N_x N_y$. The FFT-based propagation algorithm is slower than Crank–Nicolson[6] but the error in the approximation of the kinetic-energy operator $\hat{p}_x^2/2$ is with FFT of order $\mathcal{O}[(\Delta x)^{N_x}]$, i.e., extremely small. A minus on the FFT side may be the implicit assumption of periodic boundary conditions that, for some applications, is not adequate as it leads to "aliasing" [14]. The maximum momentum (or wavevector) that can be represented on a discretized spatial grid is $k_{\max} = \pi/\Delta x$. If wavevectors $k > k_{\max}$ are important, for instance, because the kinetic energy in a very deep (or even singular) potential $V < 0$ is (in atomic units) $T = k^2/2 = E - V$ such that $k = \sqrt{2(E-V)} > k_{\max}$, then the band-limited FFT will mirror these high wavevectors beyond $k_{\max}$ back into the interval $[-k_{\max}, k_{\max}]$, leading to artificial, unphysical dynamics of the wavefunction. Hence, in practice, one has to ensure that all relevant wavevectors k obey $|k| \ll \pi/\Delta x$ by choosing Δx sufficiently small. The same condition should be fulfilled in a Crank–Nicolson (or any other grid-based) scheme in order to propagate these high-k components of the wavefunction correctly. The advantage of Crank–Nicolson (besides being $\mathcal{O}(N)$) is the absence of aliasing.

6 $\mathcal{O}(N \log N)$ for FFT vs $\mathcal{O}(N)$ for Crank–Nicolson.

1.5 Expansion in spherical harmonics

The operator splitting outlined above for two Cartesian dimensions x and y is easily extended to three dimensions (3D) or more. For instance, a code for Cartesian coordinates x, y, and z could be used to study one 3D particle driven by an external field or, e.g., three 1D particles [15, 16]. However, a direct discretization in Cartesian position space is often not the most economic way to solve the TDSE. Whenever the problem at hand displays certain symmetries, one should exploit them. Consider the TDSE with a spherically symmetric potential $V(\mathbf{r}, t) = V(r, t)$ in atomic units

$$i\frac{\partial}{\partial t}\Psi(\mathbf{r}, t) = \left(-\frac{1}{2}\nabla^2 + V(r, t)\right)\Psi(\mathbf{r}, t). \tag{74}$$

Spherically symmetric time-dependent potentials $V(r, t)$ are usually not of interest, as they do not describe properly the coupling to an external electromagnetic field. However, it is advisable to introduce complications one after the other. Hence, for the moment let us stick with $V(r, t)$. Below, in Section 1.5.2, we will introduce external, time-dependent drivers that break the spherical symmetry.

Inserting the expansion in spherical harmonics $Y_{lm}(\Omega)$

$$\Psi(\mathbf{r}, t) = \frac{1}{r}\sum_{l=0}^{\infty}\sum_{m=-l}^{l}\phi_{lm}(r, t)\, Y_{lm}(\Omega), \qquad \Omega = \theta, \varphi \tag{75}$$

yields a set of uncoupled 1D TDSEs

$$i\frac{\partial}{\partial t}\phi_{lm}(r, t) = \left(-\frac{1}{2}\frac{\partial^2}{\partial r^2} + V(r, t) + \frac{l(l+1)}{2r^2}\right)\phi_{lm}(r, t) \tag{76}$$

in which the repulsive centrifugal potential $l(l+1)/2r^2$ appears. Instead of an x, y, z grid, we are now working on an r, l, m grid, on which the problem separates. In practice, the grid has to be finite, so we actually have

$$\Psi(\mathbf{r}, t) = \frac{1}{r}\sum_{l=0}^{N_l-1}\sum_{m=-l}^{l}\phi_{lm}(r, t)\, Y_{lm}(\Omega) \tag{77}$$

with N_l the number of grid points for the relevant angular momentum quantum numbers l that occur in the problem. The number of grid points for the magnetic quantum numbers m is then at most $2(N_l - 1) + 1$. The numerical grid for l and m is sketched in Figure 3.

The total size of the numerical grid is thus of order $\mathcal{O}(N_r N_l^2)$ with N_r the number of radial grid points. If the initial state is restricted to a few l and m values, the time-dependent state will be as well because the different components ϕ_{lm} remain uncoupled as long as the spherical symmetry is unbroken by the external driver.

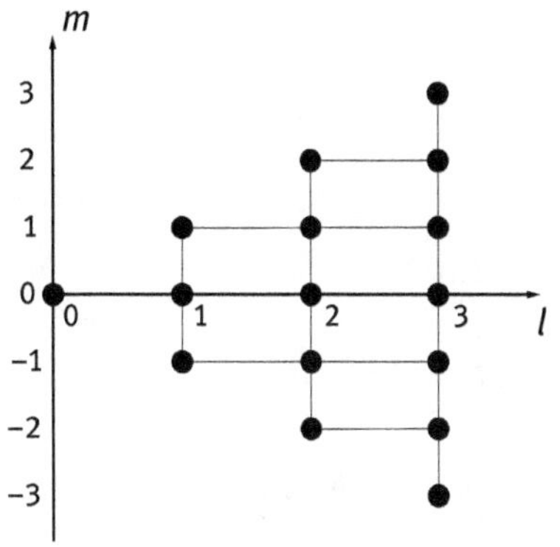

Fig. 3. Sketch of the l, m part of the numerical r, l, m grid.

We assume that the radial coordinate r is directly discretized in position space,

$$r = s\Delta r, \qquad s = 1, 2, 3, \dots, N_r. \tag{78}$$

Note that $r = 0$ is not part of the numerical grid so that we do not need to worry about how to represent, e.g., the singular Coulomb potential of hydrogen-like ions $V(r) = -Z/r$. The boundary condition of the radial wavefunction at $r = 0$ for potentials not more singular than the Coulomb potential reads as

$$\forall l, m \qquad \phi_{lm}(0, t) = 0 \tag{79}$$

at all times t. The expansion in spherical harmonics dictates the l, m grid,

$$l = 0, 1, 2, \dots, N_l - 1, \qquad m = -(N_l - 1), -(N_l - 1) + 1, \dots, N_l - 1. \tag{80}$$

In examples in Chapter II, we discuss how to estimate a big enough N_l in typical strong-field problems.

1.5.1 Propagation by the Muller method

We propagate the wavefunction using the Crank–Nicolson scheme introduced previously. We also make use of the Numerov "boost" in accuracy, as explained in Section 1.2.3. We absorb the $-1/2$ in front of the second derivative in the kinetic energy, defining

$$\mathbf{M}_2^{-1} = -\frac{1}{2}\mathbf{M}^{-1}, \tag{81}$$

with **M** given in (37), i.e.,

$$\mathbf{M}_2 = -\frac{1}{6}\begin{pmatrix} 10 & 1 & & & \\ 1 & 10 & 1 & & \\ & 1 & 10 & 1 & \\ & & & & \ddots \end{pmatrix}. \tag{82}$$

Then, with $h = \Delta r$ (indices l, m and arguments r, t suppressed),

$$\frac{\partial^2}{\partial r^2}\phi = \phi'' = -2\mathbf{M}_2^{-1}\mathbf{D}\phi + \mathcal{O}(h^4) = \left(1 + \frac{h^2}{12}\mathbf{D}\right)^{-1}\mathbf{D}\phi + \mathcal{O}(h^4). \tag{83}$$

The Hamiltonian in (76) in position-space matrix representation becomes

$$\hat{H}_l = \mathbf{M}_2^{-1}\mathbf{D} + \mathbf{V}_l, \qquad V_l(r,t) = V(r,t) + \frac{l(l+1)}{2r^2}. \tag{84}$$

The effective potential V_l depends on the angular momentum quantum number l but is diagonal in it. The Numerov-improved Crank–Nicolson approximant (41) is

$$\mathbf{U}_{\mathrm{CN}}(\Delta t) = \left(\mathbf{M}_2 + i\frac{\Delta t}{2}[\mathbf{D} + \mathbf{M}_2\mathbf{V}_l]\right)^{-1}\left(\mathbf{M}_2 - i\frac{\Delta t}{2}[\mathbf{D} + \mathbf{M}_2\mathbf{V}_l]\right), \tag{85}$$

with $\mathbf{M}_2$, $\mathbf{D}$, and $\mathbf{M}_2\mathbf{V}_l$ tridiagonal so that the efficient forward-backward substitution introduced in Section 1.2.1 can be employed for the implicit Crank–Nicolson step.

Boosting further the accuracy for Coulomb-like potentials

Consider the time-independent case $V(r,t) = V(r)$ with

$$V(r) = -\frac{Z}{r}. \tag{86}$$

For $l = m = 0$, the radial wavefunction $\phi_{00}(r,t)$ fulfills for $r \to 0$ according (76)

$$i\partial_t\phi_{00}(0) = -\frac{1}{2}\phi''_{00}(0) - \frac{Z}{r}[\phi_{00}(0) + r\phi'_{00}(0) + \cdots], \tag{87}$$

and thus, with (79),

$$\phi''_{00}(0) = -2Z\phi'_{00}(0) \neq 0. \tag{88}$$

Our discretized representation of second derivative (83) does not implement this particular boundary condition properly. However, we can incorporate it with little effort, as proposed by Muller [11], boosting further the numerical accuracy. Bringing

$\left(1+\frac{h^2}{12}\mathbf{D}\right)$ in (83) to the left-hand side, we obtain

$$\left(1+\frac{h^2}{12}\mathcal{D}\right)\phi''=\mathcal{D}\phi, \tag{89}$$

where we introduce $\mathbf{D}\to\mathcal{D}$ to be constructed in such a way that (88) is fulfilled. The idea is to modify the upper-left corner matrix element δ of $\mathcal{D}$,

$$\mathcal{D}=\frac{1}{h^2}\begin{pmatrix} h^2\delta & 1 & & & \\ 1 & -2 & 1 & & \\ & 1 & -2 & 1 & \\ & & & & \ddots \end{pmatrix}. \tag{90}$$

We try to get along with modifying the corner element only, because this is the simplest way without spoiling the Hermiticity of the Hamiltonian.

At $r=h$, we obtain from (89) and (90) for $\phi=\phi_{00}$

$$\frac{1}{12}\phi''(2h)+\left(1+\frac{h^2\delta}{12}\right)\phi''(h)=\frac{1}{h^2}\phi(2h)+\delta\phi(h).$$

Taylor expansion yields

$$\begin{aligned}&\frac{1}{12}\left[\phi''(0)+2h\phi'''(0)+\mathcal{O}(h^2)\right]+\left(1+\frac{h^2\delta}{12}\right)\left[\phi''(0)+h\phi'''(0)+\mathcal{O}(h^2)\right]\\ &=\frac{1}{h^2}\left[2h\phi'(0)+2h^2\phi''(0)+\mathcal{O}(h^3)\right]+\delta\left[h\phi'(0)+\frac{1}{2}h^2\phi''(0)+\mathcal{O}(h^3)\right],\end{aligned}$$

where $\phi(0)=0$ was used. Keeping all terms up to (excluding) $\mathcal{O}(h)$ and taking into account that δh^2 is expected to be $\mathcal{O}(1)$ because the uncorrected $(-2/h^2)h^2$ is $\mathcal{O}(1)$ as well, we have

$$\frac{1}{12}\phi''(0)+\left(1+\frac{h^2\delta}{12}\right)\phi''(0)+\mathcal{O}(h)=\phi'(0)\left(\frac{2}{h}+\delta h\right)+\phi''(0)\left(2+\frac{\delta h^2}{2}\right)+\mathcal{O}(h).$$

Making use of the boundary condition (88), i.e., replacing $\phi'(0)$ by $-\phi''(0)/2Z$, we obtain

$$0=\phi''(0)\left[-\frac{11}{12}-\frac{5}{12}\delta h^2+\frac{1}{Zh}+\frac{\delta h}{2Z}\right].$$

The square bracket must vanish. Solving it for δ gives the desired modified matrix corner element

$$\delta=-\frac{2}{h^2}\left(1-\frac{Zh}{12-10Zh}\right)\qquad\text{for}\quad l=m=0 \tag{91}$$

and $\delta = -2/h^2$ otherwise. With modified $\mathbf{D} \to \mathcal{D}$, also $\mathbf{M}_2 \to \mathcal{M}_2$ changes. The modified upper-left matrix element reads as

$$(\mathcal{M}_2)_{11} = -2\left(1 + \frac{h^2}{12}\delta\right) \qquad \text{for} \quad l = m = 0 \tag{92}$$

and $(\mathcal{M}_2)_{11} = (\mathbf{M}_2)_{11} = -10/6$ otherwise.

Let us consider as an example the ground state of atomic hydrogen where $l = m = 0$, $V(r) = -r^{-1}$, and analytically $\phi_{00}(r) = 2re^{-r}$ and $E_0 = -1/2$. Without the modification of the matrix corner element, we obtain with converged imaginary time propagation using the Crank–Nicolson time-evolution operator (85) on a (sufficiently big) numerical grid with a resolution $\Delta r = h = 0.2$ the energy $E_0^{(\text{without})} = -0.489388404$. With modification, we obtain $E_0^{(\text{with})} = -0.500151077$. The relative errors are

$$\frac{|E_0^{(\text{without})} - E_0|}{|E_0|} = 2 \cdot 10^{-2}, \qquad \frac{|E_0^{(\text{with})} - E_0|}{|E_0|} = 3 \cdot 10^{-4}.$$

Hence, with very little effort, we gain an improvement over two orders of magnitude. If a certain accuracy is prescribed, the corner-element correction allows us to work with a coarser grid. This is very valuable if grids with large spatial extension are needed to describe, e.g., the interaction with external drivers such as lasers. Figure 4 shows the two wavefunctions obtained with and without modified upper-left corner element, leading to the ground state energy values just given. While in the linear plot in (a) the two wavefunctions look innocently similar, the logarithmic plot of $\psi_0(r) = \phi_0(r)/r$ in (b) clearly shows the wrong slope without modified corner matrix element as $r \to 0$.

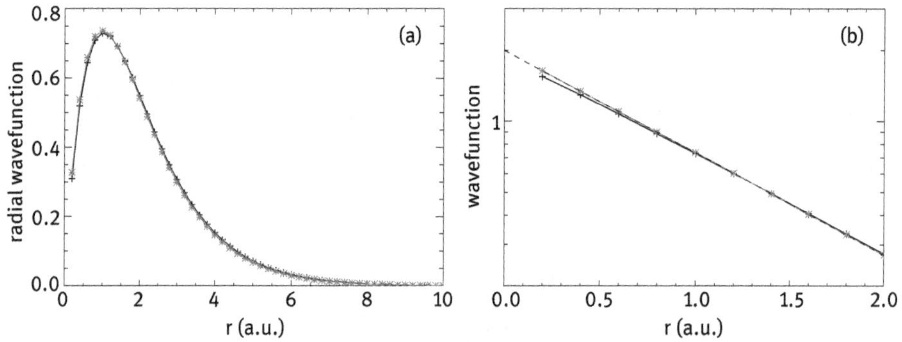

Fig. 4. (a) Radial wavefunctions $\phi_0(r)$ with (*) and without (+) corrected upper-left matrix element. (b) Logarithmic plot of the wavefunction $\psi_0(r) = \phi_0(r)/r$ with and without modified upper-left matrix element. The exact, analytical result $\psi_0(r) = 2e^{-r}$ is included (dashed).

Consistency

Note that if the Numerov boost in accuracy is employed, it is important to calculate the energy expectation value in a consistent way, i.e.,

$$E = \sum_{lm} \langle \phi_{lm} | \hat{H}_0 | \phi_{lm} \rangle = \langle \phi_{lm} | \hat{\mathcal{M}}_2^{-1} (\hat{\mathcal{D}} + \hat{\mathcal{M}}_2 \hat{V}_l) | \phi_{lm} \rangle . \tag{93}$$

Unlike in the Crank–Nicolson propagation, $\hat{\mathcal{M}}_2^{-1}$ does not cancel. Hence, an additional forward-backward substitution is required because

$$\langle \phi'_{lm} | \hat{\mathcal{M}}_2 = \langle \phi_{lm} |$$

has to be solved first for $\langle \phi'_{lm} |$, and then

$$E = \sum_{lm} \langle \phi'_{lm} | (\hat{\mathcal{D}} + \hat{\mathcal{M}}_2 \hat{V}_l) | \phi_{lm} \rangle \tag{94}$$

can be evaluated. If we propagated with the Numerov-improved scheme but still calculated the energy expectation value with the "old" $\langle \phi_{lm} | (-\frac{1}{2}\hat{D} + \hat{V}_l) | \phi_{lm} \rangle$, we would not get necessarily an improved energy value. In fact, it might be even worse.

1.5.2 Coupling to a classical external field

Electromagnetic fields can be described by a vector potential $\mathbf{A}(\mathbf{r}, t)$ and a scalar potential $\Phi(\mathbf{r}, t)$. Electric and magnetic field are given by

$$\mathbf{E}(\mathbf{r}, t) = -\nabla \Phi(\mathbf{r}, t) - \partial_t \mathbf{A}(\mathbf{r}, t), \qquad \mathbf{B}(\mathbf{r}, t) = \nabla \times \mathbf{A}(\mathbf{r}, t). \tag{95}$$

The coupling to a free particle of charge q and mass m, governed by the free TDSE $[i\hbar\partial_t - \hat{\mathbf{p}}^2/2m]\Psi(\mathbf{r}, t) = 0$, amounts to the "minimal coupling" replacement

$$i\hbar\partial_t \longrightarrow i\hbar\partial_t - q\Phi(\mathbf{r}, t), \qquad \hat{\mathbf{p}} \longrightarrow \hat{\mathbf{p}} - q\mathbf{A}(\mathbf{r}, t) \tag{96}$$

(in SI units). In the following, we specialize on an electron and use atomic units again, i.e., $q = -e = -1$, so that the TDSE reads as

$$i\partial_t \Psi(\mathbf{r}, t) = \left\{ \frac{[\hat{\mathbf{p}} + \mathbf{A}(\mathbf{r}, t)]^2}{2} - \Phi(\mathbf{r}, t) \right\} \Psi(\mathbf{r}, t). \tag{97}$$

We choose a gauge where the vector potential alone governs the electromagnetic radiation and the scalar potential $q\Phi = V$ the static potential (e.g., $-Z/r$). The Hamiltonian then is (arguments $\mathbf{r}, t$ suppressed)

$$\hat{H} = \frac{1}{2}[\hat{\mathbf{p}} + \mathbf{A}]^2 + V = \frac{1}{2}[\hat{\mathbf{p}}^2 + \hat{\mathbf{p}} \cdot \mathbf{A} + \mathbf{A} \cdot \hat{\mathbf{p}} + \mathbf{A}^2] + V. \tag{98}$$

In most of the chapters in this book (an exception is Chapter III), we make the dipole approximation where $\mathbf{A}(\mathbf{r},t) \simeq \mathbf{A}(t)$ and thus $\mathbf{B} \equiv 0$. In order to render the nonrelativistic TDSE applicable, we further require $|\mathbf{A}| \ll 137$. Otherwise (the amplitude of) the (oscillatory) electron velocity becomes comparable to the speed of light in vacuum c. In dipole approximation, the Hamiltonian (98) simplifies to

$$\hat{H} = \frac{1}{2}\hat{\mathbf{p}}^2 + \mathbf{A}\cdot\hat{\mathbf{p}} + \frac{1}{2}\mathbf{A}^2 + V, \tag{99}$$

with a purely time-dependent term $\mathbf{A}^2/2$. Now, consider a general TDSE $i\partial_t\varphi(t) = [\hat{H} + f(t)]\varphi(t)$. The transformation

$$\varphi(t) = \varphi'(t)\, e^{-i\int^t f(t')\,dt'} \tag{100}$$

yields the transformed TDSE $i\partial_t\varphi'(t) = \hat{H}\varphi'(t)$, i.e., with the $f(t)$ absent. The purely time-dependent term $\mathbf{A}^2/2$ can thus be transformed away in this manner,[7] leading to the Hamiltonian

$$\hat{H} = \frac{1}{2}\hat{\mathbf{p}}^2 + \mathbf{A}(t)\cdot\hat{\mathbf{p}} + V(\mathbf{r}). \tag{101}$$

All observables will be unaffected by the purely time-dependent phase transformation (100), which is an especially simple, global gauge transformation. Gauge-noninvariant quantities like "the energy while the laser is on" $\langle\varphi(t)|\,[\hat{H}(t) + f(t)]\,|\varphi(t)\rangle \neq \langle\varphi'(t)|\,\hat{H}\,|\varphi'(t)\rangle$, on the other hand, are *not* invariant under this transformation. Hence, when it comes to the interpretation of simulation results, one should not overinterpret gauge-noninvariant quantities. However, from the numerical point of view, a transformation may greatly reduce the numerical effort, as discussed in Section 1.5.5.

1.5.3 Linear polarization

Let us consider an external driver in dipole approximation that is polarized in z-direction,

$$\mathbf{A}(t) = A(t)\,\mathbf{e}_z, \tag{102}$$

and a spherically symmetric (binding) potential,

$$V(\mathbf{r}) = V(r). \tag{103}$$

With Hamiltonian (101), the TDSE

$$i\partial_t\Psi(\mathbf{r},t) = \left(-\frac{1}{2}\nabla^2 - iA(t)\partial_z + V(r)\right)\Psi(\mathbf{r},t) \tag{104}$$

7 Emphasizing once more: this only works in dipole approximation!

results. Insertion of expansion (75) yields a set of equations for the radial wavefunctions $\phi_{lm}(r,t)$,

$$\begin{aligned} i\partial_t \phi_{lm}(r,t) = & \left(-\frac{1}{2}\frac{\partial^2}{\partial r^2} + V_{\mathrm{eff}\,l}(r)\right)\phi_{lm}(r,t) \\ & - iA(t)\sum_{l'} \langle lm|\cos\theta|l'm\rangle \frac{\partial}{\partial r}\phi_{l'm}(r,t) \\ & + i\frac{A(t)}{r}\sum_{l'}\left(\langle lm|\cos\theta|l'm\rangle + \langle lm|\sin\theta\frac{\partial}{\partial\theta}|l'm\rangle\right)\phi_{l'm}(r,t), \end{aligned} \tag{105}$$

where

$$V_{\mathrm{eff}\,l}(r) = V(r) + \frac{l(l+1)}{2r^2}, \tag{106}$$

and $\langle lm|\cdots|l'm\rangle$ is short-hand notation for $\int d\Omega\, Y^{\star}_{lm}(\Omega)\cdots Y_{l'm}(\Omega)$. All the matrix elements in (105) can be evaluated analytically using the properties of spherical harmonics. The Hamiltonian has the form [1, 11]

$$\hat{H}(t) = \hat{H}_{\mathrm{at}} + \hat{H}_{\mathrm{mix}}(t) + \hat{H}_{\mathrm{ang}}(t), \tag{107}$$

where $\hat{H}_{\mathrm{at}}$ corresponds to the first line in (105), $\hat{H}_{\mathrm{mix}}(t)$ to the second, and $\hat{H}_{\mathrm{ang}}(t)$ to the third. As expected for linear polarization, there is no coupling of different m components, i.e., the Hamiltonian is diagonal in m, and m is fixed by the initial state. Hence, it is sufficient to discretize in r, l space, and the problem of an electron in a potential $V(r)$ plus a linearly polarized laser field in dipole approximation is effectively 2D. A clever grid representation should make use of that reduced dimensionality; cylindrical coordinates do (see Section 2), Cartesian do not.

The part $\hat{H}_{\mathrm{at}}$ is diagonal in l. With the Numerov boost in accuracy and the upper-left corner correction for Coulomb-like potentials, it becomes, discretized in r, l space,

$$\mathbf{H}_{\mathrm{at}} = \mathbf{1}_l \otimes (\mathcal{M}_2^{-1}\mathcal{D}_2 + \mathbf{V}_l), \tag{108}$$

where $\mathbf{1}_l$ indicates explicitly the diagonality in l space. We write here $\mathcal{D}_2$ instead of $\mathcal{D}$ because we will introduce a $\mathcal{D}_1$ for the first derivative soon.

$\hat{H}_{\mathrm{mix}}(t)$ is diagonal neither in r space nor in l space. How neighboring ls are coupled can be seen in matrix representation with respect to l space, acting on vectors $(\phi_{0m}(r), \phi_{1m}(r), \phi_{2m}(r), \ldots)^{\top}$,

$$\mathbf{H}_{\mathrm{mix}}(t) = -iA(t)\begin{pmatrix} 0 & c_{0m} & 0 & 0 \\ c_{0m} & 0 & c_{1m} & 0 \\ 0 & c_{1m} & 0 & c_{2m} \\ 0 & 0 & c_{2m} & \ddots \end{pmatrix}\frac{\partial}{\partial r}, \tag{109}$$

with

$$c_{lm} = \sqrt{\frac{(l+1)^2 - m^2}{(2l+1)(2l+3)}}. \tag{110}$$

After discretization in r, the first derivative will couple different r components as well, as will be discussed below.

The part $\hat{H}_{\text{ang}}(t)$ is diagonal in r space but also couples neighboring ls:

$$\mathbf{H}_{\text{ang}}(t) = -i\frac{A(t)}{r}\begin{pmatrix} 0 & c_{0m} & 0 & 0 \\ -c_{0m} & 0 & 2c_{1m} & 0 \\ 0 & -2c_{1m} & 0 & 3c_{2m} \\ 0 & 0 & -3c_{2m} & \ddots \end{pmatrix}. \tag{111}$$

We further notice that $\mathbf{H}_{\text{mix}}$ and $\mathbf{H}_{\text{ang}}$ can be written as sums over 2×2 matrices acting in neighboring, overlapping l subspaces,

$$\mathbf{H}_{\text{mix,ang}} = \sum_l \mathbf{H}^{lm}_{\text{mix,ang}}, \tag{112}$$

$$\mathbf{H}^{lm}_{\text{mix}} = -iA(t)\mathbf{L}^{lm}\frac{\partial}{\partial r}, \qquad \mathbf{H}^{lm}_{\text{ang}} = -i\frac{A(t)}{r}\mathbf{T}^{lm}, \tag{113}$$

where

$$\mathbf{L}^{lm} = \begin{pmatrix} 0 & c_{lm} \\ c_{lm} & 0 \end{pmatrix}, \qquad \mathbf{T}^{lm} = (l+1)\begin{pmatrix} 0 & c_{lm} \\ -c_{lm} & 0 \end{pmatrix}. \tag{114}$$

Discretizing ∂_r without spoiling Hermiticity

Considering also the discretization in r, we write instead of (113)

$$\mathbf{H}^{lm}_{\text{mix}} = -iA(t)\mathbf{L}^{lm} \otimes \mathbf{M}_1^{-1}\mathbf{D}_1, \qquad \mathbf{H}^{lm}_{\text{ang}} = -iA(t)\mathbf{T}^{lm} \otimes \frac{1}{r_s}\mathbf{1}_r, \tag{115}$$

where $\mathbf{1}_r$ is unity in r space, and $\partial_r = \mathbf{M}_1^{-1}\mathbf{D}_1 + \mathcal{O}(\Delta r^4)$ is the Numerov analogue of (36) for the first derivative with

$$\mathbf{M}_1 = \frac{1}{6}\begin{pmatrix} 4 & 1 & & \\ 1 & 4 & 1 & \\ & & \ddots & \\ & & 1 & 4 \end{pmatrix}, \qquad \mathbf{D}_1 = \frac{1}{2h}\begin{pmatrix} 0 & 1 & & \\ -1 & 0 & 1 & \\ & & \ddots & \\ & & -1 & 0 \end{pmatrix}. \tag{116}$$

However, in order for $\mathbf{H}^{lm}_{\text{mix}}$ to be Hermitian, $\mathbf{M}_1^{-1}\mathbf{D}_1$ needs to be anti-Hermitian. We should take all measures to keep the total Hamiltonian Hermitian. Otherwise the time propagation is not unitary, and the norm of the wavefunction is not conserved.

The problem can again be solved by modifying corner matrix elements [11]. It is straightforward (though a little cumbersome) to show that by replacing

$$\mathbf{D}_1 \longrightarrow \mathcal{D}_1 = \frac{1}{2h}\begin{pmatrix} y & 1 & & & \\ -1 & 0 & 1 & & \\ & & \ddots & & \\ & & -1 & -y \end{pmatrix}, \qquad h = \Delta r, \quad y = \sqrt{3} - 2, \tag{117}$$

$$\mathbf{M}_1 \longrightarrow \mathcal{M}_1 = \frac{1}{6}\begin{pmatrix} 4+y & 1 & & \\ 1 & 4 & 1 & \\ & & \ddots & \\ & & 1 & 4+y \end{pmatrix}. \tag{118}$$

Hermiticity is restored.

1.5.4 Propagator for linear polarization

For propagating the radial wavefunctions ϕ_{lm} by one time step Δt, we use the split operator

$$\mathbf{U}(\Delta t) = \left\{ \prod_{l=N_l-2}^{0} \exp\left(-i\frac{\Delta t}{2}\mathbf{H}_{\text{ang}}^{lm}\right) \exp\left(-i\frac{\Delta t}{2}\mathbf{H}_{\text{mix}}^{lm}\right) \right\} \exp(-i\Delta t \mathbf{H}_{\text{at}})$$
$$\times \left\{ \prod_{l=0}^{N_l-2} \exp\left(-i\frac{\Delta t}{2}\mathbf{H}_{\text{mix}}^{lm}\right) \exp\left(-i\frac{\Delta t}{2}\mathbf{H}_{\text{ang}}^{lm}\right) \right\} + \mathcal{O}(\Delta t^3). \tag{119}$$

The exponentials $\exp\left(-i\frac{\Delta t}{2}\mathbf{H}_{\text{ang}}^{lm}\right)$, being 2×2 matrices in l subspaces and diagonal in r, could be diagonalized exactly. Here, we treat them like all other exponentials in Crank–Nicolson style,

$$\mathbf{R}^{lm} = \left(\hat{1} + i\frac{\Delta t}{4}\mathbf{H}_{\text{ang}}^{lm}\right)^{-1} \left(\hat{1} - i\frac{\Delta t}{4}\mathbf{H}_{\text{ang}}^{lm}\right), \tag{120}$$

$$\mathbf{X}_{\pm}^{lm} = \hat{1} \pm i\frac{\Delta t}{4}\mathbf{H}_{\text{mix}}^{lm}, \qquad \mathbf{Q}_{\pm} = \hat{1} \pm i\frac{\Delta t}{2}\mathbf{H}_{\text{at}}, \tag{121}$$

so that

$$\mathbf{U}(\Delta t) = \left\{ \prod_{l=N_l-2}^{0} \mathbf{R}^{lm}(\mathbf{X}_{+}^{lm})^{-1}\mathbf{X}_{-}^{lm} \right\} \mathbf{Q}_{+}^{-1}\mathbf{Q}_{-} \left\{ \prod_{l=0}^{N_l-2} (\mathbf{X}_{+}^{lm})^{-1}\mathbf{X}_{-}^{lm}\mathbf{R}^{lm} \right\}. \tag{122}$$

The purely angular 2×2 parts $\mathbf{R}^{lm}$ can be applied easily to the wavefunction.[8] The purely radial Crank–Nicolson part

$$\mathbf{Q}_+^{-1}\mathbf{Q}_- = \mathbf{W}_+^{-1}\mathbf{W}_-, \qquad \mathbf{W}_\pm = \mathcal{M}_2 \pm i\frac{\Delta t}{2}(\mathcal{D}_2 + \mathcal{M}_2 \mathbf{V}_l) \tag{123}$$

is known already. It just needs to be applied to each l component ϕ_{lm} separately. The mixing part can be simplified further. Factoring out $\mathcal{M}_1^{-1}$ gives

$$(\mathbf{X}_+^{lm})^{-1}\mathbf{X}_-^{lm} = (\mathbf{Y}_+^{lm})^{-1}\mathbf{Y}_-^{lm}, \qquad \mathbf{Y}_\pm^{lm} = \mathbf{1}_l \otimes \mathcal{M}_1 \pm \frac{\Delta t}{4} A(t) \mathbf{L}^{lm} \otimes \mathcal{D}_1. \tag{124}$$

$\mathbf{Y}_\pm^{lm}$ are only block tridiagonal,

$$\mathbf{Y}_\pm^{lm} = \left(\begin{array}{cc|cc|cc|cc|cc}
\frac{4+y}{6} & \pm y g_{lm} & \frac{1}{6} & \pm g_{lm} & & & & & & \\
\pm y g_{lm} & \frac{4+y}{6} & \pm g_{lm} & \frac{1}{6} & & & & & & \\
\hline
\frac{1}{6} & \mp g_{lm} & \frac{2}{3} & & \frac{1}{6} & \pm g_{lm} & & & & \\
\mp g_{lm} & \frac{1}{6} & & \frac{2}{3} & \pm g_{lm} & \frac{1}{6} & & & & \\
\hline
 & & \frac{1}{6} & \mp g_{lm} & \frac{2}{3} & & \frac{1}{6} & \pm g_{lm} & & \\
 & & \mp g_{lm} & \frac{1}{6} & & \frac{2}{3} & \pm g_{lm} & \frac{1}{6} & & \\
\hline
 & & & & \cdot & & \cdot & & \cdot & \\
 & & & & & \cdot & & \cdot & & \cdot \\
\hline
 & & & & & & \frac{1}{6} & \mp g_{lm} & \frac{4+y}{6} & \mp y g_{lm} \\
 & & & & & & \mp g_{lm} & \frac{1}{6} & \mp y g_{lm} & \frac{4+y}{6}
\end{array}\right)$$

with $g_{lm} = \Delta t A(t) c_{lm}/8h$. It operates on vectors of the form

$$\boldsymbol{\phi}_{lm} = \left(\boxed{\phi_{lm}(r_1,t), \phi_{l+1,m}(r_1,t)}, \boxed{\phi_{lm}(r_2,t), \phi_{l+1,m}(r_2,t)}, \ldots \right)^\top.$$

However, with the help of the transform

$$\mathbf{B}\mathbf{L}^{lm}\mathbf{B}^\top = \begin{pmatrix} c_{lm} & 0 \\ 0 & -c_{lm} \end{pmatrix} = \mathbf{C}_{lm}, \qquad \mathbf{B} = \frac{1}{\sqrt{2}}\begin{pmatrix} 1 & 1 \\ -1 & 1 \end{pmatrix}, \quad \mathbf{B}^{-1} = \mathbf{B}^\top,$$

we can write for $(\mathbf{Y}_+^{lm})^{-1}\mathbf{Y}_-^{lm}$, applied to the part of the discretized wavefunction $\boldsymbol{\phi}_{lm}(r_s) = (\phi_{lm}(r_s), \phi_{l+1,m}(r_s))^\top$,

$$(\mathbf{Y}_+^{lm})^{-1}\mathbf{Y}_-^{lm}\boldsymbol{\phi}_{lm}(r_s) = \hat{B}^\top (\mathcal{Y}_+^{lm})^{-1}\mathcal{Y}_-^{lm}\,\tilde{\boldsymbol{\phi}}_{lm}(r_s)$$

with

$$\mathcal{Y}_\pm^{lm} = \mathbf{B}\mathbf{Y}_\pm^{lm}\mathbf{B}^\top, \qquad \tilde{\boldsymbol{\phi}}_{lm}(r_s) = \mathbf{B}\boldsymbol{\phi}_{lm}(r_s).$$

8 The trivial inverse of a 2×2 matrix in (120) can be calculated for each r analytically (as is done below in Section 1.5.6) or numerically.

We find that $\mathcal{Y}_{\pm}^{lm}$ are diagonal in the l subspaces:

$$\mathcal{Y}_{\pm}^{lm} = \mathbf{1}_l \otimes \mathcal{M}_1 \pm \frac{\Delta t}{4} A(t) \begin{pmatrix} c_{lm} & 0 \\ 0 & -c_{lm} \end{pmatrix} \otimes \mathcal{D}_1 .$$

Hence, applying it to $\tilde{\boldsymbol{\phi}}_{lm}(r_s) = \left(\tilde{\phi}_{lm}(r_s), \tilde{\phi}_{l+1,m}(r_s)\right)^{\top}$, we obtain for each l block

$$\mathcal{Y}_{-}^{lm} \tilde{\boldsymbol{\phi}}_{lm}(r_s) = \begin{pmatrix} -\mathcal{G}^{lm}(t) & 0 \\ 0 & \mathcal{G}^{lm}(t) \end{pmatrix} \begin{pmatrix} \tilde{\phi}_{lm}(r_s) \\ \tilde{\phi}_{l+1,m}(r_s) \end{pmatrix}$$

with

$$\mathcal{G}^{lm}(t) = \mathcal{M}_1 + \frac{\Delta t}{4} A(t) c_{lm} \mathcal{D}_1 .$$

Hence, the Crank–Nicolson step $(\mathcal{Y}_{+}^{lm})^{-1}\mathcal{Y}_{-}^{lm}$ involves tridiagonal matrices (in r space) only and can be efficiently implemented using forward-backward substitution as well.

To summarize, the time-evolution operator reads as

$$\mathbf{U}(\Delta t) = \left\{ \prod_{l=N_l-2}^{0} \mathbf{R}^{lm} \mathbf{B}^{\top} (\mathcal{Y}_{+}^{lm})^{-1} \mathcal{Y}_{-}^{lm} \mathbf{B} \right\} \mathbf{W}_{+}^{-1} \mathbf{W}_{-} \times \left\{ \prod_{l=0}^{N_l-2} \mathbf{B}^{\top} (\mathcal{Y}_{+}^{lm})^{-1} \mathcal{Y}_{-}^{lm} \mathbf{B} \mathbf{R}^{lm} \right\} . \tag{125}$$

The whole problem has been reduced to the application of 2×2 matrices in l space and tridiagonal matrices in r space, with an intermediate transformation for the rl-mixed part.

The strong-field TDSE solver `Qprop` [1] is based on the propagator (125) and is available for download at `www.qprop.de`.

1.5.5 Choice of gauge within dipole approximation

Instead of the so-called velocity gauge $\mathbf{A} \cdot \hat{\mathbf{p}}$ coupling, we may work in so-called length gauge, where the coupling of electron and external field is governed by

$$\mathbf{E}(t) \cdot \mathbf{r} = E(t) z = E(t) r \cos\theta , \tag{126}$$

and $\mathbf{E}(t) = -\partial_t \mathbf{A}(t)$ is the electric field corresponding to the vector potential. At first sight, that seems much more attractive than the velocity gauge because $E(t) r \cos\theta$ is already diagonal in r space. Hence, $\mathbf{H}_{\text{mix}}$, which required considerable extra effort above, is absent in length gauge. In fact, instead of $\mathbf{H}_{\text{mix}}(t) + \mathbf{H}_{\text{ang}}(t)$ (see (109) and

(111), respectively), we just have

$$\mathbf{H}_{\text{ang,length}}(t) = rE(t)\begin{pmatrix} 0 & c_{0m} & 0 & 0 \\ c_{0m} & 0 & c_{1m} & 0 \\ 0 & c_{1m} & 0 & c_{2m} \\ 0 & 0 & c_{2m} & \ddots \end{pmatrix}, \tag{127}$$

and instead of (115),

$$\mathbf{H}^{lm}_{\text{ang,length}} = r_s E(t)\,\mathbf{L}^{lm} \otimes \mathbf{1}_r. \tag{128}$$

The Crank–Nicolson propagator then simplifies to

$$\mathbf{U}(\Delta t) = \left\{ \prod_{l=N_l-2}^{0} \mathbf{R}^{lm}_{\text{length}} \right\} \mathbf{W}_+^{-1}\mathbf{W}_- \left\{ \prod_{l=0}^{N_l-2} \mathbf{R}^{lm}_{\text{length}} \right\} \tag{129}$$

with

$$\mathbf{R}^{lm}_{\text{length}} = \left(\hat{1} + i\frac{\Delta t}{4}\mathbf{H}^{lm}_{\text{ang,length}}\right)^{-1}\left(\hat{1} - i\frac{\Delta t}{4}\mathbf{H}^{lm}_{\text{ang,length}}\right), \tag{130}$$

$\mathbf{H}^{lm}_{\text{ang,length}}$ as defined in (128) and $\mathbf{W}_\pm$ in (123).

So why did we consider the significantly more complicated-to-implement velocity gauge in the first place? The disadvantage of the length gauge is that the energy $E(t)r$ becomes large for strong fields and large distances r and thus requires smaller Δt (and smaller Δr to resolve the corresponding momenta). Note that in velocity gauge, $\hat{\mathbf{p}}$ is the canonical momentum and does not equal the kinetic momentum $\hat{\mathbf{p}} + \mathbf{A}(t)$. In fact, considering the Hamiltonian of a free electron in a purely time-dependent external field

$$\mathbf{H} = \frac{1}{2}[\hat{\mathbf{p}} + \mathbf{A}(t)]^2,$$

we have the Heisenberg equations of motion

$$\dot{\hat{\mathbf{r}}} = \hat{\mathbf{p}} + \mathbf{A}(t), \qquad \dot{\hat{\mathbf{p}}} = \hat{\mathbf{0}},$$

i.e., the canonical momentum is a constant of the motion. It represents just a constant drift momentum since the large oscillatory part $\sim \mathbf{A}(t)$ is subtracted. Of course, in situations of interest, a binding potential is present as well, which may lead through laser-driven scattering to higher momenta. However, that happens too in length gauge. Hence, the take-home message is: use velocity gauge in numerical solutions of the TDSE for strong (laser) fields. The extra effort to handle the $\mathbf{H}_{\text{mix}}$ part in the propagator pays off very well. There are, however, situations where length gauge is more appropriate. Imagine the simulation of ionization of highly charged ions, i.e.,

$V(r) \sim -Z/r$ with high Z. In this case,[9] we have to reduce also our $\Delta r \sim Z^{-1}$, and it is beneficial that $E(t)r \to 0$ where $V(r) \to -\infty$ in order to keep the maximum absolute energy value in the problem $Z/\Delta r$ instead of making it even greater by $A(t)\hat{p}_z$, which does not vanish in the origin. In Section 1 of Chapter II, we indeed use length gauge to test Landau's ionization rate.

To summarize this little interlude, one should not only exploit the freedom to use the most efficient grid representation of the wavefunctions but also the "gauge-freedom" to choose the most efficient Hamiltonian, i.e., the one which allows for the coarsest grid spacings (and thus the smallest number of grid points) and the smallest basis sets.

1.5.6 Propagator for elliptical polarization in the *xy* plane

We now consider a vector potential of the form

$$\mathbf{A}(t) = A_x(t)\mathbf{e}_x + A_y(t)\mathbf{e}_y \tag{131}$$

so that the TDSE reads as

$$i\partial_t\Psi(\mathbf{r},t) = \left(-\frac{1}{2}\nabla^2 + V(r) - iA_x(t)\partial_x - iA_y(t)\partial_y\right)\Psi(\mathbf{r},t), \tag{132}$$

which, after inserting expansion (75), yields

$$\begin{aligned} i\partial_t\phi_{lm} = &\left(-\frac{1}{2}\frac{\partial^2}{\partial r^2} + V_{\text{eff}\,l}\right)\phi_{lm} \\ &- \frac{ir}{2}\sum_{l'm'}\langle lm|[\exp(i\varphi)\tilde{A}^* + \exp(-i\varphi)\tilde{A}]\sin\vartheta|l'm'\rangle\partial_r\frac{1}{r}\phi_{l'm'} \\ &- \frac{i}{2r}\sum_{l'm'}\langle lm|\tilde{A}^*\exp(i\varphi)\left(\cos\vartheta\partial_\vartheta + \frac{i}{\sin\vartheta}\partial_\varphi\right)|l'm'\rangle\phi_{l'm'} \\ &- \frac{i}{2r}\sum_{l'm'}\langle lm|\tilde{A}\exp(-i\varphi)\left(\cos\vartheta\partial_\vartheta - \frac{i}{\sin\vartheta}\partial_\varphi\right)|l'm'\rangle\phi_{l'm'} \end{aligned} \tag{133}$$

with $\tilde{A} = A_x + iA_y$. With the help of the ladder operators

$$\hat{L}_\pm = -\frac{1}{\sqrt{2}}\exp(\pm i\varphi)\,(\partial_\vartheta \pm i\cot\vartheta\partial_\varphi) \tag{134}$$

that act on a spherical harmonic according

$$\hat{L}_\pm|lm\rangle = \mp N^\pm_{lm}|lm \pm 1\rangle, \tag{135}$$

9 In such cases, it makes sense to switch to nonuniform radial grids, which have a finer resolution as one approaches the origin (e.g., a "logarithmic grid").

where

$$N^{\pm}_{lm} = \sqrt{\frac{l(l+1) - m(m\pm 1)}{2}} = \sqrt{\frac{(l\mp m)(l\pm m+1)}{2}},$$

we obtain

$$\begin{aligned} i\partial_t\phi_{lm} &= \left(-\frac{1}{2}\frac{\partial^2}{\partial r^2} + V_{\mathrm{eff}\,l}\right)\phi_{lm} \\ &\quad + i\sqrt{\frac{2\pi}{3}}\sum_{l'm'}\left\{\tilde{A}^*\langle lm|11|l'm'\rangle\partial_r - \tilde{A}\langle lm|1-1|l'm'\rangle\partial_r\right. \\ &\quad - \frac{\tilde{A}^*}{r}\langle lm|11|l'm'\rangle(1+m') + \frac{\tilde{A}}{r}\langle lm|1-1|l'm'\rangle(1-m') \\ &\quad \left. - \frac{\tilde{A}^*}{r}\langle lm|10|l'm'+1\rangle N^{+}_{l'm'} + \frac{\tilde{A}}{r}\langle lm|10|l'm'-1\rangle N^{-}_{l'm'}\right\}\phi_{l'm'} \end{aligned} \tag{136}$$

with $\langle lm|LM|l'm'\rangle = \int \mathrm{d}\Omega\, Y_{lm}{}^* Y_{LM} Y_{l'm'}$. Three spherical harmonics integrated over the solid angle Ω may be expressed in terms of Clebsch–Gordan coefficients $C^{c\gamma}_{a\alpha b\beta}$,

$$\langle lm|LM|l'm'\rangle = \int \mathrm{d}\Omega\, Y_{lm}{}^* Y_{LM} Y_{l'm'} = \sqrt{\frac{(2L+1)(2l'+1)}{4\pi(2l+1)}}\, C^{l0}_{l'0L0} C^{lm}_{l'm'LM}.$$

One finds

$$\begin{aligned} i\partial_t\phi_{lm} &= \left(-\frac{1}{2}\frac{\partial^2}{\partial r^2} + V_{\mathrm{eff}\,l}\right)\phi_{lm} + \frac{i}{2}\sum_{l'm'}\left\{\tilde{A}^*\delta_{m,m'+1}\delta_{l,l'+1}\sqrt{\frac{l+m}{(2l+1)(2l-1)}}\right. \\ &\quad \times\left[\left(\partial_r - \frac{m}{r}\right)\sqrt{l+m-1} - \frac{1}{r}\sqrt{(l-m)(l(l-1)-m(m-1))}\right] \\ &\quad + \tilde{A}^*\delta_{m,m'+1}\delta_{l,l'-1}\sqrt{\frac{l-m+1}{(2l+1)(2l+3)}} \\ &\quad \times\left[-\left(\partial_r - \frac{m}{r}\right)\sqrt{l-m+2} - \frac{1}{r}\sqrt{(l+m+1)((l+1)(l+2)-m(m-1))}\right] \\ &\quad + \tilde{A}\delta_{m,m'-1}\delta_{l,l'+1}\sqrt{\frac{l-m}{(2l+1)(2l-1)}} \\ &\quad \times\left[-\left(\partial_r + \frac{m}{r}\right)\sqrt{l-m-1} + \frac{1}{r}\sqrt{(l+m)(l(l-1)-m(m+1))}\right] \\ &\quad + \tilde{A}\delta_{m,m'-1}\delta_{l,l'-1}\sqrt{\frac{l+m+1}{(2l+1)(2l+3)}} \\ &\quad \left.\times\left[\left(\partial_r + \frac{m}{r}\right)\sqrt{l+m+2} + \frac{1}{r}\sqrt{(l-m+1)((l+1)(l+2)-m(m+1))}\right]\right\}\phi_{l'm'}. \end{aligned} \tag{137}$$

The Hamiltonian has again the structure

$$\hat{H} = \hat{H}_{\text{at}} + \hat{H}_{\text{mix}} + \hat{H}_{\text{ang}}, \tag{138}$$

with $\hat{H}_{\text{at}}$ being diagonal in lm space, $\hat{H}_{\text{ang}}$ diagonal in r, and the mixing part $\hat{H}_{\text{mix}}$. The matrix components in lm space[10] read as

$$\left[\hat{H}_{\text{at}}\right]_{lm}^{l'm'} = \delta_{l,l'}\delta_{m,m'}\left(-\frac{1}{2}\frac{\partial^2}{\partial r^2} + V_{\text{eff}\,l}\right), \tag{139}$$

$$\begin{aligned}\left[\hat{H}_{\text{ang}}\right]_{lm}^{l'm'} = &\Big(\mathcal{A}_{lm}\delta_{m,m'+1}\delta_{l,l'+1} + \mathcal{B}_{lm}\delta_{m,m'+1}\delta_{l,l'-1} \\ &+ \tilde{\mathcal{A}}_{lm}\delta_{m,m'-1}\delta_{l,l'+1} + \tilde{\mathcal{B}}_{lm}\delta_{m,m'-1}\delta_{l,l'-1}\Big),\end{aligned} \tag{140}$$

$$\begin{aligned}\left[\hat{H}_{\text{mix}}\right]_{lm}^{l'm'} = &\Big(\mathcal{C}_{lm}\delta_{m,m'+1}\delta_{l,l'+1} + \mathcal{D}_{lm}\delta_{m,m'+1}\delta_{l,l'-1} \\ &+ \tilde{\mathcal{C}}_{lm}\delta_{m,m'-1}\delta_{l,l'+1} + \tilde{\mathcal{D}}_{lm}\delta_{m,m'-1}\delta_{l,l'-1}\Big)\,\partial_r\end{aligned} \tag{141}$$

with

$$\mathcal{A}_{lm} = \tilde{\mathcal{A}}^\star_{l-m} = \frac{i\tilde{A}^\star}{2r}a_{lm}, \qquad \tilde{\mathcal{A}}_{lm} = \mathcal{A}^\star_{l-m} = \frac{i\tilde{A}}{2r}\tilde{a}_{lm},$$

$$\mathcal{B}_{lm} = \tilde{\mathcal{B}}^\star_{l-m} = \frac{i\tilde{A}^\star}{2r}b_{lm}, \qquad \tilde{\mathcal{B}}_{lm} = \mathcal{B}^\star_{l-m} = \frac{i\tilde{A}}{2r}\tilde{b}_{lm},$$

$$\mathcal{C}_{lm} = \tilde{\mathcal{C}}^\star_{l-m} = \frac{i\tilde{A}^\star}{2}c_{lm}, \qquad \tilde{\mathcal{C}}_{lm} = \mathcal{C}^\star_{l-m} = -\frac{i\tilde{A}}{2}\tilde{c}_{lm},$$

$$\mathcal{D}_{lm} = \tilde{\mathcal{D}}^\star_{l-m} = -\frac{i\tilde{A}^\star}{2}d_{lm}, \qquad \tilde{\mathcal{D}}_{lm} = \mathcal{D}^\star_{l-m} = \frac{i\tilde{A}}{2}\tilde{d}_{lm},$$

and

$$a_{lm} = \sqrt{\frac{l+m}{(2l+1)(2l-1)}}\left[-m\sqrt{l+m-1} - \sqrt{(l-m)(l(l-1)-m(m-1))}\right],$$

$$\tilde{a}_{lm} = \sqrt{\frac{l-m}{(2l+1)(2l-1)}}\left[-m\sqrt{l-m-1} + \sqrt{(l+m)(l(l-1)-m(m+1))}\right],$$

$$\begin{aligned}b_{lm} &= \sqrt{\frac{l-m+1}{(2l+1)(2l+3)}}\left[m\sqrt{l-m+2} - \sqrt{(l+m+1)((l+1)(l+2)-m(m-1))}\right] \\ &= -\tilde{a}_{l+1,m-1},\end{aligned}$$

$$\begin{aligned}\tilde{b}_{lm} &= \sqrt{\frac{l+m+1}{(2l+1)(2l+3)}}\left[m\sqrt{l+m+2} + \sqrt{(l-m+1)((l+1)(l+2)-m(m+1))}\right] \\ &= -a_{l+1,m+1},\end{aligned}$$

$$c_{lm} = \sqrt{\frac{(l+m)(l+m-1)}{(2l+1)(2l-1)}} = \tilde{c}_{l,-m} = \tilde{d}_{l-1,m-1},$$

10 Let us keep r nondiscretized for the moment.

$$\tilde{c}_{lm} = \sqrt{\frac{(l-m)(l-m-1)}{(2l+1)(2l-1)}} = c_{l,-m} = d_{l-1,m+1},$$

$$d_{lm} = \sqrt{\frac{(l-m+1)(l-m+2)}{(2l+1)(2l+3)}} = \tilde{d}_{l,-m},$$

$$\tilde{d}_{lm} = \sqrt{\frac{(l+m+1)(l+m+2)}{(2l+1)(2l+3)}} = d_{l,-m}.$$

The wavefunction (for one particular, fixed r) may be represented as a vector

$$\phi = \left(\boxed{\phi_{00}}, \boxed{\phi_{1-1}, \phi_{10}, \phi_{11}}, \boxed{\phi_{2-2}, \ldots, \phi_{21}, \phi_{22}}, \ldots, \boxed{\ldots, \phi_{N_l N_l -1}, \phi_{N_l N_l}} \right)^{\top},$$

where the l subblocks are indicated with boxes.[11] Both $\hat{H}_{\text{ang}}$ and $\hat{H}_{\text{mix}}$ may be written as a sum over 2×2 matrices acting in lm subspace,

$$\hat{H}_{\text{ang}} = \sum_{l=0}^{N_l-2} \sum_{m=-l}^{l} \left(\mathbf{H}_{\text{ang}}^{lm} + \tilde{\mathbf{H}}_{\text{ang}}^{lm} \right), \quad \hat{H}_{\text{mix}} = \sum_{l=0}^{N_l-2} \sum_{m=-l}^{l} \left(\mathbf{H}_{\text{mix}}^{lm} + \tilde{\mathbf{H}}_{\text{mix}}^{lm} \right), \tag{142}$$

with[12]

$$\mathbf{H}_{\text{ang}}^{lm} = \frac{i|\tilde{A}|}{2r} \mathbf{P}^{lm}, \quad \mathbf{P}^{lm} = b_{lm} \left(\begin{array}{c|cc} & lm & l+1, m-1 \\ \hline lm & 0 & \exp(-i\eta) \\ l+1, m-1 & -\exp(i\eta) & 0 \end{array} \right),$$

$$\tilde{\mathbf{H}}_{\text{ang}}^{lm} = \frac{i|\tilde{A}|}{2r} \tilde{\mathbf{P}}^{lm}, \quad \tilde{\mathbf{P}}^{lm} = \tilde{b}_{lm} \left(\begin{array}{c|cc} & lm & l+1, m+1 \\ \hline lm & 0 & \exp(i\eta) \\ l+1, m+1 & -\exp(-i\eta) & 0 \end{array} \right),$$

$$\mathbf{H}_{\text{mix}}^{lm} = -\frac{i|\tilde{A}|}{2} \mathbf{L}^{lm} \partial_r, \quad \mathbf{L}^{lm} = d_{lm} \left(\begin{array}{c|cc} & lm & l+1, m-1 \\ \hline lm & 0 & \exp(-i\eta) \\ l+1, m-1 & \exp(i\eta) & 0 \end{array} \right),$$

$$\tilde{\mathbf{H}}_{\text{mix}}^{lm} = \frac{i|\tilde{A}|}{2} \tilde{\mathbf{L}}^{lm} \partial_r, \quad \tilde{\mathbf{L}}^{lm} = \tilde{d}_{lm} \left(\begin{array}{c|cc} & lm & l+1, m+1 \\ \hline lm & 0 & \exp(i\eta) \\ l+1, m+1 & \exp(-i\eta) & 0 \end{array} \right).$$

The phase η is defined through

$$\tilde{A} = |\tilde{A}| \exp(i\eta).$$

---

11 Since $\hat{H}_{\text{ang}}$ is diagonal in r space, there is no need to indicate the value of r. $\hat{H}_{\text{ang}}$ simply must be applied to each r subblock.

12 The first column and row in the matrices indicate the l and m indices of the ϕ_{lm} components acted on.

A Crank–Nicolson propagator that advances the wavefunction over Δt is chosen as

$$\mathbf{U}(\Delta t) = \left\{ \prod_{l=0}^{N_l-2} \prod_{m=-l}^{l} \mathbf{R}^{lm} \left(\mathbf{X}_+^{lm}\right)^{-1} \mathbf{X}_-^{lm} \tilde{\mathbf{R}}^{lm} \left(\tilde{\mathbf{X}}_+^{lm}\right)^{-1} \tilde{\mathbf{X}}_-^{lm} \right\}$$
$$\times \mathbf{Q}_+^{-1}\mathbf{Q}_- \left\{ \prod_{l=0}^{N_l-2} \prod_{m=-l}^{l} \left(\tilde{\mathbf{X}}_+^{lm}\right)^{-1} \tilde{\mathbf{X}}_-^{lm} \tilde{\mathbf{R}}^{lm} \left[\mathbf{X}_+^{lm}\right]^{-1} \mathbf{X}_-^{lm} \mathbf{R}^{lm} \right\}.$$

Similar to the simpler case of linear polarization, we rewrite the mixing part

$$\left(\mathbf{X}_+^{lm}\right)^{-1} \mathbf{X}_-^{lm} = \left(\mathbf{Y}_+^{lm}\right)^{-1} \mathbf{Y}_-^{lm}, \qquad \mathbf{Y}_\pm^{lm} = \mathcal{M}_1 \pm \frac{\Delta t|\tilde{A}|}{8} \mathbf{L}^{lm} \otimes \mathcal{D}_1,$$
$$\left(\tilde{\mathbf{X}}_+^{lm}\right)^{-1} \tilde{\mathbf{X}}_-^{lm} = \left(\tilde{\mathbf{Y}}_+^{lm}\right)^{-1} \tilde{\mathbf{Y}}_-^{lm}, \qquad \tilde{\mathbf{Y}}_\pm^{lm} = \mathcal{M}_1 \mp \frac{\Delta t|\tilde{A}|}{8} \tilde{\mathbf{L}}^{lm} \otimes \mathcal{D}_1,$$

the atomic part,[13]

$$\mathbf{Q}_+^{-1}\mathbf{Q}_- = \mathbf{W}_+^{-1}\mathbf{W}_-, \qquad \mathbf{W}_\pm = \mathcal{M}_2 \pm i\frac{\Delta t}{2}(\mathcal{D}_2 + \mathcal{M}_2 \mathbf{V}_l),$$

and the purely angular part

$$\mathbf{R}^{lm} = (1 - \xi \mathbf{P}^{lm})^{-1}(1 + \xi \mathbf{P}^{lm}), \qquad \xi = \frac{\Delta t|\tilde{A}|}{8r}$$
$$= \frac{1}{1+\xi^2 b_{lm}^2} \begin{pmatrix} 1-\xi^2 b_{lm}^2 & 2\xi \exp(-i\eta) b_{lm} \\ -2\xi \exp(i\eta) b_{lm} & 1-\xi^2 b_{lm}^2 \end{pmatrix},$$
$$\tilde{\mathbf{R}}^{lm} = \frac{1}{1+\xi^2 \tilde{b}_{lm}^2} \begin{pmatrix} 1-\xi^2 \tilde{b}_{lm}^2 & 2\xi \exp(i\eta) \tilde{b}_{lm} \\ -2\xi \exp(-i\eta) \tilde{b}_{lm} & 1-\xi^2 \tilde{b}_{lm}^2 \end{pmatrix}.$$

Next is the transformation of the mixing part,

$$\mathbf{Y}_\pm^{lm} = \left(\mathcal{M}_1 \pm \zeta \mathbf{L}^{lm} \mathcal{D}_1\right), \qquad \tilde{\mathbf{Y}}_\pm^{lm} = \left(\mathcal{M}_1 \mp \zeta \tilde{\mathbf{L}}^{lm} \mathcal{D}_1\right),$$

where $\zeta = \Delta t|\tilde{A}|/8$. We note that with the unitary matrices

$$\mathbf{B} = \frac{1}{\sqrt{2}} \begin{pmatrix} -\exp(i\eta) & 1 \\ \exp(i\eta) & 1 \end{pmatrix}, \qquad \tilde{\mathbf{B}} = \frac{1}{\sqrt{2}} \begin{pmatrix} -\exp(-i\eta) & 1 \\ \exp(-i\eta) & 1 \end{pmatrix},$$

one has

$$\mathbf{B}\mathbf{L}^{lm}\mathbf{B}^\dagger = \tilde{\mathbf{B}}\tilde{\mathbf{L}}^{lm}\tilde{\mathbf{B}}^\dagger = \begin{pmatrix} -1 & 0 \\ 0 & 1 \end{pmatrix}$$

13 We suppress the explicit indication of unities in subspaces here, such as $\mathbf{1}_{lm}$ or $\mathbf{1}_r$.

and thus

$$\left(\mathbf{Y}_{+}^{lm}\right)^{-1}\mathbf{Y}_{-}^{lm}=\mathbf{B}^{\dagger}\mathbf{B}\left(\mathbf{Y}_{+}^{lm}\right)^{-1}\mathbf{B}^{\dagger}\mathbf{B}\mathbf{Y}_{-}^{lm}\mathbf{B}^{\dagger}\mathbf{B}=\mathbf{B}^{\dagger}\left(\mathcal{Y}_{+}^{lm}\right)^{-1}\mathcal{Y}_{-}^{lm}\mathbf{B}$$

with

$$\mathcal{Y}_{\pm}^{lm}=\mathbf{B}\mathbf{Y}_{\pm}^{lm}\mathbf{B}^{\dagger}=\mathcal{M}_1\pm\zeta d_{lm}\begin{pmatrix}-1 & 0\\ 0 & 1\end{pmatrix}\mathcal{D}_1, \tag{143}$$

and, analogously,

$$\left(\tilde{\mathbf{Y}}_{+}^{lm}\right)^{-1}\tilde{\mathbf{Y}}_{-}^{lm}=\tilde{\mathbf{B}}^{\dagger}\left(\tilde{\mathcal{Y}}_{+}^{lm}\right)^{-1}\tilde{\mathcal{Y}}_{-}^{lm}\tilde{\mathbf{B}}$$

with

$$\tilde{\mathcal{Y}}_{\pm}^{lm}=\tilde{\mathbf{B}}\tilde{\mathbf{Y}}_{\pm}^{lm}\tilde{\mathbf{B}}^{\dagger}=\mathcal{M}_1\mp\zeta\tilde{d}_{lm}\begin{pmatrix}-1 & 0\\ 0 & 1\end{pmatrix}\mathcal{D}_1. \tag{144}$$

As for linear polarization, we successfully diagonalized the mixing part in angular-momentum space so that simple forward-backward substitution is applicable with respect to r, as $\mathcal{M}_1$ and $\mathcal{D}_1$ are tridiagonal.

The propagator finally reads as

$$\begin{aligned}\mathbf{U}(\Delta t)=&\left\{\prod_{l=0}^{N_l-2}\prod_{m=-l}^{l}\mathbf{R}^{lm}\mathbf{B}^{\dagger}\left(\mathcal{Y}_{+}^{lm}\right)^{-1}\mathcal{Y}_{-}^{lm}\mathbf{B}\tilde{\mathbf{R}}^{lm}\tilde{\mathbf{B}}^{\dagger}\left(\tilde{\mathcal{Y}}_{+}^{lm}\right)^{-1}\tilde{\mathcal{Y}}_{-}^{lm}\tilde{\mathbf{B}}\right\}\\ &\times\mathbf{W}_{+}^{-1}\mathbf{W}_{-}\left\{\prod_{l=0}^{N_l-2}\prod_{m=-l}^{l}\tilde{\mathbf{B}}^{\dagger}\left(\tilde{\mathcal{Y}}_{+}^{lm}\right)^{-1}\tilde{\mathcal{Y}}_{-}^{lm}\tilde{\mathbf{B}}\tilde{\mathbf{R}}^{lm}\mathbf{B}^{\dagger}\left(\mathcal{Y}_{+}^{lm}\right)^{-1}\mathcal{Y}_{-}^{lm}\mathbf{B}\mathbf{R}^{lm}\right\},\end{aligned} \tag{145}$$

having the same structure as (125) but additional factors (those with tilde), modified 2×2 matrices $\mathbf{B}$, $\mathbf{R}$, $\tilde{\mathbf{B}}$, $\tilde{\mathbf{R}}$, and the products run also over the magnetic quantum number m now. As a consequence, the advancement of the wavefunction by Δt using propagator (145) scales $\sim N_l^2$ for elliptical polarization instead of $\sim N_l$ in the case of linear polarization using (125).

In the strong-field TDSE solver `Qprop` [1], propagator (145) is implemented as well.

2 Scaled cylindrical coordinates

Among the infinitely many coordinate choices that might be appropriate for the infinitely many single-active-electron TDSE problems one can think of, we present one more as an example. Consider the case of an azimuthally symmetric system such as H_2^+ with fixed nuclei. The binding potential for the electron does not depend on the azimuthal angle φ and reads as

$$V(\rho, z) = -\frac{1}{\sqrt{\rho^2 + (z - R/2)^2}} - \frac{1}{\sqrt{\rho^2 + (z + R/2)^2}}, \quad \rho = \sqrt{x^2 + y^2}. \tag{146}$$

Here, R is the internuclear distance, and ρ is the radial coordinate of the cylindrical coordinates ρ, φ, z. Assuming that the molecular ion is aligned to the linearly polarized laser field, we have in velocity gauge and dipole approximation the TDSE

$$i\frac{\partial}{\partial t}\Psi(\rho, z, \varphi, t) = \hat{H}_{\rho z \varphi}(t)\Psi(\rho, z, \varphi, t) \tag{147}$$

with

$$\hat{H}_{\rho z \varphi}(t) = -\frac{1}{2}\left(\frac{1}{\rho}\frac{\partial}{\partial \rho}\rho\frac{\partial}{\partial \rho} + \frac{\partial^2}{\partial z^2} + \frac{1}{\rho^2}\frac{\partial^2}{\partial \varphi^2}\right) - iA(t)\frac{\partial}{\partial z} + V(\rho, z). \tag{148}$$

The wavefunction is normalized such that

$$\iiint \mathrm{d}z\,\mathrm{d}\varphi\,\mathrm{d}\rho\;\rho|\Psi(\rho, z, \varphi, t)|^2 = 1. \tag{149}$$

The azimuthal angle φ is a cyclic variable, and $\Psi(\rho, z, \varphi, t)$ may be factorized as

$$\Psi(\rho, z, \varphi, t) = \frac{1}{\sqrt{2\pi}}\psi(\rho, z, t)\exp(im\varphi) \tag{150}$$

with constant azimuthal quantum number m. The TDSE for $\psi(\rho, z, t)$ reads as

$$i\frac{\partial}{\partial t}\psi(\rho, z, t) = \hat{H}_{\rho z m}(t)\psi(\rho, z, t) \tag{151}$$

with

$$\hat{H}_{\rho z m}(t) = -\frac{1}{2}\frac{\partial^2}{\partial z^2} - iA(t)\frac{\partial}{\partial z} - \frac{1}{2\rho}\frac{\partial}{\partial \rho}\rho\frac{\partial}{\partial \rho} + V_m(\rho, z), \tag{152}$$

where

$$V_m(\rho, z) = V(\rho, z) + \frac{m^2}{2\rho^2}, \tag{153}$$

and

$$\iint \mathrm{d}z\,\mathrm{d}\rho\;\rho|\psi(\rho, z, t)|^2 = 1. \tag{154}$$

We observe that $\hat{H}_{\rho z m}(t)$ has the structure $\hat{H}_{\rho z m}(t) = \hat{K}_z(t) + \hat{K}_\rho + V_m(\rho, z)$ so that, e.g., the Peaceman–Rachford scheme in Section 1.4.1 could be directly applied to $\hat{H}_x = \hat{K}_z + V_m/2$ and $\hat{H}_y = \hat{K}_\rho + V_m/2$ discretized on a numerical grid $\rho_j = j\Delta\rho$, $j = 1, 2, \ldots N_\rho$, $z_i = [i - (N_z - 1)/2]\Delta z$, $i = 0, 1, \ldots N_z - 1$. However, because of the behavior of the wavefunction for $\rho \to 0$, this is not the smartest coordinate choice.

Following Kono [7], we allow for a more general radial coordinate ξ,

$$\rho = \xi^\lambda, \qquad \lambda > \frac{1}{2}. \tag{155}$$

With the substitution

$$\Phi(\xi, z, t) = \sqrt{\lambda}\,\xi^{\lambda-1/2}\,\psi(\xi, z, t), \tag{156}$$

we obtain the TDSE

$$i\frac{\partial}{\partial t}\Phi(\xi, z, t) = \hat{H}^{(\lambda)}_{\xi z m}(t)\Phi(\xi, z, t), \tag{157}$$

where

$$\hat{H}^{(\lambda)}_{\xi z m}(t) = \hat{K}_z(t) + \hat{K}^{(\lambda)}_{\xi} + V^{(\lambda)}_m(\xi, z) \tag{158}$$

with

$$\hat{K}_z(t) = -\frac{1}{2}\frac{\partial^2}{\partial z^2} - iA(t)\frac{\partial}{\partial z}, \tag{159}$$

$$\hat{K}^{(\lambda)}_{\xi} = -\frac{1}{2\lambda^2\xi^{2\lambda}}\left[\xi^2\frac{\partial^2}{\partial \xi^2} - 2(\lambda-1)\xi\frac{\partial}{\partial \xi}\right], \tag{160}$$

$$V^{(\lambda)}_m(\xi, z) = V(\xi, z) + \frac{m^2}{2\xi^{2\lambda}} - \frac{1}{2\lambda^2\xi^{2\lambda}}\left(\lambda - \frac{1}{2}\right)^2, \tag{161}$$

and

$$\iint \mathrm{d}\xi\,\mathrm{d}z\,|\Phi(\xi, z, t)|^2 = 1. \tag{162}$$

Because $\lambda > 1/2$, substitution (156) ensures that $\Phi(\xi \to 0) = 0$ so that $\xi = 0$ does not need to be part of the numerical grid $\xi_j = j\Delta\xi$, $j = 1, 2, \ldots N_\xi$, $z_i = [i - (N_z - 1)/2]\Delta z$, $i = 0, 1, \ldots N_z - 1$, and $V^{(\lambda)}_m(\xi, z)$ stays nonsingular. The Peaceman–Rachford scheme in Section 1.4.1 can be applied with $\hat{H}_x = \hat{K}_z + V^{(\lambda)}_m/2$ and $\hat{H}_y = \hat{K}^{(\lambda)}_{\xi} + V^{(\lambda)}_m/2$.

Depending on the discretization of the derivatives and the external potential $V(\xi, z)$, there is an optimal choice of λ in the sense that, e.g., the most accurate ground-state energy is obtained for given $\Delta\xi$, Δz and number of grid points N_ξ, N_z. For Coulomb potentials like in the H_2^+ case (146) and standard 3-pt approximations for the derivatives,[14] $\lambda = 3/2$ turns out to be optimal [7, 10]. From (155) follows that uniform gridspacing $\Delta\xi$ in ξ corresponds to a nonuniform spacing in ρ, with increasing $\Delta\rho$ as ρ increases. This is a positive side effect as long as $\pi/\Delta\rho$ is large enough to allow the wavefunction to spread properly. In practice, the wavefunction is usually absorbed anyway at small enough distances so that the coarseness of the radial grid is not an issue. The large excursions occur in (equidistantly discretized) z direction.

3 Employing second-quantization notion

There should be a connection between the lattice models extensively used in condensed matter theory (Ising, Heisenberg, Hubbard, etc.) and the computational lattice

14 Note, however, that a straight-forwardly discretized $\hat{K}^{(\lambda)}_{\xi}$ will, in general, not be exactly Hermitian.

introduced just for discretization, i.e., born out of necessity. A beautiful method for solving the TDSE that highlights this analogy and employs a second-quantization–like terminology is based on the operator splitting introduced in Section 1.4.2 and "grid hopping" [3, 4].[15] One may object that as long as we solve problems with fixed particle numbers, second quantization is unnecessary. This is true. Yet, technically, creation and annihilation operators can be used instead of the matrices that arise from the spatial discretization of derivatives. This may help devise clever splitting schemes and algorithms, as we will see in the following.

Before we discuss the actual grid hopping algorithm, a brief reminder about second-quantization is indicated. Consider the general, second-quantized Hamiltonian for arbitrary many particles in an external potential V, interacting pairwise via the potential W,

$$\hat{H} = \int \mathrm{d}x\, \hat{\Psi}^\dagger(x) \left(-\frac{1}{2}\frac{\partial^2}{\partial x^2} + V(x)\right) \hat{\Psi}(x) + \frac{1}{2}\int \mathrm{d}x \int \mathrm{d}x'\, \hat{\Psi}^\dagger(x)\hat{\Psi}^\dagger(x')W(|x-x'|)\hat{\Psi}(x')\hat{\Psi}(x). \tag{163}$$

The field operators $\hat{\Psi}$, $\hat{\Psi}^\dagger$ annihilate and create particles at x and fulfill

$$[\hat{\Psi}(x), \hat{\Psi}^\dagger(x')]_\mp = \delta(x-x'), \tag{164}$$

where the upper sign (commutator) applies for Bosons and the lower (anticommutator) for Fermions.[16]

In the Heisenberg picture, the Heisenberg equation of motion for, e.g., the annihilation field operator $\hat{\Psi}(x)$ reads as (for both Bosons and Fermions) $\dot{\hat{\Psi}} = i[\hat{H}, \hat{\Psi}]_-$, from which follows

$$i\frac{\partial\hat{\Psi}(x,t)}{\partial t} = \left(-\frac{1}{2}\frac{\partial^2}{\partial x^2} + V(x) + \int \mathrm{d}x'\, \hat{\Psi}^\dagger(x',t)W(|x-x'|)\hat{\Psi}(x',t)\right) \hat{\Psi}(x,t) \tag{165}$$

(also for both Bosons and Fermions).

The advantage of second quantization is that the number of particles in the system is open. However, it is instructive to formally recover the single-particle (first-quantized) TDSE by applying equation (165) to a state $|\Phi\rangle$ (that is constant in the Heisenberg picture) and multiplying from the left by the vacuum $\langle 0|$. Because

$$\langle 0|\, \hat{\Psi}(x,t)\, |\Phi\rangle = \langle \hat{\Psi}^\dagger(x,t) 0|\Phi\rangle = \langle x(t)|\Phi\rangle = \Phi(x,t) \tag{166}$$

(note that, in the Heisenberg picture, the eigenstates $|x(t)\rangle = \hat{\Psi}^\dagger(x,t)\,|0\rangle$ of the operator $\hat{x}(t)$ are time dependent) and $\langle 0|\, \hat{\Psi}^\dagger(x',t)\, |\Phi\rangle = \langle \hat{\Psi}(x',t) 0|\Phi\rangle = 0$, we indeed obtain the

15 The authors of [3, 4] do not call it "grid hopping" though.

16 x may comprise spatial coordinates and spin, e.g., $x = (\mathbf{r}, m_s)$. In that case, $\delta(x-x') = \delta^3(\mathbf{r}-\mathbf{r}')\,\delta_{m_s m_s'}$.

TDSE

$$i\frac{\partial \Phi(x,t)}{\partial t} = \left(-\frac{1}{2}\frac{\partial^2}{\partial x^2} + V(x)\right)\Phi(x,t). \tag{167}$$

In passing, we note that in the Schrödinger picture, where $|\Phi(t)\rangle$ is time dependent and the operators are stationary, equation (166) reads as $\langle 0|\hat{\Psi}(x)|\Phi(t)\rangle = \langle x|\Phi(t)\rangle = \Phi(x,t)$, giving the same time-dependent wavefunction. Using this, one can alternatively derive the one-body TDSE (167) from $i|\dot{\Phi}(t)\rangle = \hat{H}|\Phi(t)\rangle$ with the second-quantized Hamiltonian (163), multiplying from the left by $\langle x'| = \langle 0|\hat{\Psi}(x')$.

Continuing in the Heisenberg picture, from (165) also follows

$$\begin{aligned} i\frac{\partial\left(\hat{\Psi}(y,t)\hat{\Psi}(x,t)\right)}{\partial t} &= \hat{\Psi}(y,t)\left(-\frac{1}{2}\frac{\partial^2}{\partial x^2} + V(x) + \int \mathrm{d}x'\, \hat{\Psi}^\dagger(x',t)W(|x-x'|)\hat{\Psi}(x',t)\right)\hat{\Psi}(x,t) \\ &+ \left(-\frac{1}{2}\frac{\partial^2}{\partial y^2} + V(y) + \int \mathrm{d}x'\, \hat{\Psi}^\dagger(x',t)W(|y-x'|)\hat{\Psi}(x',t)\right)\hat{\Psi}(y,t)\hat{\Psi}(x,t). \end{aligned}$$

Applying again $|\Phi\rangle$ from the right and $\langle 0|$ from the left, the two-body TDSE

$$i\frac{\partial \Phi^\pm(x,y,t)}{\partial t} = \left[-\frac{1}{2}\left(\frac{\partial^2}{\partial x^2} + \frac{\partial^2}{\partial y^2}\right) + V(x) + V(y) + W(|x-y|)\right]\Phi^\pm(x,y,t) \tag{168}$$

is derived, where $\langle 0|\hat{\Psi}(y,t)\hat{\Psi}(x,t)|\Phi\rangle = {}^\pm\langle x(t)y(t)|\Phi\rangle = \Phi^\pm(x,y,t)$ is a symmetric (+) or antisymmetric (−) two-body wavefunction. The same two-body TDSE can be derived using a different Hamiltonian,

$$\hat{H}' = \int \mathrm{d}x \int \mathrm{d}y\, \hat{\Psi}^\dagger(x,y)\left(-\frac{1}{2}\frac{\partial^2}{\partial x^2} - \frac{1}{2}\frac{\partial^2}{\partial y^2} + \mathcal{V}(x,y)\right)\hat{\Psi}(x,y) \tag{169}$$

with the one-body potential

$$\mathcal{V}(x,y) = V(x) + V(y) + W(|x-y|),$$

where we hazard the consequences of twice as many degrees of freedom for getting rid of the interaction. With that Hamiltonian, TDSE (168) is derived as a one-body TDSE, i.e., analogously to the derivation of (167). However, since the particle statistics is irrelevant for a one-body TDSE, the (anti-) symmetry of the wavefunction has to be enforced "manually" if, e.g., a 2D TDSE is used to simulate two 1D identical particles.

3.1 Grid hopping

To grasp the idea behind the grid-hopping method, it is sufficient to outline it for just one spatial dimension without interaction. The extension to more dimensions (or interaction), spin, etc. is straightforward after the preliminary notes. Hence, let us consider a second-quantized Hamiltonian

$$\hat{H} = \int \mathrm{d}x\, \hat{\Psi}^{\dagger}(x) \left(-\frac{1}{2}\frac{\partial^2}{\partial x^2} + V(x,t) - iA(t)\frac{\partial}{\partial x} \right) \hat{\Psi}(x).$$

In order to compete with the accuracy of our Numerov-improved Crank–Nicolson approach in Section 1.2.3, we boldly discretize up to order Δx^4 (time arguments suppressed again),

$$\hat{H} = \sum_{s=0}^{N_x-1} \hat{c}_s^{\dagger} \left(-\frac{1}{2}\frac{-\hat{c}_{s+2} + 16\hat{c}_{s+1} - 30\hat{c}_s + 16\hat{c}_{s-1} - \hat{c}_{s-2}}{12\Delta x^2} + V_s \hat{c}_s \right.$$
$$\left. -iA\frac{-\hat{c}_{s+2} + 8\hat{c}_{s+1} - 8\hat{c}_{s-1} + \hat{c}_{s-2}}{12\Delta x} \right).$$

The operators $\hat{c}_s$, $\hat{c}_s^{\dagger}$ are the discretized versions of $\hat{\Psi}(x)$ and $\hat{\Psi}^{\dagger}(x)$; they annihilate and create a particle at lattice site s. Collecting terms and shifting indices, we can write $\hat{H}$ as

$$\hat{H} = \hat{H}_0 + \hat{H}_1 + \hat{H}_2 \tag{170}$$

with

$$\hat{H}_0 = \frac{1}{\Delta x^2} \sum_{s=0}^{N_x-1} \left(\frac{5}{4} + V_s \Delta x^2 \right) \hat{c}_s^{\dagger} \hat{c}_s, \tag{171}$$

$$\hat{H}_1 = -\frac{2}{3\Delta x^2} \sum_{s=0}^{N_x-1} \left\{ [1 + iA\Delta x]\, \hat{c}_s^{\dagger} \hat{c}_{s+1} + [1 - iA\Delta x]\, \hat{c}_{s+1}^{\dagger} \hat{c}_s \right\}, \tag{172}$$

$$\hat{H}_2 = \frac{1}{12\Delta x^2} \sum_{s=0}^{N_x-1} \left\{ \left[\frac{1}{2} + iA\Delta x \right] \hat{c}_s^{\dagger} \hat{c}_{s+2} + \left[\frac{1}{2} - iA\Delta x \right] \hat{c}_{s+2}^{\dagger} \hat{c}_s \right\}. \tag{173}$$

This looks like a Hamiltonian for a lattice system with an on-site energy $\sim \hat{c}_s^{\dagger}\hat{c}_s$ in $\hat{H}_0$ and hopping terms between next-neighbor ($\hat{H}_1$) and second–next-neighbor ($\hat{H}_2$) sites due to the kinetic energy and, in our case, the coupling to an external field $A(t)$.[17]

For a fixed s, hopping terms annihilate at site s and create at site $s+1$ (or $s+2$), and vice versa. In other words, such a hopping term acts like a 2×2 matrix. Note that

17 If we quantized the electromagnetic field as well, we would have another type of (Bosonic) operators $\hat{b}_k$, $\hat{b}_k^{\dagger}$ that annihilate and create photons in mode k. However, quantization of the electromagnetic field is usually not necessary in strong-field physics as long as the electromagnetic field is well described by a coherent state with large photon expectation number, and spontaneous emission is not important for the process under consideration.

we can write

$$\hat{H}_1(t) = \hat{H}_{11}(t) + \hat{H}_{12}(t), \qquad \hat{H}_{1i}(t) = \sum_{s \in S_{1i}} \{\cdots\}, \tag{174}$$

where $i = 1, 2$, the two terms within the curly bracket are the same as in (172), and, e.g.,

$$S_{11} = \{0, 2, 4, \ldots\}, \qquad S_{12} = \{1, 3, 5, \ldots\}.$$

Analogously,

$$\hat{H}_2(t) = \hat{H}_{21}(t) + \hat{H}_{22}(t), \qquad \hat{H}_{2i}(t) = \sum_{s \in S_{2i}} \{\cdots\} \tag{175}$$

with, e.g.,

$$S_{21} = \{0, 1, 4, 5, 8, 9, \ldots\}, \qquad S_{22} = \{2, 3, 6, 7, 10, 11, \ldots\}.$$

Approximating the propagator with the help of some splitting scheme, e.g.,

$$\begin{aligned} \hat{U}(\Delta t) = {} & e^{-i\Delta t/2\hat{H}_{22}} e^{-i\Delta t/2\hat{H}_{21}} e^{-i\Delta t/2\hat{H}_{12}} e^{-i\Delta t/2\hat{H}_{11}} e^{-i\Delta t\hat{H}_0} \\ & \times e^{-i\Delta t/2\hat{H}_{11}} e^{-i\Delta t/2\hat{H}_{12}} e^{-i\Delta t/2\hat{H}_{21}} e^{-i\Delta t/2\hat{H}_{22}}, \end{aligned} \tag{176}$$

we notice that, e.g., all the summands in $\hat{H}_{11}(t)$ commute with each other. The same applies to the summands in $\hat{H}_{12}(t)$, $\hat{H}_{21}(t)$, and $\hat{H}_{22}(t)$. Hence, splitting further, e.g.,

$$\exp\left(-i\tau\hat{H}_{11}\right) = \prod_{s \in S_{11}} \exp\left\{-i\tau\left[\alpha(t)\hat{c}_s^\dagger\hat{c}_{s+1} + \alpha^*(t)\hat{c}_{s+1}^\dagger\hat{c}_s\right]\right\}, \tag{177}$$

with

$$\alpha(t) = -\frac{2}{3\Delta x^2}\left[1 + iA(t)\Delta x\right] \tag{178}$$

and $\tau = \Delta t/2$ does not introduce additional errors. Moreover, all the exponentials in (177) can be applied in parallel since they affect distinct lattice sites only. If the numerical representation of the state is of the form

$$\Phi = (\ldots, \Phi_s, \Phi_{s+1}, \ldots)^\top,$$

then $\alpha(t)\hat{c}_s^\dagger\hat{c}_{s+1} + \alpha^*(t)\hat{c}_{s+1}^\dagger\hat{c}_s$ acts like a 2×2 matrix,

$$\begin{pmatrix} 0 & \alpha \\ \alpha^* & 0 \end{pmatrix} \begin{pmatrix} \Phi_s \\ \Phi_{s+1} \end{pmatrix} = \begin{pmatrix} \alpha\Phi_{s+1} \\ \alpha^*\Phi_s \end{pmatrix}.$$

Applying the factorized propagator in (177) thus amounts to multiplications of the state components at the respective two lattice sites by unitary 2×2 matrices

$$\exp\left[-i\tau\begin{pmatrix} 0 & \alpha \\ \alpha^* & 0 \end{pmatrix}\right] = \begin{pmatrix} \cos\tau|\alpha| & -i\frac{|\alpha|}{\alpha^*}\sin\tau|\alpha| \\ -i\frac{|\alpha|}{\alpha}\sin\tau|\alpha| & \cos\tau|\alpha| \end{pmatrix}, \tag{179}$$

and similarly for $\hat{H}_2$. The Hamiltonian $\hat{H}_0$ is diagonal, i.e., the wavefunction component at lattice site s is simply multiplied by $\exp(-i\Delta t[5/(4\Delta x^2) + V_s])$.

4 Summary

As already emphasized in the introduction, this chapter on the propagation of wavefunctions according to the TDSE was very selective. However, the methods outlined may equip the reader to develop a powerful toolbox of self-made codes to tackle strong-field physics or TDSE problems in general. Alternatively, the chapter may help better understand how publicly available codes like `Qprop` work. We covered imaginary-time propagation, which is *the* method for finding low-lying eigenstates of the target system that are typically used as initial states for the real-time propagation with a TDSE solver. With regard to the propagation methods, the (Numerov-enhanced) Crank–Nicolson, the Feit–Fleck–Steiger, the Muller, and the grid-hopping approach were discussed. All approaches have in common that in more than 1D, Suzuki–Trotter or Peaceman–Rachford-like operator splitting is required to render the propagation efficient. While propagating wavefunctions is ethereal, artistic craftwork, the calculation of observables is earthly duty when it comes to the understanding[18] of experimental facts. This is the topic of the subsequent chapter.

Bibliography

[1] D. Bauer and P. Koval. Qprop: A Schrödinger-solver for intense laser–atom interaction. *Computer Physics Communications*, 174(5):396–421, 2006. www.qprop.de

[2] J. Crank and P. Nicolson. A practical method for numerical evaluation of solutions of partial differential equations of the heat-conduction type. *Advances in Computational Mathematics*, 6(1):207–226, 1996. Reprinted from *Proc. Camb. Phil. Soc.* 43 (1947), 50–67.

[3] H. De Raedt. Product formula algorithms for solving the time dependent Schrödinger equation. *Computer Physics Reports*, 7(1):1–72, 1987.

[4] H. De Raedt and K. Michielsen. Algorithm to solve the time-dependent Schrödinger equation for a charged particle in an inhomogeneous magnetic field: Application to the Aharonov–Bohm effect. *Computers in Physics*, 8(5):600–607, 1994.

[5] M. D. Feit, J. A. Fleck Jr., and A. Steiger. Solution of the Schrödinger equation by a spectral method. *Journal of Computational Physics*, 47:412–433, 1982.

[6] M. Frigo and S. G. Johnson. The design and implementation of FFTW3. *Proceedings of the IEEE*, 93(2):216–231, 2005. Special issue on "Program Generation, Optimization, and Platform Adaptation."

18 Not to be confused with "reproduction."

[7] H. Kono, A. Kita, Y. Ohtsuki, and Y. Fujimura. An efficient quantum mechanical method for the electronic dynamics of the three-dimensional hydrogen atom interacting with a linearly polarized strong laser pulse. *Journal of Computational Physics*, 130(1):148–159, 1997.
[8] S. K. Lele. Compact finite difference schemes with spectral-like resolution. *Journal of Computational Physics*, 103(1):16–42, 1992.
[9] C. Marante, L. Argenti, and F. Martín. Hybrid Gaussian–*B*-spline basis for the electronic continuum: Photoionization of atomic hydrogen. *Physical Review A*, 90:012506, 2014.
[10] V. Mosert and D. Bauer. Dissociative ionization of H_2^+: Few-cycle effect in the joint electron-ion energy spectrum. *Physical Review A*, 92:043414, 2015. 10.1103/PhysRevA.92.043414.
[11] H. G. Muller. An efficient propagation scheme for the time-dependent Schrödinger equation in the velocity gauge. *Laser Physics*, 9:138–148, 1999.
[12] D. W. Peaceman and H. H. Rachford, Jr. The numerical solution of parabolic and elliptic differential equations. *Journal of the Society for Industrial and Applied Mathematics*, 3(1): 28–41, 1955.
[13] R. M. Potvliege. STRFLO: A program for time-independent calculations of multiphoton processes in one-electron atomic systems. *Computer Physics Communications*, 114(1–3): 42–93, 1998.
[14] W. H. Press, S. A. Teukolsky, W. T. Vetterling, and B. P. Flannery. *Numerical Recipes 3rd Edition: The Art of Scientific Computing*. Cambridge University Press, New York, NY, 2007.
[15] J. Rapp and D. Bauer. Effects of inner electrons on atomic strong-field-ionization dynamics. *Physical Review A*, 89:033401, 2014.
[16] C. Ruiz, L. Plaja, and L. Roso. Lithium ionization by a strong laser field. *Physical Review Letters*, 94:063002, 2005.
[17] R. Santra. Why complex absorbing potentials work: A discrete-variable-representation perspective. *Physical Review A*, 74:034701, 2006.
[18] J. Sherman and W. J. Morrison. Adjustment of an inverse matrix corresponding to a change in one element of a given matrix. *Annals of Mathematical Statistics*, 21(1):124–127, 1950.
[19] M. Suzuki. Generalized Trotter's formula and systematic approximants of exponential operators and inner derivations with applications to many-body problems. *Communications in Mathematical Physics*, 51(2):183–190, 1976.
[20] H. Tal-Ezer and R. Kosloff. An accurate and efficient scheme for propagating the time dependent Schrödinger equation. *The Journal of Chemical Physics*, 81:3967–3971, 1984.
[21] H. F. Trotter. On the product of semi-groups of operators. *Proceedings of the American Mathematical Society*, 10(4):545–551, 1959. http://www.jstor.org/stable/2033649
[22] M. Weinmüller, M. Weinmüller, J. Rohland, and A. Scrinzi. Perfect absorption in Schrödinger-like problems using non-equidistant complex grids. *arXiv:1509.04947*, September 2015. http://arxiv.org/abs/1509.04947.

Dieter Bauer

II Calculation of typical strong-field observables

With the propagation methods introduced in Chapter I, we are able to prepare the target system in an initial state (often the ground state) and advance the wavefunction in time up to the end of the laser pulse. The final wavefunction contains all the possible knowledge about the system that one can measure in principle. Typical strong-field observables are ion yields, photoelectron spectra, or the emitted radiation. However, computationally, it is sometimes not feasible to keep the entire wavefunction on the numerical grid because of the rapid spreading in position space. One may switch to momentum space representation, in which the wavefunction spreads less. However, in momentum representation, the binding potential becomes nonlocal, which also poses a numerical challenge. Another option is to monitor the wavefunction during the pulse. In this way, one can compensate for the loss of information due to absorbing boundary conditions.

The strength of numerical simulations is actually the access to "nonobservables" an experimentalist cannot measure (may it be for practical or for fundamental reasons). Examples are the time-dependent probability density in position space (or any other space that might be helpful) and gauge-dependent entities such as energies and projections onto unperturbed or dressed states during the laser pulse. Although such nonobservables are not an end in itself, they often help understand *why*, e.g., certain spectral features occur.

1 Ionization rates

Let us begin with the single-ionization probability (or rate), a simple observable whose calculation does not require huge numerical grids.

The ionization rate of atomic hydrogen, initially in the ground state and subject to a static, homogeneous electric field, say, $\mathbf{E} = E\mathbf{e}_z$ is (in atomic units) [11]

$$\Gamma = \frac{4}{E} e^{-2/3E}. \tag{1}$$

The exponential factor $\sim e^{-2/3E}$ appears already in the problem of one-dimensional (1D) tunneling through a triangular barrier. The tricky part in the derivation of

Dieter Bauer: Institute of Physics, University of Rostock, 18051 Rostock, Germany; email: dieter.bauer@uni-rostock.de

De Gruyter Graduate – Computational Strong-Field Quantum Dynamics, Volume 5, 2017, pp. 45–76.
DOI 10.1515/9783110417265-002

strong-field tunneling rates is to get the potential-specific preexponential factor right.[1] The derivation of (1) involves the separation of the time-independent Schrödinger equation in parabolic coordinates, the Wentzel–Kramers–Brillouin (WKB) approximation of the wavefunction with respect to tunneling through the "downhill potential" in these coordinates, and the estimation of the inneratomic electron velocity to obtain a rate. It is not expected that (1) is valid for arbitrary high-field strengths E. This is because the WKB treatment involves a matching in the "under-barrier" (i.e., in the classically forbidden) region. Hence, at the latest, when E approaches the overbarrier ionization regime, the analytical rate is expected to become inaccurate.

Let us apply the Crank–Nicolson propagator for linear polarization in Section 1.5 of Chapter I to determine accurate ionization rates for atomic hydrogen, starting from the 1s ground state. In Section 1.5.5 of Chapter I, we discussed the role of the gauge in strong field problems. In the case of a constant electric field $E\mathbf{e}_z$, the vector potential is time dependent, $\mathbf{A}(t) = -\mathbf{E}\,t$, and of ever-increasing absolute value. Hence, we expect for long run-times large expectation values for the velocity gauge coupling term $\mathbf{A}(t)\cdot\hat{\mathbf{p}}$ in the Hamiltonian. Instead, the length gauge coupling $\mathbf{r}\cdot\mathbf{E}$ remains small if the numerical grid can be kept of reasonable size. We thus choose length-gauge and the Crank–Nicolson propagator (129) of Chapter I.

In order to determine the ionization rate from a real-time time-dependent Schrödinger equation (TDSE) simulation, we calculate the ground-state population $|\langle\Psi(0)|\Psi(t)\rangle|^2$, i.e., discretized

$$P_{1s}(t) = \left|\sum_{s=1}^{N_r} \phi_{00}^{\star}(r_s,0)\phi_{00}(r_s,t)\Delta r\right|^2, \qquad r_s = s\Delta s \tag{2}$$

and the norm $\int_0^{R_{\max}} \mathrm{d}r\, r^2 \int \mathrm{d}\Omega\, |\Psi(\mathbf{r},t)|^2$ on the grid of radius $R_{\max}$ and with absorbing boundaries

$$P_{\mathrm{grid}}(t) = \sum_{l=0}^{N_l-1}\sum_{s=1}^{N_r} |\phi_{l0}(r_s,t)|^2 \Delta r. \tag{3}$$

The following results are obtained with the publicly availabe code `Qprop` [2] in which propagators (125), (129), and (145) of Chapter I are implemented. The grid parameters were $\Delta r = 0.1$, $N_r = 1000$, $N_l = 20$ with an imaginary, absorbing potential $-100i[(s+1/2)/N_r]^8$. The time step for real-time propagation was $\Delta t = 0.025$. The field was switched-on instantaneously at time $t = 0$, which leads to transient dynamics before a steady decay of the ground-state population, and the norm on the grid can be observed and fitted by a function

$$P_{\mathrm{fit}}(t) = C_{\mathrm{fit}}\, e^{-\Gamma_{\mathrm{fit}} t} \tag{4}$$

1 Powerful analytical approaches have been developed, see, e.g., the review [17] on Keldysh theory.

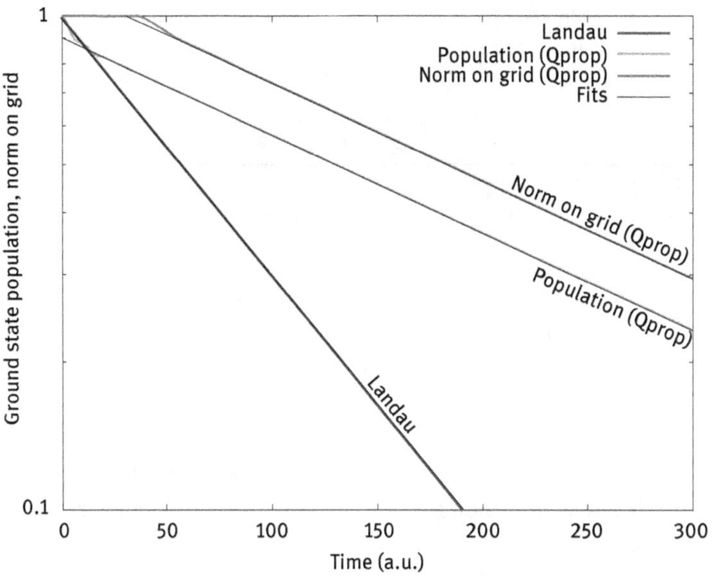

Fig. 1. Population (2) and norm on the grid (3) vs time as obtained from the numerical solution of the TDSE for an instantaneously switched-on electric field $E = 0.08$. The fits (thin black lines on top of population and norm) are $0.9\exp(-4.54 \times 10^{-3}t)$ and $1.15\exp(-4.54 \times 10^{-3}t)$, respectively. The analytical rate (labeled "Landau") is too high, i.e., it predicts a too rapid decay.

to determine the numerical rate Γ_{fit}. Figure 1 illustrates this for the run with $E = 0.08$. In the logarithmic plot, the decay curves appear linear. The longer the run time the more precisely can Γ_{fit} be determined (C_{fit} is used to shift the fit curves vertically). We see that the ground-state population and the norm on the grid have the same slope and are just shifted in time with respect to one another because the probability density takes some time to arrive at the absorbing grid boundary.

Table 1 gives numerical values for the ionization rate for electric field strengths between $E = 0.02$ and 1.0. Besides the analytical rate, we give highly accurate reference values from [27], obtained with the help of complex scaling. Our calculations reproduce these results. We are content here with two decimal places. More are possible but require longer run times so that the fitting can be done more precisely. The analytical rate increasingly overestimates the true ionization rate with increasing E. One would naively expect the geometrical overbarrier regime (i.e., where the initial energy level touches the tunneling barrier maximum) to start at $E_{\text{ob}} = 0.0625$. A more detailed investigation that takes the exceptional symmetry in parabolic coordinates and Stark shift into account predicts rather $E_{\text{ob}} = 0.147$ [1]. The analytical rate has a maximum at $E = 2/3$ after which it slowly decreases again, whereas the true rate continues to increase. However, the rate is then already of order unity in atomic units, which means that the initial state will be depleted within a few atomic time units.

Tab. 1. Ionization rates of atomic hydrogen H(1s) in a static, homogeneous electric field of strength E (first column). The columns contain our numerical results Γ_{fit} obtained with Qprop, the analytical "Landau rate" (1), and the results from [27], respectively. "-" indicate values not given or calculated.

E	Numerical Γ_{fit}	Landau	Scrinzi [27]
0.03	2.24×10^{-8}	2.98×10^{-8}	–
0.04	3.89×10^{-6}	5.78×10^{-6}	3.8927000×10^{-6}
0.06	5.15×10^{-4}	9.96×10^{-4}	$5.15077494 \times 10^{-4}$
0.08	4.54×10^{-3}	1.20×10^{-2}	$4.53965755 \times 10^{-3}$
0.1	1.45×10^{-2}	5.09×10^{-2}	$1.453811353 \times 10^{-2}$
0.5	0.56	2.11	0.5594896
1.0	1.3	2.05	1.2936418

Hence, the concept of a rate does not make sense anymore because the details matter of how such strong fields are switched on (or how the atoms enter regions in which such strong fields exists).

In the same way we have checked here the ionization rate for a static electric field, ionization rate formulas developed for oscillatory fields [17] can be tested. The numerical grid in position space should have at least the size of the excursion amplitude $\hat{E}/\omega^2$. Several benchmarking publications exist [18, 19], which might be used to test a TDSE solver. However, the situation is more involved in the case of laser pulses because of the frequency and electric field amplitude-dependent population of excited states, leading to sharp changes in the ionization rate [19].

2 Photoelectron spectra

Let the state of the system after the laser pulse be expanded in the discrete ($|E_m\rangle$) and continuum ($|E\rangle$) energy eigenstates of $\hat{H}_0$,

$$|\Psi\rangle = \sum_m c_m |E_m\rangle + \int \mathrm{d}E\, c(E)\, |E\rangle\,, \tag{5}$$

where, as usual, normalization is such that

$$\langle E_m|E_l\rangle = \delta_{ml}, \qquad \langle E|E'\rangle = \delta(E-E'), \qquad \langle E_m|E\rangle = 0. \tag{6}$$

We have in mind the typical single-active electron atomic case where the discrete part of the spectrum is restricted to energies $E_m < 0$, and the continuum starts at $E = 0$. Normalization $\langle\Psi|\Psi\rangle = 1$ requires

$$\sum_m |c_m|^2 + \int \mathrm{d}E\, |c(E)|^2 = 1. \tag{7}$$

Plotting all the $|c_m|^2$ and $|c(E)|^2$ vs energy yields a "stick spectrum" for the discrete-state part and a continuous part, respectively. However, $|c_m|^2$ is dimensionless while $|c(E)|^2$ has the dimension of an inverse energy (i.e., "per energy"). The continuous part may be "binned" according to a certain resolution γ to obtain the (dimensionless) probability to measure the electron with an energy $\in [E_\nu - \gamma, E_\nu + \gamma]$

$$P_\gamma(E_\nu) = \int_{E_\nu-\gamma}^{E_\nu+\gamma} \mathrm{d}E\,|c(E)|^2 \tag{8}$$

and then plotted in a histogram-like fashion. If $|c(E)|^2$ does not vary much over the bin-width 2γ, we have that $P_\gamma(E_\nu)/(2\gamma) \simeq |c(E_\nu)|^2$.

Alternatively, we may define the spectrum as

$$P_\gamma(E_\alpha) = \begin{cases} \mathrm{Tr}\,(|E_\alpha\rangle\langle E_\alpha|\hat{\rho}) & \text{if } E_\alpha \text{ in discrete part of spectrum,} \\ \mathrm{Tr}\left(\int_{E_\alpha-\gamma}^{E_\alpha+\gamma} \mathrm{d}E\,|E\rangle\langle E|\hat{\rho}\right) & \text{if } E_\alpha \text{ in continuum,} \end{cases} \tag{9}$$

where $\hat{\rho}$ is the density operator, and "Tr" indicates the trace. For a pure state (5) and $\hat{\rho} = |\Psi\rangle\langle\Psi|$, we obtain, using (6),

$$P(E_\alpha) = \begin{cases} |c_\alpha|^2 & \text{if } E_\alpha \text{ in discrete part of spectrum,} \\ \int_{E_\alpha-\gamma}^{E_\alpha+\gamma} \mathrm{d}E\,|c(E)|^2 \simeq 2\gamma|c(E_\alpha)|^2 & \text{if } E_\alpha \text{ in continuum.} \end{cases}$$

If we separate the E_α in the continuum by 2γ, the normalization (7) translates to

$$\sum_\alpha P(E_\alpha) = 1. \tag{10}$$

So far, we considered only total-energy-differential spectra. In strong-field laser experiments, often more differential spectra are measured, for instance,

$$P(\mathbf{p}, \Delta\mathbf{p}) = \mathrm{Tr}\left[\hat{P}(\mathbf{p}, \Delta\mathbf{p})\hat{\rho}\right], \tag{11}$$

where $\mathbf{p}$ is the photoelectron momentum at a detector of resolution $\Delta\mathbf{p}$. The projector in this case is

$$\hat{P}(\mathbf{p}, \Delta\mathbf{p}) = \int_{\mathbf{p}\pm\Delta\mathbf{p}} \mathrm{d}^3k\,|\mathbf{k}\rangle\langle\mathbf{k}|, \tag{12}$$

where the integration is over the momentum bin according to the resolution $\Delta\mathbf{p}$. The discrete states $|E_m\rangle$ will not overlap with the detector states $|\mathbf{k}\rangle$, as electrons in such states will never reach the detector.[2] We thus obtain

$$\begin{aligned} P(\mathbf{p},\Delta\mathbf{p}) &= \mathrm{Tr}\left[\int_{\mathbf{p}\pm\Delta\mathbf{p}} \mathrm{d}^3k\,|\mathbf{k}\rangle\langle\mathbf{k}| \int \mathrm{d}E\,c(E)\,|E\rangle \int \mathrm{d}E'\,c^\star(E')\,\langle E'|\right] \\ &= \int_{\mathbf{p}\pm\Delta\mathbf{p}} \mathrm{d}^3k \int \mathrm{d}E\,c(E)\,\langle\mathbf{k}|E\rangle \int \mathrm{d}E'\,c^\star(E')\,\langle E'|\mathbf{k}\rangle \\ &= \int_{\mathbf{p}\pm\Delta\mathbf{p}} \mathrm{d}^3k \left|\int \mathrm{d}E\,c(E)\Psi_E(\mathbf{k})\right|^2 , \end{aligned}$$

which is the continuum part of the probability density in momentum-space representation, integrated over the momentum bin of interest.

2.1 Energy window operator method

An elegant way to calculate the total energy-differential spectrum without the need to determine energy eigenstates on the grid is the "window operator" or resolvent approach [4, 21, 22]. Consider the operator

$$\hat{P}_{\gamma n}(E_\nu) = \frac{\gamma^{2^n}}{(\hat{H}_0 - E_\nu)^{2^n} + \gamma^{2^n}}. \tag{13}$$

Here, $E_\nu = E_{\mathrm{min}} + 2\nu\gamma$, $\nu = 0, 1, 2, \ldots$ are the energy values for which the spectrum is calculated with resolution γ. We assume that E_{min} is chosen below the ground-state energy so that the entire energy spectrum is covered by $\nu \in [0, \infty)$. The positive integer n is the "order" of the window: the higher n the more rectangular is the energy window (for $n = 1$, the window is Lorentzian shaped). If this operator acts on a state (5), we obtain the projection

$$\hat{P}_{\gamma n}(E_\nu)\,|\Psi\rangle = |\Psi_{\gamma n}(E_\nu)\rangle = \gamma^{2^n} \begin{cases} \displaystyle\sum_m \frac{c_m}{(E_m - E_\nu)^{2^n} + \gamma^{2^n}}\,|E_m\rangle & \text{discrete,} \\ \displaystyle\int \mathrm{d}E\,\frac{c(E)}{(E - E_\nu)^{2^n} + \gamma^{2^n}}\,|E\rangle & \text{continuum.} \end{cases} \tag{14}$$

The corresponding expectation value is

$$P_{\gamma n}(E_\nu) = \langle\Psi|\,\hat{P}_{\gamma n}(E_\nu)\,|\Psi\rangle = \gamma^{2^n} \begin{cases} \displaystyle\sum_m \frac{|c_m|^2}{(E_m - E_\nu)^{2^n} + \gamma^{2^n}} & \text{discrete,} \\ \displaystyle\int \mathrm{d}E\,\frac{|c(E)|^2}{(E - E_\nu)^{2^n} + \gamma^{2^n}} & \text{continuum.} \end{cases} \tag{15}$$

2 We neglect here effects such as Rydberg electrons liberated by electric detector extraction fields.

We see that when E_ν hits exactly a discrete eigenenergy E_m, a value $|c_m|^2$ is obtained, which is the correct weight. However, E_ν slightly off by ΔE will still give a weight $|c_m|^2\gamma^{2^n}/(\Delta E^{2^n}+\gamma^{2^n})$ instead of 0 in the analytically ideal stick spectrum. The sticks turn into lineshapes, determined by the window parameters γ and n.

Consider the normalization condition (10). With the window operator and with discretized, equidistant E_ν, i.e., $|E_{\nu\pm 1}-E_\nu|=2\gamma$, we have

$$\lim_{\gamma\to 0}\sum_\nu P_{\gamma n}(E_\nu)=\lim_{\gamma\to 0}\int\frac{\mathrm{d}E_\nu}{2\gamma}P_{\gamma n}(E_\nu).$$

Since

$$\int_{-\infty}^{\infty}\mathrm{d}E_\nu\,\frac{\gamma^{2^n}}{(E-E_\nu)^{2^n}+\gamma^{2^n}}=\gamma\int_{-\infty}^{\infty}\mathrm{d}u\,\frac{1}{u^{2^n}+1}=2\gamma\,\frac{\pi/2^n}{\sin(\pi/2^n)},\tag{16}$$

we find

$$\lim_{\gamma\to 0}\sum_\nu P_{\gamma n}(E_\nu)=\frac{\pi/2^n}{\sin(\pi/2^n)}\left(\sum_m|c_m|^2+\int\mathrm{d}E\,|c(E)|^2\right)=\frac{\pi/2^n}{\sin(\pi/2^n)}.\tag{17}$$

The right-hand side rapidly converges to 1 for increasing n so that we refrain from renormalizing

$$\hat{P}_{\gamma n}(E_\nu)\longrightarrow\hat{P}_{\gamma n}(E_\nu)=\frac{\sin(\pi/2^n)}{\pi/2^n}\,\frac{\gamma^{2^n}}{(\hat{H}_0-E_\nu)^{2^n}+\gamma^{2^n}}\tag{18}$$

in the following.

After these window-operator preliminaries, we now come to the actual numerical calculation of the spectrum. First, we notice that the window operator can be factorized,

$$P_{\gamma n}(E_\nu)=\langle\Psi|\,\frac{\gamma^{2^{n-1}}}{[(\hat{H}_0-E_\nu)^{2^{n-1}}+i\gamma^{2^{n-1}}]}\,\frac{\gamma^{2^{n-1}}}{[(\hat{H}_0-E_\nu)^{2^{n-1}}-i\gamma^{2^{n-1}}]}\,|\Psi\rangle\,.\tag{19}$$

We can thus calculate $P_{\gamma n}(E_\nu)$ from

$$|\tilde{\Psi}^{n-1}_{E_\nu,\gamma}\rangle=\frac{\gamma^{2^{n-1}}}{[(\hat{H}_0-E_\nu)^{2^{n-1}}-i\gamma^{2^{n-1}}]}\,|\Psi\rangle\tag{20}$$

as

$$P_{\gamma n}(E_\nu)=\langle\tilde{\Psi}^{n-1}_{E_\nu,\gamma}|\tilde{\Psi}^{n-1}_{E_\nu,\gamma}\rangle\,.\tag{21}$$

This means for $n=1$, we have to solve

$$[(\hat{H}_0-E_\nu)-i\gamma]\,|\tilde{\Psi}^0_{E_\nu,\gamma}\rangle=\gamma\,|\Psi\rangle\tag{22}$$

for $|\tilde{\Psi}^0_{E_\nu,\gamma}\rangle$ and are done. Equation (22) translates—with $|\Psi\rangle$ discretized—to the standard problem “matrix times unknown equals known right-hand-side.” The discretized $\hat{H}_0$ is, e.g., tridiagonal with the usual 3-pt stencil in position space so that forward-backward substitution from Section 1.2.1 of Chapter I can be applied.

For higher window orders n, intermediate steps can be easily added. For $n = 2$, we have $P_{\gamma 2}(E_\nu) = \langle \tilde{\Psi}^1_{E_\nu,\gamma} | \tilde{\Psi}^1_{E_\nu,\gamma} \rangle$ and

$$[(\hat{H}_0 - E_\nu)^2 - i\gamma^2] \, |\tilde{\Psi}^1_{E_\nu,\gamma}\rangle = \gamma^2 \, |\Psi\rangle \, . \tag{23}$$

Factorizing once more,

$$[(\hat{H}_0 - E_\nu) + \sqrt{i}\gamma][(\hat{H}_0 - E_\nu) - \sqrt{i}\gamma] \, |\tilde{\Psi}^1_{E_\nu,\gamma}\rangle = \gamma^2 \, |\Psi\rangle \, , \tag{24}$$

we find $|\tilde{\Psi}^1_{E_\nu,\gamma}\rangle$ by first solving

$$[(\hat{H}_0 - E_\nu) + \sqrt{i}\gamma] \, |\tilde{\Psi}^{1'}_{E_\nu,\gamma}\rangle = \gamma^2 \, |\Psi\rangle \tag{25}$$

for $|\tilde{\Psi}^{1'}_{E_\nu,\gamma}\rangle$ and then

$$[(\hat{H}_0 - E_\nu) - \sqrt{i}\gamma] \, |\tilde{\Psi}^1_{E_\nu,\gamma}\rangle = |\tilde{\Psi}^{1'}_{E_\nu,\gamma}\rangle \tag{26}$$

for $|\tilde{\Psi}^1_{E_\nu,\gamma}\rangle$. Similarly, higher orders n can be implemented. The general factorization formula reads as

$$(\hat{H}_0 - E_\nu)^{2^n} + \gamma^{2^n} = \prod_{k=1}^{2^{n-1}} [(\hat{H}_0 - E_\nu) + e^{iq_{nk}}\gamma][(\hat{H}_0 - E_\nu) - e^{iq_{nk}}\gamma] \, , \tag{27}$$

where $q_{nk} = (2k-1)\pi/2^n$.

In three dimensions (3D), with an expansion of the wavefunction in spherical harmonics as in Section 1.5 of Chapter I, we have

$$P_{\gamma n}(E) = \int \mathrm{d}\Omega \int \mathrm{d}r \left| \sum_{lm} \phi_{\gamma n,lm}(E,r) Y_{lm}(\Omega) \right|^2 = \sum_{lm} \int \mathrm{d}r \, |\phi_{\gamma n,lm}(E,r)|^2 \, , \tag{28}$$

where $\phi_{\gamma n,lm}(E,r)$ are the radial wavefunctions in the projection

$$\langle \mathbf{r} | \Psi_{\gamma n}(E) \rangle = \frac{1}{r} \sum_{l=0}^{\infty} \sum_{m=-l}^{l} \phi_{\gamma n,lm}(E,r) Y_{lm}(\Omega) \, , \tag{29}$$

and $\int \mathrm{d}\Omega Y^*_{lm}(\Omega) Y_{l'm'}(\Omega) = \delta_{ll'}\delta_{mm'}$ has been used. We see that the total spectrum $P_{\gamma n}(E)$ is simply the sum of the partial spectra $P^{(lm)}_{\gamma n}(E)$,

$$P_{\gamma n}(E) = \sum_{lm} P^{(lm)}_{\gamma n}(E) \tag{30}$$

with

$$P^{(lm)}_{\gamma n}(E) = \int \mathrm{d}r \, |\phi_{\gamma n,lm}(E,r)|^2 \, . \tag{31}$$

A slightly modified window operator is used in the Qprop code [2] where the energy spectrum is calculated as

$$P^{\mathrm{Q}}_{\gamma n}(E_\nu) = \langle \Psi_{\gamma n}(E_\nu) | \Psi_{\gamma n}(E_\nu) \rangle \tag{32}$$

with $|\Psi_{\gamma n}(E_\nu)\rangle$ defined in (14). This is equivalent to working with the window operator $\hat{P}^2_{\gamma n}(E_\nu)$ instead of $\hat{P}_{\gamma n}(E_\nu)$.

Example for a total photoelectron spectrum

Figure 2 shows total photoelectron spectrum (32) for the case of atomic hydrogen, starting from the 1s ground state, after the interaction with an $n_\mathrm{c} = 20$-cycle, linearly polarized laser pulse, described by the vector potential

$$\mathbf{A}(t) = \mathbf{e}_z A_z(t), \qquad A_z(t) = \hat{A} \sin^2\left(\frac{\omega t}{2 n_\mathrm{c}}\right) \sin \omega t \tag{33}$$

with $\omega = 0.085$ (wavelength $\lambda = 535$ nm) and an electric field amplitude $\hat{E} = \hat{A}\omega = 0.02387$ (i.e., a peak intensity $I = 2 \times 10^{13}$ W/cm^2). The wavefunction was propagated according (125) of Chapter I using Qprop [2]. The grid parameters were $N_r = 20000$, $N_l = 15$, $\Delta r = 0.2$, and the time step $\Delta t = 0.05$. The window operator parameters were $n = 3$ and $\gamma = 0.001$.

A typical above-threshold ionization (ATI) spectrum [15] is observed, i.e., photoelectron peaks at

$$E_k = E_0 - U_p + k\omega, \tag{34}$$

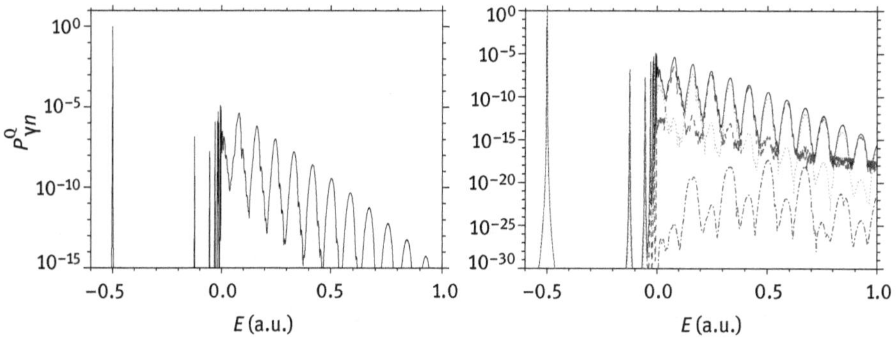

Fig. 2. ATI spectrum for H(1s) in a linearly polarized, $n_\mathrm{c} = 20$-cycle, $\sin^2$-shaped, $\lambda = 535$-nm laser pulse of peak intensity $I = 2 \times 10^{13}$ W/cm^2. Left: total spectrum. Right: total spectrum (solid) and partial spectra for $l = 0$ (dotted), for $l = 1$ (dashed), and for $l = 14$ (dashed-dotted).

where $E_0 = -I_p < 0$ is the electron binding energy, $U_p = I/4\omega^2$ is the ponderomotive energy, and $k = n_{\min}, n_{\min}+1, n_{\min}+2, \ldots$ is the number of photons absorbed, which must be at least $n_{\min}$ to overcome the net ionization potential $I_p + U_p$. In this particular example, we are within the realm of perturbation theory because U_p is small compared with both ω and $I_p = 0.5$. However, already $n_{\min} = 7$ photons are required for ionization, and peaks up to $k = 17$ are visible, although dropping exponentially in probability.

The window-operator method provides the bound, "stick-like" part of the spectrum as well. We observe that the ground state is still populated with high probability ($|\langle 1\mathrm{s}|\Psi\rangle|^2 = 0.999882$) after the pulse. Excited states up to the continuum threshold are populated as well.

The availability of partial spectra (31) for the different orbital angular momentum quantum numbers l helps analyze the numerical results further and check convergence with respect to N_l. From the right panel in Figure 2, we infer that both the 2s and the 2p states are populated after the pulse and that $l = 0$ and $l = 1$ dominate the ATI peaks in an alternating fashion. This is expected because each additional photon absorption changes l by ± 1. Convergence with respect to N_l can be ensured by proving that the partial spectrum for the maximum $l = N_l - 1$ contributes negligibly in the entire energy interval of interest. As seen in the right panel of Figure 2, this is the case: the partial spectrum $P_{\gamma n}^{(14,0)}$ is at least five orders of magnitude below the total spectrum. With increasing laser intensity or decreasing laser frequency, N_l has to be increased, making the numerical simulation more demanding.

2.1.1 More differential spectra with the window-operator method

Imagine we solved the 1D TDSE in position space. In that case, (21) or (32) is of the form (suppressing indices ν, γ, n)

$$P(E) = \int_{-\infty}^{\infty} \mathrm{d}x\, |\Psi(E,x)|^2 = P_{\mathrm{left}}(E) + P_{\mathrm{right}}(E) \tag{35}$$

with

$$P_{\mathrm{left}}(E) = \int_{-\infty}^{0} \mathrm{d}x\, |\Psi(E,x)|^2, \quad P_{\mathrm{right}}(E) = \int_{0}^{\infty} \mathrm{d}x\, |\Psi(E,x)|^2. \tag{36}$$

One may argue that the relation between momentum and energy is $p = -\sqrt{2E}$ with respect to P_{left} and $p = +\sqrt{2E}$ for P_{right}. One can then construct a momentum-differential spectrum as

$$P(p) = |p| \begin{cases} P_{\mathrm{right}}(E = p^2/2) & p \geq 0 \\ P_{\mathrm{left}}(E = p^2/2) & p < 0 \end{cases}, \tag{37}$$

in that way separating left- and right-going electrons.[3] However, this procedure is not rigorous, as is easily illustrated by the extreme case of a wave packet located in the left spatial region $x < 0$ but moving with positive momentum to the right. Clearly, (37) maps to the wrong momentum sign in this case. On the other hand, if one propagates long enough so that whatever travels to the right actually arrives in the region $x > 0$ (37) works. The smaller $|p|$ the longer one needs to postpropagate the wavefunction in order to ensure that $P(p)$ is converged and independent of time (after the laser pulse).

Similar to the 1D case above, we may employ position space information by *not* performing the angle integration in (28),

$$P(E,\Omega) = \int \mathrm{d}r \left| \sum_{lm} \phi_{lm}(E,r) Y_{lm}(\Omega) \right|^2 . \tag{38}$$

Again, this is not a rigorous energy-angle–resolved spectrum because the solid-angle element $\mathrm{d}\Omega$ in position space is not equivalent to the solid-angle element $\mathrm{d}\Omega_p$ in momentum space. Imagine a free electron that is still close to the nucleus at the end of the pulse so that is deflected by the Coulomb potential. Its momentum vector then is not yet equal to the asymptotic momentum at the detector. Hence, one should postpropagate until the angle-resolved energy spectrum is converged down to the smallest energy $E > 0$ of interest.

The conversion to a momentum spectrum may be performed noticing that, with the assumption $\Omega \simeq \Omega_p$, $P(E,\Omega)\mathrm{d}E\,\mathrm{d}\Omega$ must equal $P(\mathbf{p})\mathrm{d}^3p = P(\mathbf{p})p^2\mathrm{d}p\,\mathrm{d}\Omega_p$. Hence, with $\mathrm{d}E = p\mathrm{d}p$,

$$P(\mathbf{p}) = P(E,\Omega)\frac{\mathrm{d}E}{p^2\mathrm{d}p} = \frac{1}{p}P(p^2/2,\Omega). \tag{39}$$

Modern graphics and visualization tools (e.g., gnuplot) are capable of plotting a function $f(p,\Omega)$ that is sampled over spherical coordinates (such as $f(p,\Omega) = P(p^2/2,\Omega)/p$) vs Cartesian coordinates p_x, p_y, p_z.

Figure 3 shows such a photoelectron momentum spectrum $\log P(\mathbf{p})$ in the p_xp_z plane. In dipole approximation, there is azimuthal symmetry about the laser polarization axis $\mathbf{e}_z$ so that the spectrum looks the same in, e.g., the p_xp_z plane and the p_yp_z plane. Because $p = \sqrt{2E}$, the momentum difference between ATI peaks decreases with increasing momentum $p = |\mathbf{p}|$. The lowest-order ATI peak is a ring of radius $\sqrt{2[-(I_p + U_p) + 7\omega]} = 0.39$ in the momentum plane. Each ATI ring is modulated, i.e., θ dependent. In perpendicular direction $\theta = \pi/2$ ($p_z = 0$), every other ATI ring has a node.

3 The factor $|p|$ comes from the transformation of variables and $P(p)\,dp = P(E)\,dE$, see also the 3D version below, equation (39).

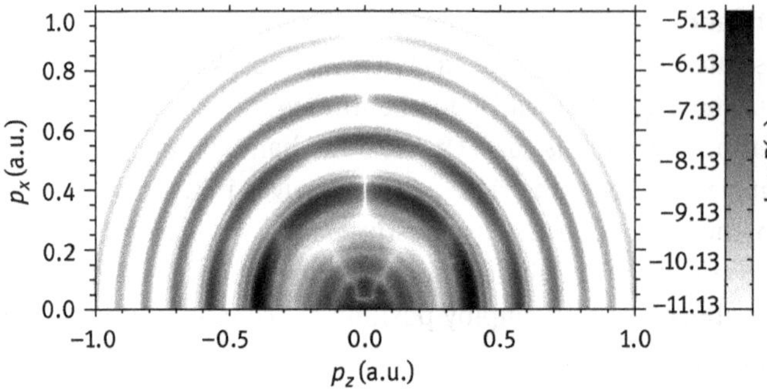

Fig. 3. Photoelectron momentum spectrum $\log P(\mathbf{p})$ in the $p_x p_z$ plane according (38) and (39) for the laser and target parameters of Figure 2.

2.2 Spectral method

Let

$$|\Psi(0)\rangle = \sum_m c_m |E_m\rangle + \int \mathrm{d}E\, c(E)\, |E\rangle \tag{40}$$

be the state of the system after the interaction (say, at time $t = 0$, without loss of generality) expanded in energy eigenstates of $\hat{H}_0$, as above in (5). The overlap of this state with the field-free propagated state at a later time $t > 0$,

$$|\Psi(t)\rangle = \sum_m c_m e^{-iE_m t} |E_m\rangle + \int \mathrm{d}E\, c(E) e^{-iEt} |E\rangle\,, \tag{41}$$

is called autocorrelation function,

$$A(t) = \langle \Psi(0)|\Psi(t)\rangle = \sum_m |c_m|^2 e^{-iE_m t} + \int \mathrm{d}E\, |c(E)|^2 e^{-iEt}. \tag{42}$$

The Fourier-transformed autocorrelation function with the initial time pushed to $-\infty$ and field-free postpropagation until ∞ would be

$$A(E) := \frac{1}{2\pi} \int_{-\infty}^{\infty} A(t)\, e^{iEt}\, \mathrm{d}t = \sum_m |c_m|^2\, \delta(E - E_m) + \int \mathrm{d}E'\, |c(E')|^2\, \delta(E - E'). \tag{43}$$

If E hits one of the discrete eigenenergies, $A(E)$ gives a δ peak (weighted by $|c_m|^2$). In the continuum, it gives $|c(E)|^2$. In practice, we postpropagate only from $t = 0$ to a finite time τ. Let us calculate how the finiteness in time affects the line shape of the peaks in the spectrum [6]. Instead of (43), we have

$$
\begin{aligned}
A^{(0)}(E) &= \frac{1}{\tau}\int_0^{\tau} \sum_m |c_m|^2 e^{i(E-E_m)t}\,\mathrm{d}t + \frac{1}{\tau}\int_0^{\tau}\mathrm{d}t \int \mathrm{d}E'\,|c(E')|^2\,e^{i(E-E')t} \\
&= \sum_m |c_m|^2 \int_{-\infty}^{\infty} w_\tau^{(0)}(t) e^{i(E-E_m)t}\,\mathrm{d}t + \int \mathrm{d}E'\,|c(E')|^2 \int_{-\infty}^{\infty}\mathrm{d}t\, w_\tau^{(0)}(t) e^{i(E-E')t} \\
&= \sum_m |c_m|^2 \mathcal{L}^{(0)}(E-E_m) + \int \mathrm{d}E'\,|c(E')|^2 \mathcal{L}^{(0)}(E-E'),
\end{aligned}
$$

where

$$
w_\tau^{(0)}(t) = \frac{1}{\tau}\Theta(t)\Theta(\tau - t) \tag{44}
$$

is a square-shaped time window whose Fourier transform gives the line-shape function

$$
\mathcal{L}^{(0)}(E-E') = \int_{-\infty}^{\infty} e^{i(E-E')t} w_\tau^{(0)}(t)\,\mathrm{d}t = e^{iz}\frac{\sin z}{z}, \qquad z = \frac{(E-E')\tau}{2}, \tag{45}
$$

which replaces the δ-like lines that occured above in the case of an infinite-time Fourier transform in the discrete part of the spectrum. The line shape is such that there is a central peak of height $\mathcal{L}^{(0)}(0) = 1$ at the position $E = E'$ but also wings with higher-order maxima (like in single-slit diffraction). For E in the discrete part of the spectrum, $A^{(0)}(E = E_m) = |c_m|^2$, as it should. In the continuum, we obtain $|c(E')|^2$, integrated over the line shape centered at E.

Clearly, an energy spectrum is an observable and should be real. However, we see from (45) that the finite integral limits introduce a τ-dependent complex phase factor. In order to eliminate that unphysical dependence on the finite upper integration limit, we consider $|A^{(0)}(E)|$ as the total energy–differential spectrum.

2.2.1 Increasing the dynamic range with the Hanning window

The slowly off-rolling "sinc-wings" in (45) may mask other peaks corresponding to less-populated states nearby. Hence, in order to increase the dynamic range over which peaks can be identified (say, from populations close to one down to 10^{-20}), a window function with faster decreasing wings than those for the square window is desirable. One option is the so-called Hanning window

$$
w_\tau^{(\mathrm{H})}(t) = \frac{1}{2\tau}\left(1 - \cos\frac{2\pi t}{\tau}\right), \tag{46}
$$

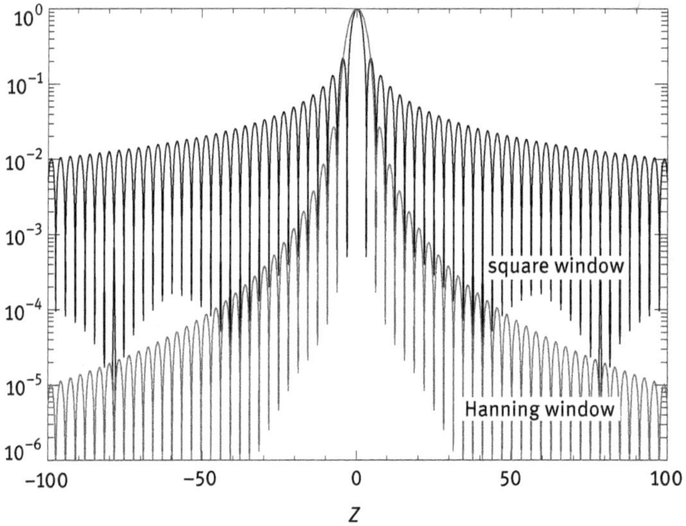

Fig. 4. $|\mathcal{L}^{(0)}|$ for the square window and $|\mathcal{L}^{(\mathrm{H})}|$ for the Hanning window vs z. The Hanning window leads to a broader central peak but a faster drop of the wings thus allowing for a higher dynamic range.

which leads to the line shape

$$\mathcal{L}^{(\mathrm{H})}(E-E') = e^{iz} \sin z \left(\frac{1}{z} - \frac{z}{z^2 - \pi^2} \right) \tag{47}$$

with z as in (45). The line shapes obtained with the square and the Hanning window are compared in Figure 4.

2.2.2 Actual numerical treatment

During the field-free postpropagation for even N_t time steps, the vector

$$\mathbf{A}_w = \langle \Psi(0) | \Big(| \Psi(0) \rangle \, w(0), | \Psi(\Delta t) \rangle \, w(\Delta t), \ldots, | \Psi(N_t \Delta t) \rangle \, w(N_t \Delta t) \Big)^\top \tag{48}$$

is stored. Here, $w(t)$ is the window of choice, for instance, the Hanning window (46) (with $\tau = N_t \Delta t$). The fast Fourier transform (FFT) $\mathrm{FFT}(\mathbf{A}_w) = \tilde{\mathbf{A}}_w$ gives back a vector of the same size,

$$\tilde{\mathbf{A}} = \left(\tilde{A}_0, \ldots, \tilde{A}_{N_t} \right)^\top, \tag{49}$$

typically ordered such that the corresponding frequencies are

$$\boldsymbol{\omega} = \Delta\omega \left(0, 1, \ldots, N_t/2 - 1, -N_t/2, -N_t/2 + 1, \ldots, -1\right)^{\top}, \tag{50}$$

where

$$\Delta\omega = \frac{2\pi}{N_t \Delta t}. \tag{51}$$

In atomic units, $\omega = E$. Hence we can assemble and plot the spectrum $|\tilde{A}(E)|$.

2.2.3 Resolution and bandwidth

By fitting the peaks in the numerically obtained spectrum to the analytical line shape, the discrete eigenenergies E_m can be determined to high accuracy. The accuracy is the higher the higher is the energy resolution

$$\Delta E = \frac{2\pi}{\tau}. \tag{52}$$

The latter is determined by the total postpropagation time τ. Moreover, the energies of interest must fall into the energy range covered,

$$\max |E| = \frac{\pi}{\Delta t}, \tag{53}$$

which is determined by the temporal resolution Δt used to generate the time series on which the FFT is performed. Equations (52) and (53) reflect the usual energy-time uncertainty.

2.2.4 Local analysis

Equation (41) reads in position-space representation

$$\Psi(\mathbf{r}, t) = \sum_m c_m e^{-iE_m t} \Psi_{E_m}(\mathbf{r}) + \int \mathrm{d}E' \, c(E') e^{-iE' t} \Psi_{E'}(\mathbf{r}). \tag{54}$$

Performing the finite-time Fourier transform as above for a time interval $[0, \tau]$ after the interaction with the laser pulse, we obtain

$$\Psi(E, \mathbf{r}) = \sum_m c_m \Psi_{E_m}(\mathbf{r}) \mathcal{L}(E - E_m) + \int \mathrm{d}E' \, c(E') \Psi_{E'}(\mathbf{r}) \mathcal{L}(E - E'). \tag{55}$$

If we pick a particular point $\mathbf{r}_\mathrm{d}$ ("d" for "detector"), $|\Psi(E, \mathbf{r}_\mathrm{d})|^2$ plotted vs E will show peaks of shape $\mathcal{L}$ for all energies whose eigenstates (i) contribute at $\mathbf{r}$ and (ii) were populated during the interaction. Close to the origin (where the binding potential is centered), only the bound, discrete states will contribute. Far away (e.g., close to the grid boundary but before the absorbing potential), only the continuum contributes. This “virtual-detector” method of obtaining spatially resolved spectra is very instructive but far from rigorous. It has been improved and made quantitatively correct in the semiclassical regime by considering the local current density (instead of directly the wavefunction), “collecting” the flux, and binning according momentum [7].

2.2.5 Determination of eigenstates

Although this section is about the calculation of photoelectron spectra, we note in passing how to extract eigenstates in a similar way. Given a “target energy” E, we can distill during field-free propagation a state

$$|\Psi_E(\tau)\rangle = \int_0^\tau |\Psi(t)\rangle\, w_\tau(t)\, e^{iEt}\, \mathrm{d}t. \tag{56}$$

Here, $w_\tau(t)$ is one of the above window functions. With the corresponding line shape function $\mathcal{L}(E) = \int \mathrm{d}t\, e^{iEt} w_\tau(t)$ and after insertion of the expansion (41), we obtain

$$|\Psi_E(\tau)\rangle = \sum_m c_m\, |E_m\rangle\, \mathcal{L}(E-E_m) + \int \mathrm{d}E'\, c(E')\, |E'\rangle\, \mathcal{L}(E-E'). \tag{57}$$

In the limit $\tau \to \infty$, the line shape becomes δ-like, and a state $|E_m\rangle$ with an eigenenergy E_m closest to the targeted energy E will be singled out (if E falls into the discrete part of the spectrum, otherwise $|E\rangle$).

In an actual field-free TDSE run, the distilled wavefunction $\Psi_E(\mathbf{r},\tau)$ is calculated via numerical integration $\Psi_E(\mathbf{r},\tau) \to \Psi_E(\mathbf{r},\tau) + \Psi(\mathbf{r},t) w_\tau(t) e^{iEt} \Delta t$ during the propagation of an, e.g., randomly initialized wavefunction $\Psi(\mathbf{r}, t = 0)$. Because of rapidly rotating phases $e^{i(E-E_m)t}$ (or $e^{i(E-E')t}$), all components in $\Psi(\mathbf{r}, t)$ will be integrated away apart from the components for which the phase vanishes, i.e., $E = E_m$ or $E = E'$. As a consequence, $\Psi_E(\mathbf{r},\tau)$ will be more and more “purified” toward an eigenfunction $\Psi_E(\mathbf{r})$, $\hat{H}_0 \Psi_E(\mathbf{r}) = E\Psi_E(\mathbf{r})$ the longer the integration time is. $\Psi_E(\mathbf{r},\tau)$ should be normalized to unity if normalized eigenstates are sought. In case of degeneracy, one finds in this way one state in the degenerate subspace, which one depends on the initial $\Psi(\mathbf{r}, 0)$. Another one of the degenerate states can be found by projecting out the previously found state in each numerical integration step. What happens if the initial wavefunction does not contain the component one is looking for? This may happen if $\Psi(\mathbf{r}, 0)$ has the parity opposite to the desired $\Psi_E(\mathbf{r})$, for instance. In this case, the procedure should converge to a state closest in energy to E and incorporated in the guess $\Psi(\mathbf{r}, 0)$. However, because of the limited machine precision states may

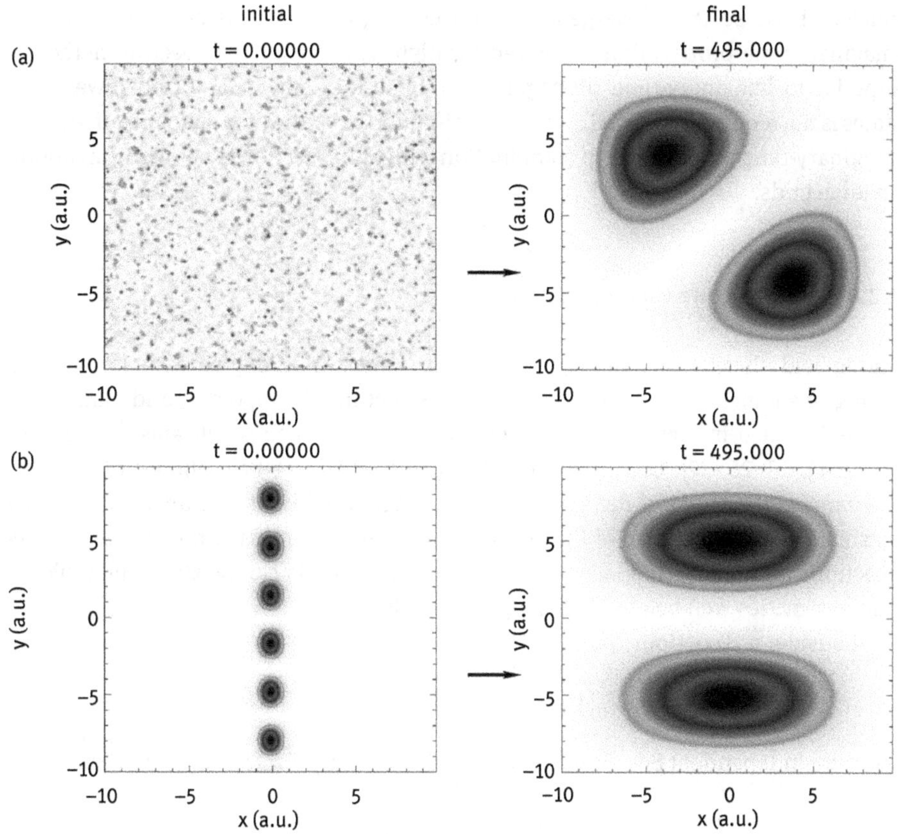

Fig. 5. Probability densities for two noninteracting 1D particles in a 1D square well of width $L = 20$ (or one 2D particle in a 20 × 20 2D box). Two initial guess wavefunctions $\Psi(x, y, 0)$ (left) and the obtained eigenstates that emerge after $\tau = 495$ for the targeted (first excited state) energy $E = 0.0605$ (right). In (a), $\Psi(x, y, 0)$ was initialized randomly; in (b), $\Psi(x, y, 0) \sim e^{-x^2} \sin y$. Different states in the two-fold degenerate subspace for E are found. Numerical parameters are as follows: $N_x = N_y = 100$, $\Delta x = \Delta y = 0.2$, $\Delta t = 0.05$, propagation with split time-evolution operator, and Crank–Nicolson.

be distilled that are actually not (supposed to be) present in $\Psi(\mathbf{r}, 0)$. To test this, the reader may initialize with the ground state and try to find an excited state.

Figure 5 shows an example where for two noninteracting 1D particles in a 1D box (or one 2D particle in a 2D box), one of two linearly independent degenerate states belonging to the second lowest eigenenergy emerges out of two different guesses for the initial wavefunction.

In Section 1.3.2 of Chapter I, we learned how excited states can be determined by imaginary propagation in combination with outprojection of the lower-lying states. In the case of first excited states as in Figure 5, that would be much more efficient because the convergence of procedure (56) with τ is slow compared to imaginary-time propagation. However, for high-lying states, the spectral method is superior to imaginary-time propagation because the knowledge of all lower-lying states is not required. Especially when quasi-bound or continuum states are of interest, imaginary-time propagation in combination with outprojection is not really an option on large grids.

2.2.6 Band structure via propagation

Strong-field physics in solids is of increasing interest, not only because of the fascinating phenomena that arise from the band structure, the topology, and many-body effects but also in view of technological applications such as ultrafast electronics [8, 12, 14, 23–26, 32]. Of course, in this context, the laser field should not exceed the damage threshold of the solid, i.e., the target should not be transformed into a plasma. However, because of the possibly lower effective mass of the charge carriers (electrons and holes), the laser-driven dynamics in solids can be strong-field-like at laser intensities well below the damage threshold.

Consider a 1D periodic potential[4]

$$V(x) = -V_0[1+\cos(2\pi x/a)], \qquad V(x+a) = V(x). \tag{58}$$

The time-independent Schrödinger equation for a single electron in that potential

$$E|\psi\rangle = \left[\frac{\hat{p}^2}{2} + V(\hat{x})\right]|\psi\rangle \tag{59}$$

has solutions of the form

$$\psi_{nk}(x) = \langle x|\psi_{nk}\rangle = e^{ikx}u_{nk}(x), \qquad u_{nk}(x+a) = u_{nk}(x) \tag{60}$$

(Bloch theorem) with n as the band index. As a consequence,

$$E_{nk}u_{nk}(x) = \left[\frac{(\hat{p}+k)^2}{2} + V(x)\right]u_{nk}(x) \tag{61}$$

or

$$\left(E_{nk} - \frac{k^2}{2}\right)u_{nk}(x) = \left[-\frac{1}{2}\frac{\partial^2}{\partial x^2} - ik\frac{\partial}{\partial x} + V(x)\right]u_{nk}(x). \tag{62}$$

4 This potential was used in [34] to simulate high-harmonic generation in solids.

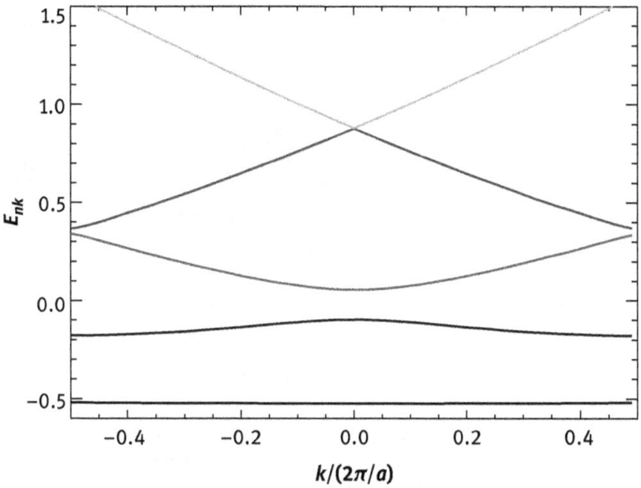

Fig. 6. Five lowest bands in the first Brillouin zone of potential (58) with $V_0 = 0.37$ and $a = 8$, calculated by solving (62) through imaginary-time propagation of (63) for $A(t) = k$. Energy E_{nk} in atomic units.

As we know from Chapter I how to solve a TDSE of the form

$$i\frac{\partial}{\partial t}\Psi(x,t) = \left[-\frac{1}{2}\frac{\partial^2}{\partial x^2} - iA(t)\frac{\partial}{\partial x} + V(x)\right]\Psi(x,t), \tag{63}$$

we can find with $A(t) = k$ the eigenenergies $\epsilon_{nk} = E_{nk} - k^2/2$ through imaginary-time propagation. With periodic boundary conditions, it is enough to solve (62) for each Bloch wavenumber k of interest on a spatial grid that covers $[0, a)$. The implementation of periodic boundary conditions can be achieved by amending the forward-backward substitution for tridiagonal matrices applying the Sherman–Morrison formula [20, 29] to the additional upper-right and lower-left corner matrix elements. The result of such a calculation for $V_0 = 0.37$, $a = 8$ (as in [34]), $N_x = 100$, $\Delta x = a/N_x$ for the first five bands is shown in Figure 6.

If one is anyway interested in solids in laser fields, one may follow an alternative approach to obtain the band structure. We assume the same periodic potential (58) with $V_0 = 0.37$ and $a = 8$ and periodic boundary conditions but this time include N_{pc} primitive cells, i.e., the spatial grid covers $[0, N_{\text{pc}}a)$. The smallest represented momentum thus is $2\pi/N_{\text{pc}}a$. We initialize $\Psi(x, t = 0)$ randomly and propagate (63) in real time for vanishing external driver $A(t) \equiv 0$. By Fourier-transforming in space and time $\Psi(x,t)$ and plotting $|\Psi(k,\omega)|^2$ logarithmically over five orders of magnitude, we obtain the band structure shown in Figure 7(a). The momentum range covered is determined by $|k_{\max}| = \pi/\Delta x$ (we used $\Delta x = 0.2$) and the momentum resolution by $N_x\Delta x = N_{\text{pc}}a$ (we used $N_{\text{pc}} = 80$ so that $\Delta k = 2\pi/(N_{\text{pc}}a) \simeq 0.01$). The frequency

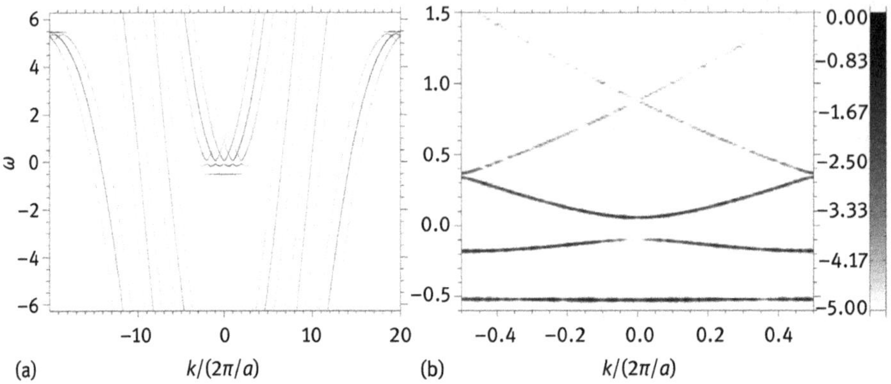

Fig. 7. Logarithm of $|\Psi(k,\omega)|^2$, obtained by an FFT of $\Psi(x,t)$, initialized randomly, and propagated in real time according (63) (with $A(t) \equiv 0$). Panel (a) shows $\log\left(|\Psi(k,\omega)|^2\right)$ in the entire frequency (energy) and momentum range; panel (b) is a zoom into the first Brillouin zone, to be compared with Figure 6.

(i.e., energy) range is determined by the $|\omega_{\max}| = \pi/\Delta t_{\text{output}}$ where Δt_{output} is the time step used for storing the wavefunction $\Psi(x,t)$ on which the Fourier transform is subsequently performed (usually larger than the time step for the Crank–Nicolson propagation). The energy resolution is determined by the total propagation time (see Section 2.2.3). A Hanning window was applied with respect to time prior to the FFT. It is important to choose momentum and energy range large enough so that aliasing does not contaminate the first Brillouin zone shown in Figure 7(b), where we observe the same band structure as in Figure 6 but obtained by completely different means. Because of the initialization with a random wavefunction, we see all bands, although a bit noisy. The method based on equation (62), leading to Figure 6, provides "cleaner" results. In particular, it provides with little effort accurate numerical values for the band energy, given a certain k. The strength of the latter, spectral method based on real-time propagation lies in the very simple determination of the actual population of the bands after the interaction with an external driver. To that end, one would propagate further the wavefunction after the interaction (instead of the randomly initialized one).

2.3 Time-dependent surface flux method

The window-operator method for calculating photoelectron spectra outlined in Section 2.1 requires as input the wavefunction at the end of the pulse. Because the wavefunction rapidly spreads, the necessary grid size might be huge. If the fastest spreading parts of the wavefunction have been in contact with the absorbing boundary, the corresponding high-energy wing of the spectrum will be missing or spoiled. Using the autocorrelation method in Section 2.2, the energy spectrum is extracted from a time series $\langle\Psi(0)|\Psi(t)\rangle$. But still, absorbed parts of the wavefunction will spoil the spectrum. And worse, because of the field-free postpropagation, the wavefunction spreads further, and an even larger grid is required.

The time-dependent surface flux method (t-SURFF) [31] trades the large grids for temporal information on a surface enclosing a much smaller spatial region where the interesting interaction physics takes place. Let this region, in the 1D case, be

$$-X < x < X, \qquad X > 0. \tag{64}$$

Narrow grid regions beyond $\pm X$ can be used to absorb outgoing flux via an imaginary potential or a mask function acting there, or something fancier like exterior complex scaling [28, 33] or perfectly transparent boundary conditions [3]. After a while, the bound and the free part of the system's state,

$$|\Psi(t)\rangle = |\Psi_{\text{bound}}(t)\rangle + |\Psi_{\text{free}}(t)\rangle\,, \tag{65}$$

will separate spatially, i.e.,

$$\langle x|\Psi_{\text{bound}}(t)\rangle \simeq 0 \quad \text{for} \quad |x| > X, \qquad \langle x|\Psi_{\text{free}}(t)\rangle \simeq 0 \quad \text{for} \quad |x| < X. \tag{66}$$

The free part

$$|\Psi_{\text{free}}(t)\rangle = \int \mathrm{d}k\, b(k)\, |\chi_k(t)\rangle \tag{67}$$

is expanded in Volkov states $|\chi_k(T)\rangle$ fulfilling the TDSE for a free electron in a field of vector potential $A(t)$,

$$i\partial_t |\chi_k(t)\rangle = \hat{H}_{\text{V}}(t)\, |\chi_k(t)\rangle\,. \tag{68}$$

We use velocity gauge with the A^2 term transformed away so that

$$\hat{H}_{\text{V}}(t) = \frac{1}{2}\hat{k}^2 + A(t)\hat{k} = -\frac{1}{2}\frac{\partial^2}{\partial x^2} - iA(t)\frac{\partial}{\partial x} \tag{69}$$

and

$$|\chi_k(t)\rangle = e^{-iS_k(t)}\,|k\rangle\,, \qquad S_k(t) = \int^t \mathrm{d}t' \left(\frac{k^2}{2} + A(t')k\right). \tag{70}$$

We can write this in position space as

$$\chi_k(x,t) = \frac{1}{\sqrt{2\pi}} e^{-itk^2/2}\, e^{ik[x-\alpha(t)]}\,, \tag{71}$$

where

$$\alpha(t) = \int^t \mathrm{d}t'\, A(t') \tag{72}$$

is the excursion. In the following, we assume that $A(t)$ vanishes for times $t < 0$.

We wish to calculate the momentum spectrum $P(k) = |b(k)|^2$. The Volkov expansion coefficients $b(k)$ can be approximately obtained from

$$\begin{aligned} b(k) &\simeq \langle \chi_k(\tau)| \{\Theta(x-X) + \Theta(-x-X)\} |\Psi(\tau)\rangle \\ &= \int_0^\tau \mathrm{d}t \frac{\mathrm{d}}{\mathrm{d}t} \langle \chi_k(t)| \{\Theta(x-X) + \Theta(-x-X)\} |\Psi(t)\rangle . \end{aligned} \tag{73}$$

We consider the first term and rewrite with the help of the TDSEs (68) and $i\partial_t |\Psi(t)\rangle = \hat{H}(t) |\Psi(t)\rangle$

$$\begin{aligned} b_1(k) &= \int_0^\tau \mathrm{d}t \frac{\mathrm{d}}{\mathrm{d}t} \langle \chi_k(t)| \Theta(x-X) |\Psi(t)\rangle \\ &= \int_0^\tau \mathrm{d}t \left(\frac{\mathrm{d}}{\mathrm{d}t} \langle \chi_k(t)| \right) \Theta(x-X) |\Psi(t)\rangle + \int_0^\tau \mathrm{d}t\, \langle \chi_k(t)| \Theta(x-X) \frac{\mathrm{d}}{\mathrm{d}t} |\Psi(t)\rangle \\ &= i \int_0^\tau \mathrm{d}t\, \langle \chi_k(t)| \{\hat{H}_\mathrm{V}(t)\Theta(x-X) - \Theta(x-X)\hat{H}(t)\} |\Psi(t)\rangle . \end{aligned}$$

The full Hamiltonian is

$$\hat{H}(t) = \hat{H}_\mathrm{V}(t) + \hat{V}, \tag{74}$$

but in the region $x > X$, we assume the potential $\hat{V}$ to vanish so that $\Theta(x-X)\hat{H}(t) \simeq \hat{H}_\mathrm{V}(t)$ and thus

$$b_1(k) = i \int_0^\tau \mathrm{d}t\, \langle \chi_k(t)| [\hat{H}_\mathrm{V}(t), \Theta(x-X)] |\Psi(t)\rangle , \tag{75}$$

and in position space

$$b_1(k) = i \int_0^\tau \mathrm{d}t \int \mathrm{d}x \chi_k^*(x,t) \left[-\frac{1}{2}\frac{\partial^2}{\partial x^2} - iA(t)\frac{\partial}{\partial x}, \Theta(x-X) \right] \Psi(x,t). \tag{76}$$

The commutator is

$$\begin{aligned} &\left[-\frac{1}{2}\frac{\partial^2}{\partial x^2} - iA(t)\frac{\partial}{\partial x}, \Theta(x-X) \right] \Psi(x,t) \\ &= -\left\{ \frac{1}{2}\delta'(x-X) + iA(t)\delta(x-X) + \delta(x-X)\frac{\partial}{\partial x} \right\} \Psi(x,t) \end{aligned}$$

so that

$$b_1(k) = \int_0^\tau \mathrm{d}t \chi_k^*(x,t) \left\{ \frac{1}{2}k + A(t) - \frac{i}{2}\frac{\partial}{\partial x} \right\} \Psi(x,t) \Bigg|_X$$
$$= \frac{1}{\sqrt{2\pi}} \int_0^\tau \mathrm{d}t \, e^{itk^2/2} \, e^{-ik[x-\alpha(t)]} \left\{ \frac{1}{2}k + A(t) - \frac{i}{2}\frac{\partial}{\partial x} \right\} \Psi(x,t) \Bigg|_X . \qquad (77)$$

For $b_2(k)$, one finds analogously

$$b_2(k) = -\frac{1}{\sqrt{2\pi}} \int_0^\tau \mathrm{d}t \, e^{itk^2/2} \, e^{-ik[x-\alpha(t)]} \left\{ \frac{1}{2}k + A(t) - \frac{i}{2}\frac{\partial}{\partial x} \right\} \Psi(x,t) \Bigg|_{-X} . \qquad (78)$$

The momentum-differential spectrum is then calculated as

$$P(k) \simeq |b_1(k) + b_2(k)|^2 . \qquad (79)$$

In order to calculate $b_1(k)$, $b_2(k)$, the wavefunction at the t-SURFF boundaries $\pm X$ and its derivative need to be stored with a sufficient time resolution such that the time integrations in (77) and (78) for each momentum k of interest can be performed numerically. One could also perform these integrations during the actual propagation of the wavefunction. However, this is less flexible than a postprocessing because the k-range of interest and the resolution Δk might not be clear beforehand.

In the case of a laser pulse of vector potential amplitude $\hat{A}$ and frequency ω, the boundary distance X should be larger than the excursion amplitude $\hat{\alpha} = \hat{A}/\omega$.[5] The potential $V(x)$ should *really* vanish at $|x| = X$. If it is asymptotically Coulombic, it is a good idea to "manually" bring it smoothly down to zero. Further, multiplication of a Hanning window

$$h(t) = \frac{1}{2}\left[1 - \cos\left(\frac{2\pi t}{\tau}\right)\right] \qquad (80)$$

to the time integrands in (77) and (78) removes unphysical dependencies on the choice of τ and increases the dynamic range. Note that in practicable TDSE simulations, one cannot choose τ so large that the flux through the boundary ceases completely.

Figure 8 shows photoelectron spectra for the widely used 1D "fruit-fly" model atom (electron in soft-core potential $V(x) = (x^2 + 1)^{-1/2}$ [9]) in an $N_c = 6$-cycle $\sin^2$-envelope laser pulse of vector potential $A(t) = A_0 \sin^2\left[\omega t/(2N_c)\right] \sin \omega t$ for $0 < t < T_p = 2N_c\pi/\omega$ (and zero otherwise), with $\omega = 0.057$ (800 nm) and $A_0 = 1$ (1.14 ×

5 In principle, the t-SURFF method allows for flux back into the interior region, in practice discretization errors and absorption of the wavefunction beyond the t-SURFF boundary spoil the spectra though.

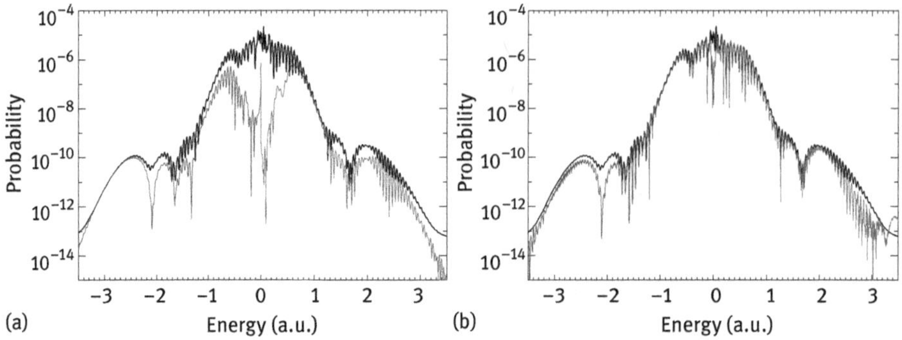

Fig. 8. Ionization probability vs kinetic energy $\frac{k^2}{2}\frac{\mathbf{k}}{k}$ as obtained by the t-SURFF (gray) and the window-operator (black) method. Laser parameters are given in the text. Numerical parameters: $\Delta x = 0.2$, $\Delta t = 0.05$; $N_x = 3600$ (including absorbing boundary), $X = 200$ for t-SURFF, $N_x = 32000$ for window operator. Panel (a) shows the t-SURFF spectrum at time $T_c = 661$ (with Hanning window (80) applied for $\tau = T_c$) and panel (b) at $2T_c = \tau$. The window-operator spectrum was calculated at time T_c on the bigger grid.

10^{14} Wcm^{-2}). The spectra are plotted vs "directional" kinetic energy $\frac{k^2}{2}\mathbf{k}/k$. In the t-SURFF calculation, right- and left-going electrons were distinguished according (77) and (78), respectively. For comparison, the window-operator results, calculated according (37) on a ten times bigger grid, are included (numerical parameters are specified in the figure caption). In panel (a), the t-SURFF spectrum was calculated up to the end of the laser pulse at $T_p = 661$. A substantial fraction of the low-energy part had not yet arrived at the t-SURFF boundaries $\pm X$ by that time. Twice the time later, in panel (b), the t-SURFF spectrum is already converged down to very low kinetic energies, as can be inferred from the excellent agreement with the window-operator spectrum.

In [16] (`Qprop 2.0`), t-SURFF has been implemented for the three-dimensional case with the wavefunction expanded in spherical harmonics (cf. Section 1.5 of Chapter I). The strong-field, low-frequency photoelectron spectrum on the book cover was calculated this way.

2.4 Pros and cons of the various methods for photoelectron spectra

The window-operator method has the advantage that, at least for total energy-differential spectra, no postpropagation is required. The wavefunction immediately after the interaction with the laser pulse is sufficient to determine the energy spectrum. Instead, with the spectral method based on the autocorrelation function and with t-SURFF, one needs to propagate further for a time τ according to the field-free time-evolution operator $\exp(-i\hat{H}_0\tau)$. The time τ can be large if a high energy resolution is required (autocorrelation method) or very slow electrons are of interest (t-SURFF). In fact, with increasing τ, the t-SURFF spectrum builds up from high to low $|k|$ because low-$|k|$ electrons take longer to arrive at the t-SURFF boundary, as illustrated in Figure 8. The t-SURFF method is, in this respect, complementary to the window operator where the fast electrons disappear first upon absorption at the grid boundary but the slow electrons do not have to arrive at a boundary to be captured. Hence, a combination of t-SURFF (for the fast) and window operator (for the slow electrons) is optimal.

There are circumstances where the autocorrelation method or variations of it are useful, in particular for total-energy spectra for systems where $\hat{H}_0$ has not much symmetry. Note that the window operator is inefficient if the discretized $\hat{H}_0$ cannot be made tridiagonal in a cleverly chosen basis. t-SURFF becomes inefficient in high dimensions because of the necessity to sample and store the wavefunction (and its gradient in normal direction) at the surface. Lacking symmetry, the surface may become difficult to chose.

Another useful application of the autocorrelation method is the calculation of dressed states. Imagine we perform the autocorrelation analysis while the external driver of frequency ω (e.g., a laser) is on. The spectrum calculated in this way yields directly so-called dressed states, also known as Floquet states or light-induced states [10].

3 Emitted radiation and high-harmonics spectra

A fundamental description of the radiation that is generated during the interaction of intense laser light with matter and actually measured at a detector is intricate and falls under the cognizance of Quantum Optics. Feeding the current density, evaluated from the quantum mechanical dipole or acceleration expectation value of a single atom into Maxwell's equations is a bold approximation, which, however, seems to give reasonable results under many circumstances. The following brief arguments may at least help understand why such a bold approach works for rarefied gas targets and strong laser fields [30].

Consider the single-active-electron Hamiltonian with quantized electromagnetic field in dipole approximation and length gauge,

$$\hat{H} = \hat{H}_\mathrm{A} + \hat{H}_\mathrm{F} + \hat{\mathbf{r}} \cdot \hat{\mathbf{E}} \tag{81}$$

with $\hat{H}_\mathrm{A} = \hat{\mathbf{p}}^2/2 + V(\hat{\mathbf{r}})$ the atomic Hamiltonian,

$$\hat{H}_\mathrm{F} = \sum_{\mathbf{k},\lambda} \omega_k \hat{a}^\dagger_{\mathbf{k},\lambda} \hat{a}_{\mathbf{k},\lambda} \tag{82}$$

the field Hamiltonian composed of modes $(\mathbf{k},\lambda)$ with $\lambda = 1,2$ indicating the possible polarizations, $\hat{a}^{(\dagger)}_{\mathbf{k},\lambda}$ annihilation (creation) operators fulfilling $[\hat{a}_{\mathbf{k},\lambda}, \hat{a}^{\dagger}_{\mathbf{k}',\lambda'}] = \delta_{\mathbf{k},\mathbf{k}'}\delta_{\lambda,\lambda'}\hat{1}$, and

$$\hat{\mathbf{E}} = i\sum_{\mathbf{k},\lambda} \xi_k(\hat{a}_{\mathbf{k},\lambda} - \hat{a}^{\dagger}_{\mathbf{k},\lambda})\mathbf{e}_{\mathbf{k},\lambda}, \qquad \xi_k = \left(\frac{2\pi\omega_k}{V}\right)^{1/2} \tag{83}$$

with polarization vector $\mathbf{e}_{\mathbf{k},\lambda}$ and quantization volume V. In the Heisenberg picture,

$$\dot{\hat{a}}_{\mathbf{k},\lambda}(t) = i[\hat{H}, \hat{a}_{\mathbf{k},\lambda}(t)] = i[\hat{H}_{\mathrm{F}} + \hat{\mathbf{r}}\cdot\hat{\mathbf{E}}, \hat{a}_{\mathbf{k},\lambda}(t)] = -i\omega_k\hat{a}_{\mathbf{k},\lambda}(t) - \xi_k\hat{\mathbf{r}}(t)\cdot\mathbf{e}_{\mathbf{k},\lambda} \tag{84}$$

and thus

$$\hat{a}_{\mathbf{k},\lambda}(t) = \hat{a}_{\mathbf{k},\lambda}(0)e^{-i\omega_k t} - \xi_k\mathbf{e}_{\mathbf{k},\lambda}\cdot\int^t \mathrm{d}t'\,\hat{\mathbf{r}}(t')\,e^{i\omega_k(t'-t)}. \tag{85}$$

We are interested in the expectation value for the number of photons in mode $(\mathbf{k},\lambda)$, which is given by

$$\left\langle \hat{a}^{\dagger}_{\mathbf{k},\lambda}(t)\hat{a}_{\mathbf{k},\lambda}(t)\right\rangle = \left\langle \hat{a}^{\dagger}_{\mathbf{k},\lambda}(0)\hat{a}_{\mathbf{k},\lambda}(0)\right\rangle \tag{86}$$

$$- 2\xi_k \mathrm{Re}\,\mathbf{e}^{*}_{\mathbf{k},\lambda}\cdot\int^t \mathrm{d}t'\,\langle\hat{\mathbf{r}}(t')\hat{a}_{\mathbf{k},\lambda}(0)\rangle\, e^{-i\omega_k t'} \tag{87}$$

$$+ \xi_k^2\int^t \mathrm{d}t'\int^t \mathrm{d}t''\left\langle \mathbf{e}_{\mathbf{k},\lambda}\cdot\hat{\mathbf{r}}(t')\mathbf{e}^{*}_{\mathbf{k},\lambda}\cdot\hat{\mathbf{r}}(t'')\right\rangle e^{i\omega_k(t'-t'')}. \tag{88}$$

The three terms can be easily interpreted. The first term (86) is simply the expectation value for the initial number of photons in the mode of interest. The second term describes stimulated absorption and emission. In fact, assuming the closest-to-classical situation where the electromagnetic field is in a coherent state $|\alpha_{\mathbf{k},\lambda}\rangle$ for which $\hat{a}_{\mathbf{k},\lambda}|\alpha_{\mathbf{k},\lambda}\rangle = \alpha_{\mathbf{k},\lambda}|\alpha_{\mathbf{k},\lambda}\rangle$, the second term becomes [30]

$$(87) \quad\rightarrow\quad -2\xi_k \mathrm{Re}\,\alpha_{\mathbf{k},\lambda}\mathbf{e}^{*}_{\mathbf{k},\lambda}\cdot\int^t \mathrm{d}t'\,\langle\hat{\mathbf{r}}(t')\rangle\, e^{-i\omega_k t'}, \tag{89}$$

i.e., the stimulated absorption or emission spectrum is proportional to the real part of the (finite-time) Fourier-transformed dipole expectation value $-\langle\hat{\mathbf{r}}(t)\rangle$ in polarization direction.

The third term is the only one that contributes to the population of modes that were initially not populated, i.e., to scattering and spontaneous emission. In particular, this term must be responsible for high-harmonic generation (HHG) because the nth-harmonic mode where $k = 2\pi/\omega_k = 2\pi/(n\omega)$, with ω the fundamental frequency (of the incoming laser), is initially unpopulated. Abbreviating $\hat{x}_{\mathbf{k},\lambda}(t) =$

$\mathbf{e}_{\mathbf{k},\lambda} \cdot \hat{\mathbf{r}}(t)$ and assuming real polarization vectors, we see from

$$(88) \quad \rightarrow \quad \langle n_{\mathbf{k},\lambda}(t) \rangle_{\mathrm{HHG}} = \xi_k^2 \int^t \mathrm{d}t' \int^t \mathrm{d}t'' \, \langle \hat{x}_{\mathbf{k},\lambda}(t') \hat{x}_{\mathbf{k},\lambda}(t'') \rangle \, e^{i\omega_k(t'-t'')} \tag{90}$$

that the single-atom source for HHG is actually to be calculated from the two-time dipole-dipole correlation function $\langle \hat{x}_{\mathbf{k},\lambda}(t') \hat{x}_{\mathbf{k},\lambda}(t'') \rangle$. Only if we assume an uncorrelated sample of N_A equal single-active-electron atoms (and thus the interaction term in the Hamiltonian (81) replaced by $\sum_{j=1}^{N_\mathrm{A}} \hat{\mathbf{r}}_j \cdot \hat{\mathbf{E}}$) distributed within a volume over which the dipole approximation is still valid, i.e., all atoms "see" the same field, we may write

$$\begin{aligned} \langle n_{\mathbf{k},\lambda}(t) \rangle_{\mathrm{HHG}} &= \xi_k^2 \sum_{j=1}^{N_\mathrm{A}} \sum_{k=1}^{N_\mathrm{A}} \int^t \mathrm{d}t' \int^t \mathrm{d}t'' \, \left\langle \hat{x}^{(j)}_{\mathbf{k},\lambda}(t') \hat{x}^{(k)}_{\mathbf{k},\lambda}(t'') \right\rangle \, e^{i\omega_k(t'-t'')} \\ &\simeq \xi_k^2 \sum_{j=1}^{N_\mathrm{A}} \sum_{k=1}^{N_\mathrm{A}} \int^t \mathrm{d}t' \int^t \mathrm{d}t'' \, \left\langle \hat{x}^{(j)}_{\mathbf{k},\lambda}(t') \right\rangle \left\langle \hat{x}^{(k)}_{\mathbf{k},\lambda}(t'') \right\rangle \, e^{i\omega_k(t'-t'')} \\ &= \xi_k^2 \left| \int^t \mathrm{d}t' \sum_{j=1}^{N_\mathrm{A}} \left\langle \hat{x}^{(j)}_{\mathbf{k},\lambda}(t') \right\rangle \, e^{i\omega_k t'} \right|^2 \end{aligned} \tag{91}$$

with $\hat{x}^{(j)}_{\mathbf{k},\lambda}(t) = \mathbf{e}_{\mathbf{k},\lambda} \cdot \hat{\mathbf{r}}_j(t)$, justifying that HHG spectra from rarefied gas targets are commonly calculated from the modulus square of the respective Fourier-transformed (projected) dipole expectation value $\mu_{\mathbf{k},\lambda}(t) = -\sum_{j=1}^{N_\mathrm{A}} \left\langle \hat{x}^{(j)}_{\mathbf{k},\lambda}(t) \right\rangle$. The same holds true for a single N_e-electron atom where $\mu_{\mathbf{k},\lambda}(t) = -\sum_{j=1}^{N_\mathrm{e}} \left\langle \hat{x}^{(j)}_{\mathbf{k},\lambda}(t') \right\rangle$. However, the assumption of uncorrelated electrons within the same atom might be hard to justify.

Let us simplify (91) a bit by assuming a single electron and a certain polarization, say, $\lambda = 1$. Further, we adopt the continuum limit

$$\sum_{\mathbf{k}} \rightarrow \frac{V}{(2\pi)^3} \int \mathrm{d}^3 k = \frac{V}{(2\pi)^3} \int \mathrm{d}k \, k^2 \int \mathrm{d}\Omega_{\hat{\mathbf{k}}} \tag{92}$$

so that, with $\omega_k = kc$, $t \rightarrow \infty$, and the initial time $-\infty$,

$$\sum_{\mathbf{k}} \langle n_{\mathbf{k},\lambda} \rangle_{\mathrm{HHG}} \rightarrow \frac{1}{(2\pi)^2} \int \mathrm{d}\omega_k \int \mathrm{d}\Omega_{\hat{\mathbf{k}}} \, \frac{\omega_k^3}{c^3} \left| \int_{-\infty}^{\infty} \mathrm{d}t' \, \langle \hat{x}_{\mathbf{k}}(t') \rangle \, e^{i\omega_k t'} \right|^2 . \tag{93}$$

We identify the spectral distribution

$$\frac{\mathrm{d}\mathcal{N}(\omega_k, \Omega_{\hat{\mathbf{k}}})}{\mathrm{d}\omega_k \, \mathrm{d}\Omega_{\hat{\mathbf{k}}}} = \frac{\omega_k^3}{(2\pi)^2 c^3} \left| \langle \hat{x}_{\mathbf{k}} \rangle (\omega_k) \right|^2 , \tag{94}$$

where

$$\langle \hat{x}_{\mathbf{k}} \rangle (\omega_k) = \int_{-\infty}^{\infty} \mathrm{d}t' \, \langle \hat{x}_{\mathbf{k}}(t') \rangle \, e^{i\omega_k t'} \tag{95}$$

is the Fourier transform. As $\langle \hat{x}_{\mathbf{k}}(t) \rangle$ vanishes for $t \to \pm\infty$, we can, upon integration by parts of the right hand side of (93), write as well

$$\frac{\mathrm{d}\mathcal{N}(\omega_k, \Omega_{\hat{\mathbf{k}}})}{\mathrm{d}\omega_k \, \mathrm{d}\Omega_{\hat{\mathbf{k}}}} = \frac{\omega_k}{(2\pi)^2 c^3} \left| \left\langle \dot{\hat{x}}_{\mathbf{k}} \right\rangle (\omega_k) \right|^2 \tag{96}$$

or

$$\frac{\mathrm{d}\mathcal{N}(\omega_k, \Omega_{\hat{\mathbf{k}}})}{\mathrm{d}\omega_k \, \mathrm{d}\Omega_{\hat{\mathbf{k}}}} = \frac{1}{(2\pi)^2 \omega_k c^3} \left| \left\langle \ddot{\hat{x}}_{\mathbf{k}} \right\rangle (\omega_k) \right|^2 , \tag{97}$$

with $\left\langle \dot{\hat{x}}_{\mathbf{k}} \right\rangle (\omega_k)$ and $\left\langle \ddot{\hat{x}}_{\mathbf{k}} \right\rangle (\omega_k)$ defined analogously to (95).

The last expression is particularly suited for numerical purposes. Let us assume that the polarization vector is $\mathbf{e}_{\mathbf{k}} = \mathbf{e}_z$ and that we are interested in photons emitted inside a narrow cone in forward direction, say, $\mathbf{k} = k\mathbf{e}_x$,

$$\frac{\mathrm{d}\mathcal{N}_x(\omega_k)}{\mathrm{d}\omega_k} = \frac{1}{(2\pi)^2 \omega_k c^3} \left| \left\langle \ddot{\hat{z}} \right\rangle (\omega_k) \right|^2 . \tag{98}$$

In order to obtain $\langle \ddot{\hat{z}} \rangle (\omega_k)$ via Fourier transformation, we need $\langle \ddot{\hat{z}} \rangle (t)$ as input. With the Heisenberg equations of motion for the electron

$$\dot{\hat{z}} = i[\hat{H}, \hat{z}] = \hat{v}_z, \qquad \dot{\hat{p}}_z = i[\hat{H}, \hat{p}_z] \tag{99}$$

and the Hamiltonian (81), we obtain, using $[\hat{z}, \hat{p}_z] = i\hat{1}$ and $[f(\hat{z}), \hat{p}_z] = i\partial f / \partial \hat{z}$,

$$\langle \ddot{\hat{z}} \rangle = -\left\langle \frac{\partial V}{\partial \hat{z}} \right\rangle - \langle \hat{E}_z \rangle . \tag{100}$$

If we use the velocity-gauge Hamiltonian for a single active electron

$$\hat{H}_{\mathrm{vg}} = \frac{1}{2} (\hat{\mathbf{p}} + \hat{\mathbf{A}})^2 + V(\hat{\mathbf{r}}) \tag{101}$$

with the quantized vector potential $\hat{\mathbf{A}}$, we have

$$\dot{\hat{z}} = i[\hat{H}_{\mathrm{vg}}, \hat{z}] = \hat{v}_z = \hat{p}_z + \hat{A}_z, \qquad \dot{\hat{p}}_z = i[\hat{H}_{\mathrm{vg}}, \hat{p}_z] = -\frac{\partial V}{\partial \hat{z}} \tag{102}$$

and thus, with $\hat{\mathbf{E}} = -\dot{\hat{\mathbf{A}}}$, the same as (100) because

$$\langle \ddot{\hat{z}} \rangle = \langle \dot{\hat{p}}_z + \dot{\hat{A}}_z \rangle = -\left\langle \frac{\partial V}{\partial \hat{z}} \right\rangle - \langle \hat{E}_z \rangle . \tag{103}$$

This just illustrates that $\langle\ddot{\hat{z}}\rangle$, from which we calculate the observable $\mathrm{d}\mathcal{N}_x(\omega_k)/\mathrm{d}\omega_k$, is gauge invariant.

As all the numerical approaches introduced in Chapter I are based on the Schrödinger picture, but the calculations in this section are, for convenience, in the Heisenberg picture, we briefly remind the reader that any observable's (time-dependent) expectation value $\langle\hat{O}\rangle(t)$ is independent of the picture because

$$\langle\hat{O}\rangle(t) = \langle\Psi_{\mathrm{H}}|\,\hat{O}_{\mathrm{H}}(t)\,|\Psi_{\mathrm{H}}\rangle = \langle\Psi_{\mathrm{S}}(t)|\,\hat{O}_{\mathrm{S}}\,|\Psi_{\mathrm{S}}(t)\rangle\,, \tag{104}$$

where the subscript "H" ("S") refers to the Heisenberg (Schrödinger) picture. Moreover, we usually treat the electromagnetic field classically in strong-field TDSE simulations so that (100) becomes

$$\langle\ddot{\hat{z}}\rangle(t) = -\left\langle \Psi(t) \left| \frac{\partial V}{\partial \hat{z}} \right| \Psi(t) \right\rangle - E_z(t). \tag{105}$$

In order to calculate $\langle\hat{z}\rangle(t)$ or $\langle\ddot{\hat{z}}\rangle(t)$ with $\Psi(t)$ expanded in spherical harmonics, one has to evaluate integrals of the form

$$I(t) = \int dr\, r^2 \int d\Omega\, f(r) \cos\theta\, |\Psi(\mathbf{r}, t)|^2. \tag{106}$$

With the c_{lm} defined in (110) of Chapter I, $c_{lm} = \sqrt{[(l+1)^2 - m^2]/[(2l+1)(2l+3)]}$, we obtain

$$\begin{aligned} I(t) &= \sum_{lm}\sum_{l'm'} \int dr\, f(r) \phi^*_{lm}(r,t) \phi_{l'm'}(r,t) \langle Y_{lm}|\cos\theta|Y_{l'm'}\rangle \\ &= \sum_{lm} \int dr\, f(r) \phi^*_{lm}(r,t) \left(c_{l-1,m}\phi_{l-1,m}(r,t) + c_{lm}\phi_{l+1,m}(r,t)\right). \end{aligned} \tag{107}$$

In the case $I(t) = -[\langle\ddot{\hat{z}}\rangle(t) + E_z(t)]$, we have $f(r) = \frac{\partial V}{\partial r}$ and for $I(t) = \langle\hat{z}\rangle(t)$ simply $f(r) = r$.

Figure 9 shows an example for a high-harmonic spectrum. With the TDSE solver `Qprop` outlined in Section 1.5 of Chapter I, H(1s) in a linearly polarized, 800 nm, 6-cycle $\sin^2$-envelope laser pulse of intensity $3.51 \times 10^{14}\,\mathrm{Wcm}^{-2}$ was propagated. During the propagation, the acceleration (105) was calculated and stored. An FFT of the acceleration yields a complex $\langle\ddot{\hat{z}}\rangle(\omega)$. Figure 9(a) shows $|\langle\ddot{\hat{z}}\rangle(\omega)|^2$. As expected [13], the cutoff is at harmonic order

$$n_{\text{cut-off}} \simeq \frac{1}{\omega}\,(I_p + 3.17\,U_p)\,,$$

which is at $n_{\text{cut-off}} \simeq 52$ is this case. By taking windowed Fourier transforms (or a wavelet transform), a time-frequency analysis can be performed, i.e., one may investigate at which time which harmonic is predominantly emitted. Because of the time-frequency uncertainty, a narrower window in time will lead to a broader distribution in frequency and *vice versa*. Time-frequency analysis is a very powerful

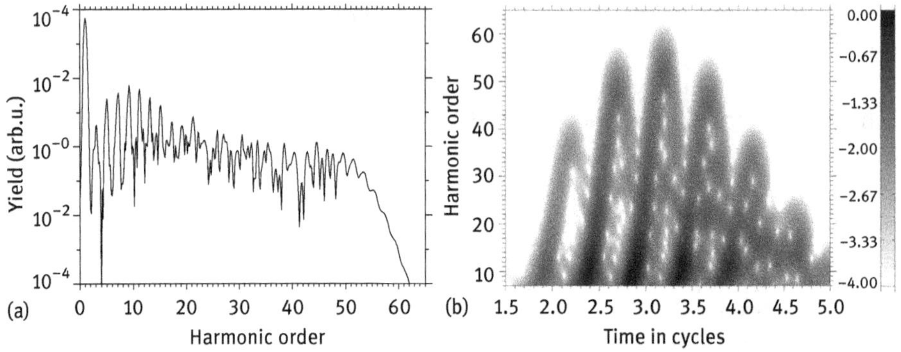

Fig. 9. High-harmonic spectrum (a) and time-frequency analysis (b) for a 6-cycle $\sin^2$-envelope laser pulse of frequency $\omega = 0.057$ (800 nm) and field amplitude $E_0 = 0.1$ (3.51×10^{14} Wcm^{-2}).

tool for identifying the physical mechanism behind harmonic emission. In Figure 9(b), the logarithm of the modulus square of the windowed Fourier transform of $\langle\ddot{\hat{z}}\rangle(t)$ is plotted over four orders of magnitude. The frequency-resolved emission times are in good agreement with "simple man's theory" (see, e.g., [5, 35]).

Bibliography

[1] D. Bauer. Ejection energy of photoelectrons in strong-field ionization. *Physical Review A*, 55: 2180–2185, 1997.

[2] D. Bauer and P. Koval. Qprop: A Schrödinger-solver for intense laser–atom interaction. *Computer Physics Communications*, 174(5):396–421, 2006. www.qprop.de.

[3] K. Boucke, H. Schmitz, and H.-J. Kull. Radiation conditions for the time-dependent Schrödinger equation: Application to strong-field photoionization. *Physical Review A*, 56:763–771, 1997.

[4] F. Catoire and H. Bachau. Extraction of the absolute value of the photoelectron spectrum probability density by means of the resolvent technique. *Physical Review A*, 85:023422, 2012.

[5] C. C. Chirilă, I. Dreissigacker, E. V. van der Zwan, and M. Lein. Emission times in high-order harmonic generation. *Physical Review A*, 81:033412, 2010.

[6] M. D. Feit, J. A. Fleck, Jr., and A. Steiger. Solution of the Schrödinger equation by a spectral method. *Journal of Computational Physics*, 47:412–433, 1982.

[7] B. Feuerstein and U. Thumm. On the computation of momentum distributions within wavepacket propagation calculations. *Journal of Physics B: Atomic, Molecular and Optical Physics*, 36(4):707, 2003.

[8] S. Ghimire, A. D. DiChiara, E. Sistrunk, P. Agostini, L. F. DiMauro, and D. A. Reis. Observation of high-order harmonic generation in a bulk crystal. *Nature Physics*, 7(2):138–141, 2011.

[9] J. Javanainen, J. H. Eberly, and Q. Su. Numerical simulations of multiphoton ionization and above-threshold electron spectra. *Physical Review A*, 38:3430–3446, 1988.

[10] V. Kapoor and D. Bauer. Floquet analysis of real-time wave functions without solving the Floquet equation. *Physical Review A*, 85:023407, 2012.

[11] L. D. Landau and L. M. Lifshitz. *Quantum Mechanics Non-Relativistic Theory, Third Edition*, vol. 3. Butterworth-Heinemann, Oxford, 1981.
[12] F. Langer, M. Hohenleutner, C. P. Schmid, C. Poellmann, P. Nagler, T. Korn, C. Schüller, M. S. Sherwin, U. Huttner, J. T. Steiner, S. W. Koch, M. Kira, and R. Huber. Lightwave-driven quasiparticle collisions on a subcycle timescale. *Nature*, 533(7602):225–229, 2016.
[13] M. Lewenstein, Ph. Balcou, M. Yu. Ivanov, A. L'Huillier, and P. B. Corkum. Theory of high-harmonic generation by low-frequency laser fields. *Physical Review A*, 49:2117–2132, 1994.
[14] T. T. Luu, M. Garg, S. Y. Kruchinin, A. Moulet, M. Th. Hassan, and E. Goulielmakis. Extreme ultraviolet high-harmonic spectroscopy of solids. *Nature*, 521(7553):498–502, 2015.
[15] D. B. Milošević, G. G. Paulus, D. Bauer, and W. Becker. Above-threshold ionization by few-cycle pulses. *Journal of Physics B: Atomic, Molecular and Optical Physics*, 39(14):R203, 2006.
[16] V. Mosert and D. Bauer. Photoelectron spectra with Qprop and t-SURFF. *Computer Physics Communications*, 2016.
[17] S. V. Popruzhenko. Keldysh theory of strong field ionization: History, applications, difficulties and perspectives. *Journal of Physics B: Atomic, Molecular and Optical Physics*, 47(20):204001, 2014.
[18] S. V. Popruzhenko, V. D. Mur, V. S. Popov, and D. Bauer. Strong field ionization rate for arbitrary laser frequencies. *Physical Review Letters*, 101:193003, 2008.
[19] R. M. Potvliege and R. Shakeshaft. Multiphoton processes in an intense laser field: Harmonic generation and total ionization rates for atomic hydrogen. *Physical Review A*, 40:3061–3079, 1989.
[20] W. H. Press, S. A. Teukolsky, W. T. Vetterling, and B. P. Flannery. *Numerical Recipes 3rd Edition: The Art of Scientific Computing*. Cambridge University Press, New York, NY, 2007.
[21] K. J. Schafer and K. C. Kulander. Energy analysis of time-dependent wave functions: Application to above-threshold ionization. *Physical Review A*, 42:5794–5797, 1990.
[22] K. J. Schafer. The energy analysis of time-dependent numerical wave functions. *Computer Physics Communications*, 63(1):427–434, 1991.
[23] A. Schiffrin, T. Paasch-Colberg, N. Karpowicz, V. Apalkov, D. Gerster, S. Muhlbrandt, M. Korbman, J. Reichert, M. Schultze, S. Holzner, J. V. Barth, R. Kienberger, R. Ernstorfer, V. S. Yakovlev, M. I. Stockman, and F. Krausz. Optical-field-induced current in dielectrics. *Nature*, 493(7430):70–74, 2013.
[24] O. Schubert, M. Hohenleutner, F. Langer, B. Urbanek, C. Lange, U. Huttner, D. Golde, T. Meier, M. Kira, S. W. Koch, and R. Huber. Sub-cycle control of terahertz high-harmonic generation by dynamical Bloch oscillations. *Nature Photonics*, 8(2):119–123, 2014.
[25] M. Schultze, E. M. Bothschafter, A. Sommer, S. Holzner, W. Schweinberger, M. Fiess, M. Hofstetter, R. Kienberger, V. Apalkov, V. S. Yakovlev, M. I. Stockman, and F. Krausz. Controlling dielectrics with the electric field of light. *Nature*, 493(7430):75–78, 2013.
[26] M. Schultze, K. Ramasesha, C. D. Pemmaraju, S. A. Sato, D. Whitmore, A. Gandman, J. S. Prell, L. J. Borja, D. Prendergast, K. Yabana, D. M. Neumark, and S. R. Leone. Attosecond band-gap dynamics in silicon. *Science*, 346(6215):1348–1352, 2014.
[27] A. Scrinzi. Ionization of multielectron atoms by strong static electric fields. *Physical Review A*, 61:041402, 2000.
[28] A. Scrinzi. Infinite-range exterior complex scaling as a perfect absorber in time-dependent problems. *Physical Review A*, 81:053845, 2010.
[29] J. Sherman and W. J. Morrison. Adjustment of an inverse matrix corresponding to a change in one element of a given matrix. *Annals of Mathematical Statistics*, 21(1):124–127, 1950.
[30] B. Sundaram and P. W. Milonni. High-order harmonic generation: Simplified model and relevance of single-atom theories to experiment. *Physical Review A*, 41:6571–6573, 1990.

[31] L. Tao and A. Scrinzi. Photo-electron momentum spectra from minimal volumes: The time-dependent surface flux method. *New Journal of Physics*, 14(1):013021, 2012.

[32] G. Vampa, T. J. Hammond, N. Thire, B. E. Schmidt, F. Legare, C. R. McDonald, T. Brabec, and P. B. Corkum. Linking high harmonics from gases and solids. *Nature*, 522(7557):462–464, 2015.

[33] M. Weinmüller, M. Weinmüller, J. Rohland, and A. Scrinzi. Perfect absorption in Schrödinger-like problems using non-equidistant complex grids. *arXiv:1509.04947*, 2015. http://arxiv.org/abs/1509.04947.

[34] M. Wu, S. Ghimire, D. A. Reis, K. J. Schafer, and M. B. Gaarde. High-harmonic generation from Bloch electrons in solids. *Physical Review A*, 91:043839, 2015.

[35] V. S. Yakovlev and A. Scrinzi. High harmonic imaging of few-cycle laser pulses. *Physical Review Letters*, 91:153901, 2003.

Heiko Bauke

III Time-dependent relativistic wave equations: Numerics of the Dirac and the Klein–Gordon equation

Computational methods are indispensable to study quantum dynamics in the realm of relativistic light-matter interactions. This chapter provides a short introduction to relativistic quantum mechanics and the Dirac and the Klein–Gordon equations. Numerical methods for solving the time-dependent Dirac equation are presented together with some exemplary implementations. These methods allow us to study Klein tunneling, the interaction of electrons with strong short laser pulses, and ionization from hydrogen-like highly charged ions in strong laser fields.

1 From nonrelativistic to relativistic quantum mechanics

In the previous chapters, we have considered strong-field quantum dynamics in the nonrelativistic regime, which is governed by the time-dependent Schrödinger equation. For sufficiently strong fields, however, charged particles can be accelerated to velocities on the order of the speed of light within a single laser half-cycle, which demands a relativistic quantum theory.

1.1 Relativistic quantum mechanical equations of motion – a naive attempt

In order to derive a relativistic quantum mechanical equation of motion, one may apply a correspondence principle. This approach is motivated by the fact that the nonrelativistic time-dependent Schrödinger equation

$$\mathrm{i}\hbar\frac{\partial\Psi(\mathbf{r},t)}{\partial t} = \hat{H}\Psi(\mathbf{r},t) = \left\{\frac{1}{2m_0}\left[-\mathrm{i}\hbar\boldsymbol{\nabla} - q\mathbf{A}(\mathbf{r},t)\right]^2 + q\phi(\mathbf{r},t)\right\}\Psi(\mathbf{r},t) \tag{1}$$

Heiko Bauke: Max-Planck-Institut für Kernphysik, Saupfercheckweg 1, 69117 Heidelberg, Germany, email: heiko.bauke@mpi-hd.mpg.de

De Gruyter Graduate – Computational Strong-Field Quantum Dynamics, Volume 5, 2017, pp. 77–109.
DOI 10.1515/9783110417265-003

for a charged particle of charge q and mass m_0 and electromagnetic potentials $\mathbf{A}(\mathbf{r}, t)$ and $\phi(\mathbf{r}, t)$ can be "derived" from the classical Hamilton function

$$H(\mathbf{r}, \mathbf{p}, t) = \frac{1}{2m_0}\left[\mathbf{p} - q\mathbf{A}(\mathbf{r}, t)\right]^2 + q\phi(\mathbf{r}, t) \tag{2}$$

by replacing the canonical momentum $\mathbf{p}$ by the canonical momentum operator $\hat{\mathbf{p}} = -\mathrm{i}\hbar\boldsymbol{\nabla}$.

The classical, relativistic Hamilton function for a charged particle is

$$H(\mathbf{r}, \mathbf{p}, t) = \sqrt{c^2\left[\mathbf{p} - q\mathbf{A}(\mathbf{r}, t)\right]^2 + m_0^2c^4} + q\phi(\mathbf{r}, t)\,. \tag{3}$$

Thus, one might propose the Hamiltonian

$$\hat{H} = \sqrt{c^2\left[\hat{\mathbf{p}} - q\mathbf{A}(\mathbf{r}, t)\right]^2 + m_0^2c^4} + q\phi(\mathbf{r}, t) \tag{4}$$

as the quantum mechanical version of (3), which yields the equation of motion

$$\mathrm{i}\hbar\frac{\partial\Psi(\mathbf{r}, t)}{\partial t} = \left\{\sqrt{c^2\left[-\mathrm{i}\hbar\boldsymbol{\nabla} - q\mathbf{A}(\mathbf{r}, t)\right]^2 + m_0^2c^4} + q\phi(\mathbf{r}, t)\right\}\Psi(\mathbf{r}, t) \tag{5}$$

for the wavefunction $\Psi(\mathbf{r}, t)$. The operator (4), however, suffers from the defect that it is very problematic to give the square-root operator, which appears in (4), a well-defined meaning. The operator $\sqrt{\left[\hat{\mathbf{p}} - q\mathbf{A}(\mathbf{r}, t)\right]^2 + m_0^2c^2}$ is a nonlinear function of three operators, the three components of $\hat{\mathbf{p}} - q\mathbf{A}(\mathbf{r}, t)$.

If two linear operators $\hat{O}_1$ and $\hat{O}_2$ commute, we can define the action of some operator function $f(\hat{O}_1, \hat{O}_2)$ in terms of its action on the common eigenfunctions of $\hat{O}_1$ and $\hat{O}_2$. Let u_{o_1,o_2} denote a common eigenfunction of $\hat{O}_1$ and $\hat{O}_2$ with

$$\hat{O}_1 u_{o_1,o_2} = o_1 u_{o_1,o_2}\,, \qquad \hat{O}_2 u_{o_1,o_2} = o_2 u_{o_1,o_2}\,; \tag{6}$$

then, one can show via a formal Taylor expansion of $f(\hat{O}_1, \hat{O}_2)$ that

$$f(\hat{O}_1, \hat{O}_2)u_{o_1,o_2} = f(o_1, o_2)u_{o_1,o_2}\,. \tag{7}$$

Thus, the operator function $f(\hat{O}_1, \hat{O}_2)$ reduces to a function of real-valued arguments in the space of common eigenfunctions of $\hat{O}_1$ and $\hat{O}_2$. If both $\hat{O}_1$ and $\hat{O}_2$ are Hermitian, then the set of their common eigenfunctions forms a basis. Hence, the action of $f(\hat{O}_1, \hat{O}_2)$ on some arbitrary function can be calculated by expanding it into this basis and employing (7). This result can be generalized to operator functions of more than two commuting operators.

However, in general, the components of $\hat{\mathbf{p}} - q\mathbf{A}(\mathbf{r}, t)$ do not commute pairwise (unless the vector potential $\mathbf{A}(\mathbf{r}, t)$ is homogeneous), and a set of common eigenfunctions does not exist. Consequently, the given recipe to apply operator functions cannot be employed to render equation (5) well defined, and the square-root equation

(5) is usually employed in contexts without a magnetic vector potential [19, 29]. In the following, we will see that there are two possible approaches that successfully avoid the square root in relativistic quantum equations of motion, which lead to the Klein–Gordon equation and the Dirac equation.

1.2 The Klein–Gordon equation

In order to remove the square root in (5), one may rearrange and square (4), which yields

$$\left[i\hbar\frac{\partial}{\partial t} - q\phi(\mathbf{r},t)\right]^2 \varphi(\mathbf{r},t) = \left\{c^2\left[-i\hbar\nabla - q\mathbf{A}(\mathbf{r},t)\right]^2 + m_0^2c^4\right\}\varphi(\mathbf{r},t), \tag{8}$$

where we have used $\varphi(\mathbf{r},t)$ to denote the wavefunction. Equation (8) is often called Klein–Gordon equation. The discovery of the Klein–Gordon equation can be attributed to various physicists. Depending on who is credited, the equation is called Klein–Gordon equation, Klein–Fock–Gordon equation, or Klein–Gordon–Schrödinger equation; see [20] for the history of the Klein–Gordon equation. Solutions $\varphi(\mathbf{r},t)$ of equation (8) satisfy the continuity equation

$$\frac{\partial\rho(\mathbf{r},t)}{\partial t} + \nabla\cdot\mathbf{j}(\mathbf{r},t) = 0 \tag{9}$$

with the density

$$\rho(\mathbf{r},t) = \frac{i\hbar}{2mc^2}\left[\varphi(\mathbf{r},t)^*\frac{\partial\varphi(\mathbf{r},t)}{\partial t} - \frac{\partial\varphi(\mathbf{r},t)^*}{\partial t}\varphi(\mathbf{r},t)\right] - \frac{q\phi(\mathbf{r},t)}{mc^2}\varphi(\mathbf{r},t)^*\varphi(\mathbf{r},t) \tag{10}$$

and the current

$$\mathbf{j}(\mathbf{r},t) = -\frac{i\hbar}{2m}\left[\varphi(\mathbf{r},t)^*\nabla\varphi(\mathbf{r},t) - \nabla\varphi(\mathbf{r},t)^*\varphi(\mathbf{r},t)\right] - \frac{q\mathbf{A}(\mathbf{r},t)}{m}\varphi(\mathbf{r},t)^*\varphi(\mathbf{r},t), \tag{11}$$

where $\varphi(\mathbf{r},t)^*$ denotes the complex conjugate of $\varphi(\mathbf{r},t)$.

Klein–Gordon equation (8) differs from the Schrödinger-type Hamiltonian form

$$i\hbar\frac{\partial}{\partial t}\Psi(\mathbf{r},t) = \hat{H}(t)\Psi(\mathbf{r},t) \tag{12}$$

of quantum mechanical equations of motion, as the Klein–Gordon equation is of second order in time. Consequently, unique initial conditions require the specification of both the wavefunction and its time derivative. Furthermore, the density $\rho(\mathbf{r},t)$ is not positive definite. Therefore, $\rho(\mathbf{r},t)$ cannot be interpreted as a probability density, and $\mathbf{j}(\mathbf{r},t)$ is not a probability density current.

However, these difficulties can be circumvented if one interprets the presence of a second-order time derivative in the Klein–Gordon equation as a second degree of freedom of the wavefunction. The wavefunction's two degrees of freedom correspond to two different charge states, and $q\rho(\mathbf{r},t)$ may be interpreted as a charge density and $q\mathbf{j}(\mathbf{r},t)$ as a charge current, see [8, 13, 36, 39, 41] for details. In the following, we will show how to transform Klein–Gordon equation (8) into the Hamiltonian form (12).

It is possible to transform (8) into a Hamiltonian form (12) [8] by introducing a two-component wavefunction $\Psi(\mathbf{r},t)$ with

$$\Psi(\mathbf{r},t) = \begin{pmatrix} \Psi_1(\mathbf{r},t) \\ \Psi_2(\mathbf{r},t) \end{pmatrix} = \begin{pmatrix} \frac{1}{2}\left\{\varphi(\mathbf{r},t) + \frac{1}{m_0c^2}\left[\mathrm{i}\hbar\frac{\partial}{\partial t} - q\phi(\mathbf{r},t)\right]\varphi(\mathbf{r},t)\right\} \\ \frac{1}{2}\left\{\varphi(\mathbf{r},t) - \frac{1}{m_0c^2}\left[\mathrm{i}\hbar\frac{\partial}{\partial t} - q\phi(\mathbf{r},t)\right]\varphi(\mathbf{r},t)\right\} \end{pmatrix}. \tag{13}$$

With the ansatz (13), it follows

$$\varphi(\mathbf{r},t) = \Psi_1(\mathbf{r},t) + \Psi_2(\mathbf{r},t)\,, \tag{14}$$

$$\frac{1}{m_0c^2}\left[\mathrm{i}\hbar\frac{\partial}{\partial t} - q\phi(\mathbf{r},t)\right]\varphi(\mathbf{r},t) = \Psi_1(\mathbf{r},t) - \Psi_2(\mathbf{r},t)\,, \tag{15}$$

and with (8) the equation of motion for the new two-component wavefunction

$$\begin{aligned}\mathrm{i}\hbar\frac{\partial\Psi(\mathbf{r},t)}{\partial t} &= \hat{H}(t)\Psi(\mathbf{r},t) \\ &= \left\{\frac{\sigma_3 + \mathrm{i}\sigma_2}{2m_0}\left[-\mathrm{i}\hbar\nabla - q\mathbf{A}(\mathbf{r},t)\right]^2 + q\phi(\mathbf{r},t) + \sigma_3 m_0c^2\right\}\Psi(\mathbf{r},t)\,, \end{aligned} \tag{16}$$

which has the desired Hamiltonian form (12). The wavefunction's components are coupled by the Pauli matrices σ_2 and σ_3 of the set

$$\sigma_1 = \begin{pmatrix} 0 & 1 \\ 1 & 0 \end{pmatrix}, \quad \sigma_2 = \begin{pmatrix} 0 & -\mathrm{i} \\ \mathrm{i} & 0 \end{pmatrix}, \quad \sigma_3 = \begin{pmatrix} 1 & 0 \\ 0 & -1 \end{pmatrix}. \tag{17}$$

The Pauli matrices are complex-valued Hermitian unitary matrices that obey the Pauli algebra

$$\sigma_i\sigma_j = \mathrm{i}\epsilon_{i,j,k}\sigma_k + \delta_{i,j}\,, \tag{18a}$$

$$\left[\sigma_i, \sigma_j\right] = 2\mathrm{i}\epsilon_{i,j,k}\sigma_k\,, \tag{18b}$$

$$\left\{\sigma_i, \sigma_j\right\} = 2\delta_{i,j}\,, \tag{18c}$$

with $i, j, k \in \{1, 2, 3\}$, and where $\epsilon_{i,j,k}$ denotes the permutation symbol (also known as the Levi–Civita symbol), and $\delta_{i,j}$ is the Kronecker delta. The density (10) in terms of the two-component wavefunction $\Psi(\mathbf{r},t)$ is given by

$$\rho(\mathbf{r},t) = \Psi^*(\mathbf{r},t)\sigma_3\Psi(\mathbf{r},t). \tag{19}$$

The operator

$$\hat{K} = \frac{\sigma_3 + \mathrm{i}\sigma_2}{2m_0}\left[-\mathrm{i}\hbar\boldsymbol{\nabla} - q\mathbf{A}(\mathbf{r},t)\right]^2 + \sigma_3 m_0 c^2 \tag{20}$$

that appears in (16) is the kinetic-energy operator of the Klein–Gordon equation.

Note that the Klein–Gordon Hamiltonian (16) is not Hermitian ($\hat{H}(t) \neq \hat{H}(t)^\dagger$) because $\sigma_3 + \mathrm{i}\sigma_2$ is not a Hermitian matrix. As a consequence, the Klein–Gordon Hamiltonian can have complex eigenvalues if the external fields become sufficiently strong [34]. Furthermore, the time-evolution operator of the Klein–Gordon Hamiltonian is not unitary ($\hat{U}^{-1} \neq \hat{U}^\dagger$). However, the Klein–Gordon Hamiltonian is a σ_3-pseudo-Hermitian operator and its time-evolution operator is σ_3-pseudo-unitary [26, 27]. A linear operator $\hat{H}$ acting in a Hilbert space $\mathcal{H}$ is called $\hat{\eta}$-pseudo-Hermitian if there is a Hermitian operator $\hat{\eta}$ such that

$$\hat{\eta}^{-1}\hat{H}^\dagger\hat{\eta} = \hat{H}\,. \tag{21}$$

The operator $\hat{H}^\# = \hat{\eta}^{-1}\hat{H}^\dagger\hat{\eta}$ is named the $\hat{\eta}$-pseudo-adjoint of $\hat{H}$. Let the wavefunctions Ψ_1 and Ψ_2 be two elements of the Hilbert space $\mathcal{H}$, and let $\langle\Psi_1|\Psi_2\rangle$ denote the inner product in $\mathcal{H}$. The operator $\hat{\eta}$ defines the pseudo–inner product

$$\langle\Psi_1|\Psi_2\rangle_{\hat{\eta}} = \langle\Psi_1|\hat{\eta}\Psi_2\rangle\,. \tag{22}$$

The pseudo–inner product (22) is conjugate symmetric, linear in its second argument, but in contrast to usual inner products, it is not necessarily positive definite. The operator $\hat{H}$ is Hermitian with respect to the pseudo-inner-product (22)

$$\langle\hat{H}\Psi_1|\Psi_2\rangle_{\hat{\eta}} = \langle\Psi_1|\hat{H}\Psi_2\rangle_{\hat{\eta}}\,. \tag{23}$$

A linear invertible operator $\hat{U}$ acting on $\mathcal{H}$ is called $\hat{\eta}$-pseudo-unitary if

$$\hat{\eta}^{-1}\hat{U}^\dagger\hat{\eta} = \hat{U}^{-1}\,. \tag{24}$$

The inner product (22) is invariant under $\hat{\eta}$-pseudo-unitary transforms

$$\langle\Psi_1|\Psi_2\rangle_{\hat{\eta}} = \langle\hat{U}\Psi_1|\hat{U}\Psi_2\rangle_{\hat{\eta}}\,. \tag{25}$$

Expectation values $\langle O\rangle$ of observables represented by the operator $\hat{O}$ are given by the pseudo-inner product (22), that is,

$$\langle O\rangle = \langle\Psi_1|\hat{O}\Psi_1\rangle_{\hat{\eta}}\,. \tag{26}$$

The Klein–Gordon equation is gauge invariant. Gauge invariance means that if we introduce via the gauge function $g(\mathbf{r},t)$, the new potentials

$$\mathbf{A}'(\mathbf{r},t) = \mathbf{A}(\mathbf{r},t) + \frac{1}{q}\frac{\partial g(\mathbf{r},t)}{\partial\mathbf{r}}\,, \tag{27a}$$

$$\phi'(\mathbf{r},t)=\phi(\mathbf{r},t)-\frac{1}{q}\frac{\partial g(\mathbf{r},t)}{\partial t}, \tag{27b}$$

which leave the electromagnetic fields $\mathbf{E}(\mathbf{r},t)$ and $\mathbf{B}(\mathbf{r},t)$ invariant; then, the wavefunction

$$\Psi'(\mathbf{r},t)=\exp\left[\mathrm{i}g(\mathbf{r},t)/\hbar\right]\Psi(\mathbf{r},t) \tag{28}$$

fulfills the Klein–Gordon equation with new potentials (27), that is,

$$\mathrm{i}\hbar\frac{\partial\Psi'(\mathbf{r},t)}{\partial t}=\hat{H}\left(\mathbf{A}'(\mathbf{r},t),\phi'(\mathbf{r},t)\right)\Psi'(\mathbf{r},t). \tag{29}$$

This can be proven by showing that

$$\begin{aligned}\hat{H}\left(\mathbf{A}'(\mathbf{r},t),\phi'(\mathbf{r},t)\right)\Psi'(\mathbf{r},t)-\mathrm{i}\hbar\frac{\partial\Psi'(\mathbf{r},t)}{\partial t}&=\\ \left[\hat{H}\left(\mathbf{A}(\mathbf{r},t),\phi(\mathbf{r},t)\right)\Psi(\mathbf{r},t)-\mathrm{i}\hbar\frac{\partial\Psi(\mathbf{r},t)}{\partial t}\right]&\exp\left[\mathrm{i}g(\mathbf{r},t)/\hbar\right].\end{aligned} \tag{30}$$

The right-hand side is zero because $\Psi(\mathbf{r},t)$ fulfills the Klein–Gordon equation.

1.3 The Dirac equation

Another possible approach to avoid the square root in (5) is to assume that the relativistic wave equation has form (12) and make a particular ansatz for the Hamiltonian. As (5) contains first-order time derivatives and special relativity puts time and space on an equal footing, it appears natural to make for free particles the linear ansatz

$$\hat{H}_0=c\boldsymbol{\alpha}\cdot(-\mathrm{i}\hbar\nabla)+\beta m_0c^2 \tag{31}$$

with the coefficients $\boldsymbol{\alpha}=(\alpha_1,\alpha_2,\alpha_3)^\mathsf{T}$ and β to be determined. In order to conform with the classical energy-momentum relation

$$E(\mathbf{p})=\sqrt{c^2\mathbf{p}^2+m_0^2c^4}, \tag{32}$$

we require that the square of the Hamiltonian is

$$\hat{H}_0^2=c^2(-\mathrm{i}\hbar\nabla)^2+m_0^2c^4 \tag{33}$$

or more explicitly

$$\begin{aligned}&-c^2\hbar^2\sum_i\sum_j\frac{\alpha_i\alpha_j+\alpha_j\alpha_i}{2}\partial_{r_i}\partial_{r_j}-\mathrm{i}\hbar m_0c^2\sum_i(\alpha_i\beta+\beta\alpha_i)\partial_{r_i}+\beta^2m_0^2c^4\\ &=-c^2\hbar^2\sum_i\partial_{r_i}^2+m_0^2c^4.\end{aligned} \tag{34}$$

Comparing the left-hand side and the right-hand side of this equation yields that the coefficients α_1, α_2, α_3, and β must obey the so-called Dirac algebra

$$\alpha_i^2 = \beta^2 = 1, \qquad \alpha_i\alpha_k + \alpha_k\alpha_i = 2\delta_{i,k}, \qquad \alpha_i\beta + \beta\alpha_i = 0. \tag{35}$$

This algebra cannot be fulfilled by ordinary numbers but by noncommuting Hermitian matrices. Hermiticity of these matrices ensures that the Hamiltonian is Hermitian, too. From $\alpha_i^2 = \beta^2 = 1$, it follows that the matrices α_i and β have eigenvalues ± 1. Furthermore, the traces vanish, because

$$\mathrm{Tr}(\alpha_i) = \mathrm{Tr}(\beta^2\alpha_i) = \mathrm{Tr}(\beta\alpha_i\beta) = \mathrm{Tr}(\beta(-\beta\alpha_i)) = -\mathrm{Tr}(\alpha_i), \tag{36}$$

$$\mathrm{Tr}(\beta) = \mathrm{Tr}(\alpha_i^2\beta) = \mathrm{Tr}(\alpha_i\beta\alpha_i) = \mathrm{Tr}(\alpha_i(-\alpha_i\beta)) = -\mathrm{Tr}(\beta_i) \tag{37}$$

and therefore $\mathrm{Tr}(\alpha_i) = \mathrm{Tr}(\beta) = 0$. As the trace equals the sum of all eigenvalues, each matrix must have the same number of positive and negative eigenvalues ± 1 and thus must have an even dimension. The lowest matrix dimension that allows to fulfill (35) in three space dimensions is four. The algebra (35) determines the matrices α_i and β only up to unitary transforms. Whenever a concrete representation is required, we adopt the so-called Dirac representation with

$$\begin{aligned}
\alpha_1 &= \begin{pmatrix} 0 & 0 & 0 & 1 \\ 0 & 0 & 1 & 0 \\ 0 & 1 & 0 & 0 \\ 1 & 0 & 0 & 0 \end{pmatrix}, \qquad
\alpha_2 = \begin{pmatrix} 0 & 0 & 0 & -\mathrm{i} \\ 0 & 0 & \mathrm{i} & 0 \\ 0 & -\mathrm{i} & 0 & 0 \\ \mathrm{i} & 0 & 0 & 0 \end{pmatrix}, \\
\alpha_3 &= \begin{pmatrix} 0 & 0 & 1 & 0 \\ 0 & 0 & 0 & -1 \\ 1 & 0 & 0 & 0 \\ 0 & -1 & 0 & 0 \end{pmatrix}, \qquad
\beta = \begin{pmatrix} 1 & 0 & 0 & 0 \\ 0 & 1 & 0 & 0 \\ 0 & 0 & -1 & 0 \\ 0 & 0 & 0 & -1 \end{pmatrix}.
\end{aligned} \tag{38}$$

Note that in this representation, the matrices α_i may also be written as

$$\alpha_i = \begin{pmatrix} 0 & \sigma_i \\ \sigma_i & 0 \end{pmatrix}. \tag{39}$$

Furthermore, only one or two α_i matrices are required in one-dimensional or two-dimensional systems, which reduces the number of constraints as given by the Dirac algebra (35). Therefore, the two-dimensional Pauli matrices (17) are sufficient to fulfill the Dirac algebra (35) in one or two dimensions, where the standard representation of Dirac matrices is given by

$$\alpha_1 = \sigma_1, \qquad \beta = \sigma_3, \tag{40}$$

and

$$\alpha_1 = \sigma_1, \qquad \alpha_2 = \sigma_2, \qquad \beta = \sigma_3, \tag{41}$$

respectively.

The equation

$$\mathrm{i}\hbar \frac{\partial \Psi(\mathbf{r},t)}{\partial t} = \hat{H}_0 \Psi(\mathbf{r},t) = \left(c\boldsymbol{\alpha}\cdot(-\mathrm{i}\hbar\boldsymbol{\nabla}) + \beta m_0 c^2\right)\Psi(\mathbf{r},t) \tag{42}$$

with the matrices α_1, α_2, α_3, and β is a quantum mechanical equation of motion for free particles that obey the relativistic energy-momentum relation. Because the matrices α_i and β are four-dimensional, the wavefunction $\Psi(\mathbf{r},t)$ has four components in case of the Dirac equation. The interaction with electromagnetic potentials may be introduced in analogy to the Schrödinger equation or the Klein–Gordon equation via the minimal-coupling principle, which yields

$$\mathrm{i}\hbar \frac{\partial \Psi(\mathbf{r},t)}{\partial t} = \hat{H}\Psi(\mathbf{r},t) = \left\{c\boldsymbol{\alpha}\cdot\left[-\mathrm{i}\hbar\boldsymbol{\nabla} - q\mathbf{A}(\mathbf{r},t)\right] + q\phi(\mathbf{r},t) + \beta m_0 c^2\right\}\Psi(\mathbf{r},t). \tag{43}$$

Like the Schrödinger and the Klein–Gordon equation, also the Dirac equation (43) is gauge invariant, i.e., the wavefunction

$$\Psi'(\mathbf{r},t) = \exp\left[\mathrm{i}g(\mathbf{r},t)/\hbar\right]\Psi(\mathbf{r},t) \tag{44}$$

fulfills the Dirac equation with the new potentials (27),

$$\mathrm{i}\hbar \frac{\partial \Psi'(\mathbf{r},t)}{\partial t} = \hat{H}\left(\mathbf{A}'(\mathbf{r},t), \phi'(\mathbf{r},t)\right)\Psi'(\mathbf{r},t) \tag{45}$$

with $\hat{H}$ being the Dirac-Hamiltonian defined in (43). As for the Klein–Gordon equation, this can be proven by showing that

$$\begin{aligned}
&\hat{H}\left(\mathbf{A}'(\mathbf{r},t), \phi'(\mathbf{r},t)\right)\Psi'(\mathbf{r},t) - \mathrm{i}\hbar\frac{\partial \Psi'(\mathbf{r},t)}{\partial t} \\
&\quad = \left\{\hat{H}\left(\mathbf{A}(\mathbf{r},t), \phi(\mathbf{r},t)\right)\Psi(\mathbf{r},t) - \mathrm{i}\hbar\frac{\partial \Psi(\mathbf{r},t)}{\partial t}\right\}\exp\left[\mathrm{i}g(\mathbf{r},t)/\hbar\right].
\end{aligned} \tag{46}$$

The right-hand side of the last equation is zero because $\Psi(\mathbf{r},t)$ fulfills the Dirac equation.

Solutions $\Psi(\mathbf{r},t)$ of equation (43) satisfy the continuity equation

$$\frac{\partial \rho(\mathbf{r},t)}{\partial t} + \boldsymbol{\nabla}\cdot\mathbf{j}(\mathbf{r},t) = 0 \tag{47}$$

with the density

$$\rho(\mathbf{r},t) = \Psi(\mathbf{r},t)^\dagger\Psi(\mathbf{r},t) \tag{48}$$

and the current

$$\mathbf{j}(\mathbf{r},t) = \Psi(\mathbf{r},t)^{\dagger} c\boldsymbol{\alpha}\Psi(\mathbf{r},t), \tag{49}$$

where $\Psi(\mathbf{r},t)^{\dagger}$ denotes the complex conjugate transpose of $\Psi(\mathbf{r},t)$.

2 Free particles and wave packets

Systems with vanishing electromagnetic potentials belong to the rare cases where the Klein–Gordon and the Dirac equations can be solved analytically [1]. These solutions can be employed to construct wave packets for initialization of the wavefunction in numerical solutions. Furthermore, analytic solutions give us insight into general properties of the Klein–Gordon and the Dirac equations.

2.1 Free-particle solution of the Klein–Gordon equation

The Klein–Gordon equation for a free particle reduces to

$$\mathrm{i}\hbar\frac{\partial\Psi(\mathbf{r},t)}{\partial t} = \left[\frac{\sigma_3+\mathrm{i}\sigma_2}{2m_0}(-\mathrm{i}\hbar\boldsymbol{\nabla})^2 + \sigma_3 m_0 c^2\right]\Psi(\mathbf{r},t). \tag{50}$$

Making the ansatz

$$\Psi(\mathbf{r},t) = \Psi(\mathbf{r})\,\mathrm{e}^{-\mathrm{i}Et/\hbar} \tag{51}$$

yields the eigenvalue equation

$$E\Psi(\mathbf{r}) = \left[\frac{\sigma_3+\mathrm{i}\sigma_2}{2m_0}(-\mathrm{i}\hbar\boldsymbol{\nabla})^2 + \sigma_3 m_0 c^2\right]\Psi(\mathbf{r}). \tag{52}$$

In Fourier space, where the differential operator $-\mathrm{i}\hbar\boldsymbol{\nabla}$ becomes a real vector $\mathbf{p}$, the eigenvalue equation reads as

$$0 = \left[\frac{\sigma_3+\mathrm{i}\sigma_2}{2m_0}\mathbf{p}^2 + \sigma_3 m_0 c^2 - E\right]\tilde{\Psi}(\mathbf{p}), \tag{53}$$

where $\tilde{\Psi}(\mathbf{p})$ denotes the Fourier transform of $\Psi(\mathbf{r})$. For a nontrivial solution of (53), the condition

$$0 = \det\left(\frac{\sigma_3+\mathrm{i}\sigma_2}{2m_0}\mathbf{p}^2 + \sigma_3 m_0 c^2 - E\right) = E^2 - \mathbf{p}^2c^2 - m_0^2c^4 \tag{54}$$

has to be fulfilled. This yields the two energy eigenvalues

$$E = \pm\sqrt{\mathbf{p}^2c^2 + m_0^2c^4} = \pm E(\mathbf{p}). \tag{55}$$

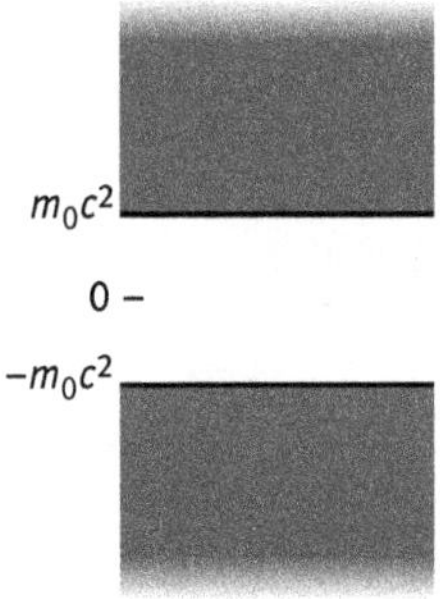

Fig. 1. The energy spectrum of the free-particle Klein–Gordon and Dirac Hamiltonians splits into two continua with positive and negative energy values.

This means that a free Klein–Gordon particle may have positive or negative kinetic energy, unlike in classical relativistic mechanics or nonrelativistic quantum mechanics where the kinetic energy is always positive, see Figure 1. Inserting the two eigenvalues (55) into (53) gives the two orthogonal eigenvectors

$$\tilde{\Psi}^{+}(\mathbf{p}) = \frac{1}{2\sqrt{p_0(\mathbf{p})/(m_0c)}}\begin{pmatrix}1+p_0(\mathbf{p})/(m_0c)\\ 1-p_0(\mathbf{p})/(m_0c)\end{pmatrix}, \tag{56a}$$

$$\tilde{\Psi}^{-}(\mathbf{p}) = \frac{1}{2\sqrt{p_0(\mathbf{p})/(m_0c)}}\begin{pmatrix}1-p_0(\mathbf{p})/(m_0c)\\ 1+p_0(\mathbf{p})/(m_0c)\end{pmatrix}, \tag{56b}$$

where

$$p_0(\mathbf{p}) = \frac{E(\mathbf{p})}{c} = \sqrt{m_0^2c^2+\mathbf{p}^2}. \tag{57}$$

Both vectors have been normalized such that

$$\tilde{\Psi}^{+}(\mathbf{p})^{\dagger}\sigma_3\tilde{\Psi}^{+}(\mathbf{p}) = 1, \tag{58a}$$

$$\tilde{\Psi}^{-}(\mathbf{p})^{\dagger}\sigma_3\tilde{\Psi}^{-}(\mathbf{p}) = -1, \tag{58b}$$

$$\tilde{\Psi}^{+}(\mathbf{p})^{\dagger}\sigma_3\tilde{\Psi}^{-}(\mathbf{p}) = 0. \tag{58c}$$

Going back to position space, the full solutions of the free Klein–Gordon equation read as

$$\Psi_{\mathbf{p}}^{+}(\mathbf{r},t) = \frac{1}{(2\pi\hbar)^{3/2}}\frac{1}{2\sqrt{p_0(\mathbf{p})/(m_0c)}}\begin{pmatrix}1+p_0(\mathbf{p})/(m_0c)\\ 1-p_0(\mathbf{p})/(m_0c)\end{pmatrix}\mathrm{e}^{\mathrm{i}(\mathbf{p}\cdot\mathbf{r}-E(\mathbf{p})t)/\hbar}, \tag{59a}$$

$$\Psi_{\mathbf{p}}^{-}(\mathbf{r},t) = \frac{1}{(2\pi\hbar)^{3/2}}\frac{1}{2\sqrt{p_0(\mathbf{p})/(m_0c)}}\begin{pmatrix}1-p_0(\mathbf{p})/(m_0c)\\ 1+p_0(\mathbf{p})/(m_0c)\end{pmatrix}\mathrm{e}^{\mathrm{i}(\mathbf{p}\cdot\mathbf{r}+E(\mathbf{p})t)/\hbar}. \tag{59b}$$

These functions are orthonormalized, i.e.,

$$\langle \Psi^+_{\mathbf{p}_1} | \Psi^+_{\mathbf{p}_2} \rangle_{\sigma_3} = \delta(\mathbf{p}_1 - \mathbf{p}_2), \tag{60a}$$

$$\langle \Psi^-_{\mathbf{p}_1} | \Psi^-_{\mathbf{p}_2} \rangle_{\sigma_3} = -\delta(\mathbf{p}_1 - \mathbf{p}_2), \tag{60b}$$

$$\langle \Psi^+_{\mathbf{p}_1} | \Psi^-_{\mathbf{p}_2} \rangle_{\sigma_3} = 0, \tag{60c}$$

where $\delta(x)$ denotes the Dirac δ distribution. The two functions (59) are also eigenfunctions of the momentum operator $-i\hbar\nabla$ with the same eigenvalue $\mathbf{p}$. Because of the nondefinite scalar product and (58), the relations

$$\Psi^+_{\mathbf{p}}(\mathbf{r},t)^\dagger \sigma_3(-i\hbar\nabla)\Psi^+_{\mathbf{p}}(\mathbf{r},t) = \mathbf{p}, \tag{61a}$$

$$\Psi^-_{\mathbf{p}}(\mathbf{r},t)^\dagger \sigma_3(-i\hbar\nabla)\Psi^-_{\mathbf{p}}(\mathbf{r},t) = -\mathbf{p} \tag{61b}$$

follow. An arbitrary solution of the free Klein–Gordon equation may be built by superimposing plane waves of different momenta, viz,

$$\Psi(\mathbf{r},t) = \int \left[g^+(\mathbf{p})\Psi^+_{\mathbf{p}}(\mathbf{r},t) + g^-(\mathbf{p})\Psi^-_{\mathbf{p}}(\mathbf{r},t) \right] \mathrm{d}^3 p \tag{62}$$

with weight functions $g^+(\mathbf{p})$ and $g^-(\mathbf{p})$. For a proper normalization,

$$\int \left[|g^+(\mathbf{p})|^2 + |g^-(\mathbf{p})|^2 \right] \mathrm{d}^3 p = 1 \tag{63}$$

is required. Figure 2 shows two one-dimensional wave packets that are superpositions of only positive-energy states and only negative-energy states, respectively, at two different times. The functions $g^\pm(\mathbf{p})$ are Gaussians with both having the mean

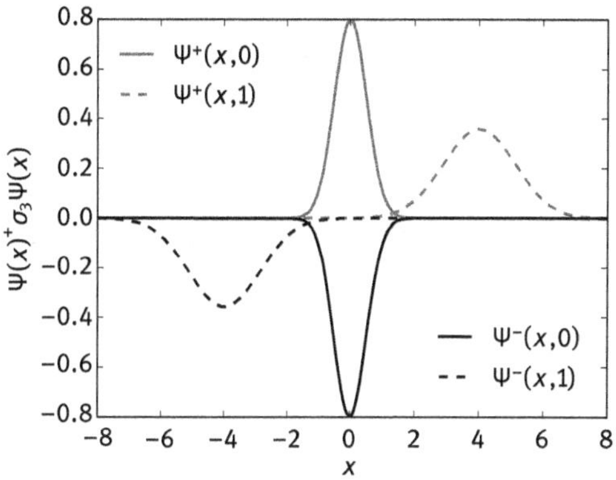

Fig. 2. The densities of free Klein–Gordon wave packets composed of plane waves having positive ($\Psi^+(x,t)$) or negative ($\Psi^-(x,t)$) energy only at two different times. Although the weight function has its maximum at a positive momentum in both cases, the negative-energy wave packet moves into the negative direction.

momentum at $p = 4$ a.u. Note that because of normalization (58), the wave packet that is a superposition of negative-energy states has a negative density. Furthermore, it features a negative mean momentum $\langle \Psi^- | -\mathrm{i}\hbar\nabla | \Psi^- \rangle_{\sigma_3} = -4$ a.u. as a consequence of (61b), while $\langle \Psi^+ | -\mathrm{i}\hbar\nabla | \Psi^+ \rangle_{\sigma_3} = 4$ a.u. Note that the wave packet that is a superposition of negative-energy states moves to the left. This is a direct consequence of the energy being negative, as it can be seen by considering a wave packet

$$\Psi(\mathbf{r}, t) \sim \int g(\mathbf{p})\, \mathrm{e}^{\mathrm{i}(\mathbf{p}\cdot\mathbf{r} \mp E(\mathbf{p})t)/\hbar} . \tag{64}$$

Let us assume that $g(\mathbf{p})$ is peaked around $\mathbf{p}^*$. We can then linearize

$$E(\mathbf{p}) \sim E(\mathbf{p}^*) + \nabla E(\mathbf{p})|_{\mathbf{p}=\mathbf{p}^*} \cdot (\mathbf{p} - \mathbf{p}^*) \tag{65}$$

so that

$$\Psi(\mathbf{r}, t) \sim \mathrm{e}^{\pm\mathrm{i}(E(\mathbf{p}^*) + \nabla E(\mathbf{p})|_{\mathbf{p}=\mathbf{p}^*} \cdot \mathbf{p}^*)t/\hbar} \int g(\mathbf{p})\, \mathrm{e}^{\mathrm{i}\mathbf{p}\cdot(\mathbf{r} \mp \nabla E(\mathbf{p})|_{\mathbf{p}=\mathbf{p}^*} t)/\hbar} , \tag{66}$$

and the group velocity equals

$$\mathbf{v} = \pm\nabla E(\mathbf{p})|_{\mathbf{p}=\mathbf{p}^*} = \pm\frac{\mathbf{p}^*}{m_0} \frac{1}{\sqrt{1 + \mathbf{p}^{*2}/(m_0 c)^2}} . \tag{67}$$

Note that this is, up to the sign, exactly the velocity of a classical, relativistic point-like particle with momentum $\mathbf{p}^*$.

2.2 Free-particle solution of the Dirac equation

Let us solve the free-particle Dirac equation (42) before we consider more complicated setups later on. In analogy to the Schrödinger and the Klein–Gordon equation, we utilize the ansatz

$$\Psi(\mathbf{r}, t) = \Psi(\mathbf{r})\, \mathrm{e}^{-\mathrm{i}Et/\hbar} \tag{68}$$

with the four-component function $\Psi(\mathbf{r})$ and the energy E to be determined. The ansatz (68) leads to the eigenvalue equation

$$E\Psi(\mathbf{r}) = \left[c\boldsymbol{\alpha} \cdot (-\mathrm{i}\hbar\nabla) + \beta m_0 c^2\right] \Psi(\mathbf{r}), \tag{69}$$

and in Fourier space

$$0 = \left(c\boldsymbol{\alpha} \cdot \mathbf{p} + \beta m_0 c^2 - E\right) \tilde{\Psi}(\mathbf{p}) . \tag{70}$$

For a nontrivial solution of (70), the condition

$$0 = \det\left(c\boldsymbol{\alpha} \cdot \mathbf{p} + \beta m_0 c^2 - E\right) = (E^2 - \mathbf{p}^2 c^2 - m_0^2 c^4)^2 \tag{71}$$

has to be fulfilled, where we utilized the Dirac algebra (35). This yields the two doubly degenerate energy eigenvalues

$$E = \pm\sqrt{\mathbf{p}^2 c^2 + m_0^2 c^4} = \pm E(\mathbf{p})\,. \tag{72}$$

A free Dirac particle may thus have positive or negative kinetic energy, which agrees with the energy spectrum of the Klein–Gordon equation. Furthermore, the doubly degenerate energy eigenvalues indicate that momentum and energy are not sufficient to characterize an eigenstate of the free Dirac equation. Dirac particles carry some additional internal degree of freedom that is required to characterize the particle completely. The eigenfunctions of the free Dirac equation (42) can be chosen as simultaneous eigenfunctions of the free Dirac Hamiltonian, the momentum operator $-i\hbar\boldsymbol{\nabla}$, and some further operator that characterizes this internal degree of freedom, which can be related to the spin of the electron [3]. One can show that a possible set of orthonormal eigenfunctions is given by

$$\tilde{\Psi}^{+}_{1,2}(\mathbf{p}) = \sqrt{\frac{m_0 c + p_0(\mathbf{p})}{2p_0(\mathbf{p})}} \begin{pmatrix} \chi_{1,2} \\ \dfrac{\boldsymbol{\sigma}\cdot\mathbf{p}}{m_0 c + p_0(\mathbf{p})}\chi_{1,2} \end{pmatrix}, \tag{73a}$$

$$\tilde{\Psi}^{-}_{1,2}(\mathbf{p}) = \sqrt{\frac{m_0 c + p_0(\mathbf{p})}{2p_0(\mathbf{p})}} \begin{pmatrix} -\dfrac{\boldsymbol{\sigma}\cdot\mathbf{p}}{m_0 c + p_0(\mathbf{p})}\chi_{1,2} \\ \chi_{1,2} \end{pmatrix}, \tag{73b}$$

where χ_1 and χ_2 denote two arbitrary orthogonal two-component unit-vectors. We will choose

$$\chi_1 = \begin{pmatrix} 1 \\ 0 \end{pmatrix}, \qquad \chi_2 = \begin{pmatrix} 0 \\ 1 \end{pmatrix}. \tag{74}$$

Another common choice for $\chi_{1,2}$ is such that both are eigenvectors of the operator $\boldsymbol{\sigma}\cdot\mathbf{p}$. In this case, χ_1 and χ_2 are momentum dependent, and the vectors (73) are also eigenstates of the so-called helicity operator $\Sigma\cdot\mathbf{p}/|\mathbf{p}|$ with $\Sigma = (\Sigma_1, \Sigma_2, \Sigma_3)^{\mathsf{T}}$ and

$$\Sigma_i = \begin{pmatrix} \sigma_i & 0 \\ 0 & \sigma_i \end{pmatrix}. \tag{75}$$

Both vectors (73) have been normalized such that

$$\tilde{\Psi}^{+}_{i}(\mathbf{p})^{\dagger}\tilde{\Psi}^{+}_{j}(\mathbf{p}) = \delta_{i,j}, \tag{76a}$$

$$\tilde{\Psi}^{-}_{i}(\mathbf{p})^{\dagger}\tilde{\Psi}^{-}_{j}(\mathbf{p}) = \delta_{i,j}, \tag{76b}$$

$$\tilde{\Psi}^{+}_{i}(\mathbf{p})^{\dagger}\tilde{\Psi}^{-}_{j}(\mathbf{p}) = 0. \tag{76c}$$

In position space, the full solutions of the free Dirac equation are given by

$$\Psi^{+}_{\mathbf{p},i}(\mathbf{r},t)=\frac{1}{(2\pi\hbar)^{3/2}}\sqrt{\frac{m_0c+p_0(\mathbf{p})}{2p_0(\mathbf{p})}}\begin{pmatrix}\chi_i\\ \dfrac{\boldsymbol{\sigma}\cdot\mathbf{p}}{m_0c+p_0(\mathbf{p})}\chi_i\end{pmatrix}\mathrm{e}^{\mathrm{i}(\mathbf{p}\cdot\mathbf{r}-E(\mathbf{p})t)/\hbar}, \tag{77a}$$

$$\Psi^{-}_{\mathbf{p},i}(\mathbf{r},t)=\frac{1}{(2\pi\hbar)^{3/2}}\sqrt{\frac{m_0c+p_0(\mathbf{p})}{2p_0(\mathbf{p})}}\begin{pmatrix}-\dfrac{\boldsymbol{\sigma}\cdot\mathbf{p}}{m_0c+p_0(\mathbf{p})}\chi_i\\ \chi_i\end{pmatrix}\mathrm{e}^{\mathrm{i}(\mathbf{p}\cdot\mathbf{r}+E(\mathbf{p})t)/\hbar}. \tag{77b}$$

The upper index indicates the sign of the energy eigenvalue of these solutions and the lower index $\mathbf{p}$ the momentum eigenvalue. The two-fold degeneracy is indicated by the index $i\in\{1,2\}$. The functions in (77) are orthonormalized, i.e.,

$$\langle\Psi^{+}_{\mathbf{p}_1,i}|\Psi^{+}_{\mathbf{p}_2,j}\rangle=\delta_{i,j}\delta(\mathbf{p}_1-\mathbf{p}_2), \tag{78a}$$

$$\langle\Psi^{-}_{\mathbf{p}_1,i}|\Psi^{-}_{\mathbf{p}_2,j}\rangle=\delta_{i,j}\delta(\mathbf{p}_1-\mathbf{p}_2), \tag{78b}$$

$$\langle\Psi^{+}_{\mathbf{p}_1,i}|\Psi^{-}_{\mathbf{p}_2,j}\rangle=0. \tag{78c}$$

The four functions (77) are also eigenfunctions of the momentum operator $-\mathrm{i}\hbar\boldsymbol{\nabla}$ with the eigenvalue $\mathbf{p}$. Furthermore, they form a complete basis. Thus, an arbitrary solution of the free Dirac equation may be built by superimposing plane waves of different momenta, viz,

$$\begin{aligned}\Psi(\mathbf{r},t)=\int\big[&g^{+}_{1}(\mathbf{p})\Psi^{+}_{\mathbf{p},1}(\mathbf{r},t)+g^{+}_{2}(\mathbf{p})\Psi^{+}_{\mathbf{p},2}(\mathbf{r},t)\\ &+g^{-}_{1}(\mathbf{p})\Psi^{-}_{\mathbf{p},1}(\mathbf{r},t)+g^{-}_{2}(\mathbf{p})\Psi^{-}_{\mathbf{p},2}(\mathbf{r},t)\big]\,\mathrm{d}^3p\end{aligned} \tag{79}$$

with the weight functions $g^{+}_{1}(\mathbf{p})$, $g^{+}_{1}(\mathbf{p})$, $g^{-}_{1}(\mathbf{p})$, and $g^{-}_{2}(\mathbf{p})$. The quantity $|g^{+}_{1}(\mathbf{p})|^2+|g^{+}_{2}(\mathbf{p})|^2+|g^{-}_{1}(\mathbf{p})|^2+|g^{-}_{2}(\mathbf{p})|^2$ is the wave packet's density in momentum space. Hence,

$$\int|g^{+}_{1}(\mathbf{p})|^2+|g^{+}_{2}(\mathbf{p})|^2+|g^{-}_{1}(\mathbf{p})|^2+|g^{-}_{2}(\mathbf{p})|^2\,\mathrm{d}^3p=1 \tag{80}$$

is required for a proper normalization.

Equation (77) presents the solution of the free Dirac equation in three dimensions. As mentioned before, in one and two dimensions, the Dirac matrices can be represented by 2×2 matrices, and the wavefunctions then have only two components. By a similar calculation as for the three-dimensional case, the solutions

$$\Psi^{+}_{p}(x,t)=\frac{1}{(2\pi\hbar)^{1/2}}\sqrt{\frac{m_0c+p_0(p)}{2p_0(p)}}\begin{pmatrix}1\\ \dfrac{p}{m_0c+p_0(p)}\end{pmatrix}\mathrm{e}^{\mathrm{i}(px-E(p)t)/\hbar}, \tag{81a}$$

$$\Psi^{-}_{p}(x,t)=\frac{1}{(2\pi\hbar)^{1/2}}\sqrt{\frac{m_0c+p_0(p)}{2p_0(p)}}\begin{pmatrix}-\dfrac{p}{m_0c+p_0(p)}\\ 1\end{pmatrix}\mathrm{e}^{\mathrm{i}(px+E(p)t)/\hbar} \tag{81b}$$

for the one-dimensional free Dirac equation follow. Here, p denotes the momentum eigenvalue of these functions. Analogously, the solutions of the two-dimensional free Dirac equation are

$$\Psi_{\mathbf{p}}^{+}(\mathbf{r},t) = \frac{1}{2\pi\hbar}\sqrt{\frac{m_0c+p_0(\mathbf{p})}{2p_0(\mathbf{p})}}\begin{pmatrix} 1 \\ \dfrac{p_x+\mathrm{i}p_y}{m_0c+p_0(\mathbf{p})}\end{pmatrix}\mathrm{e}^{\mathrm{i}(\mathbf{p}\cdot\mathbf{r}-E(\mathbf{p})t)/\hbar}, \tag{82a}$$

$$\Psi_{\mathbf{p}}^{-}(\mathbf{r},t) = \frac{1}{2\pi\hbar}\sqrt{\frac{m_0c+p_0(\mathbf{p})}{2p_0(\mathbf{p})}}\begin{pmatrix} \dfrac{-p_x+\mathrm{i}p_y}{m_0c+p_0(\mathbf{p})} \\ 1\end{pmatrix}\mathrm{e}^{\mathrm{i}(\mathbf{p}\cdot\mathbf{r}+E(\mathbf{p})t)/\hbar}, \tag{82b}$$

with the two-dimensional position vector $\mathbf{r} = (x, y)$ and the two-dimensional momentum eigenvalue $\mathbf{p} = (p_x, p_y)$.

The wavefunctions in (77) are four-fold degenerate for a fixed momentum. The sign of the energy and the spin are needed as further quantum numbers to specify an eigenstate completely. The four-fold degeneracy is also reflected in the need to employ four-component wavefunctions. The physical reason why the Dirac equation wavefunction in one-dimensional systems has only two components and not four as in three dimensions is that a one-dimensional vector potential cannot describe a magnetic field to which a spin could couple. Consequently, the solutions of the one-dimensional free Dirac equation do not feature the two-fold degeneracy due to the electron spin and can be expressed as two-component wavefunctions. As the functions (82) have only two components, they feature no spin too, although vector potentials can lead to a nonvanishing magnetic field in this case. However, this field is embedded into a three-dimensional space. Therefore, to describe electrons including their spin in two-dimensional systems, one has to employ a four-component Dirac equation, which can be obtained by ignoring any dependency on the third dimension in the three-dimensional Dirac equation.

Figure 3 shows two one-dimensional wave packets that are superpositions of only positive-energy states and only negative-energy states, respectively,

$$\Psi^{+}(x,t) = \int g(p)\Psi_{p}^{+}(x,t)\,\mathrm{d}p\,, \tag{83}$$

$$\Psi^{-}(x,t) = \int g(p)\Psi_{p}^{-}(x,t)\,\mathrm{d}p\,, \tag{84}$$

with

$$g(p) = \frac{1}{(2\pi\Delta p^2)^{1/4}}\exp\left(-\frac{(p-p^{*})^2}{4\Delta p^2}\right) \tag{85}$$

at two different times. For Figure 3, the function $g(p)$ is a Gaussian with mean at $p^{*} = 4$ a.u. Note that the wave packet $\Psi^{-}(x,t)$ moves to the left, analogously to the Klein–Gordon case above.

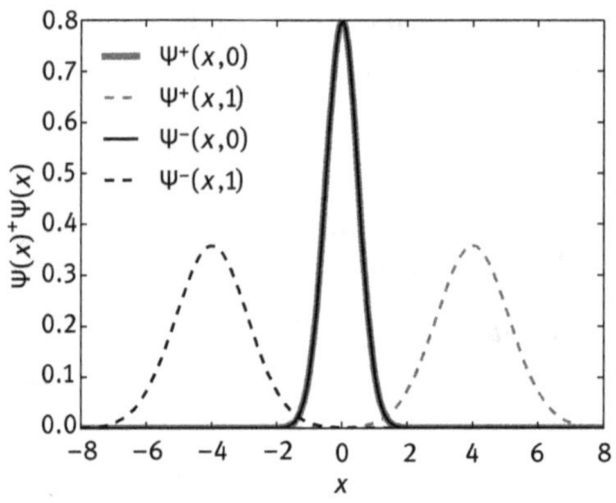

Fig. 3. The densities of free one-dimensional Dirac wave packets at two times. The wave packets are composed of plane waves having positive ($\Psi^+(x,t)$) or negative ($\Psi^-(x,t)$) energy only. Although the weight function has its maximum at a positive momentum in both cases, the negative-energy wave packet moves into the negative direction.

In Figure 3, we also observe wave packet spreading. Because at low momenta ($|p^*| \ll m_0c$), the relation between momentum and velocity is linear, the free Gaussian wave packets spread symmetrically as in the nonrelativistic Schrödinger theory. However, if a wave packet has mean momentum larger than m_0c and a broad momentum distribution, the relativistic nonlinear relation between momentum and velocity causes an asymmetric velocity distribution, resulting in an asymmetrical spreading. The wave packet forms a shock front that moves with velocity close to c, see

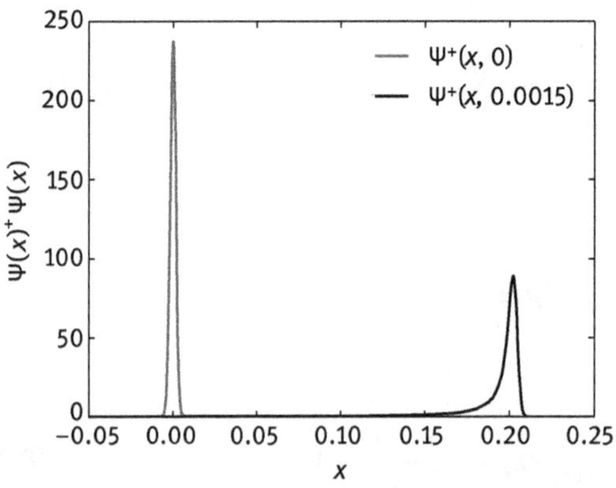

Fig. 4. A free Dirac wave packet composed of plane waves having positive energy only with mean momentum $p^* = 4m_0c$ and a broad momentum width $\Delta p = 2.5m_0c$. As a consequence of the nonlinear relation between momentum and velocity, the wave packet forms a shock front that moves with velocity close to c.

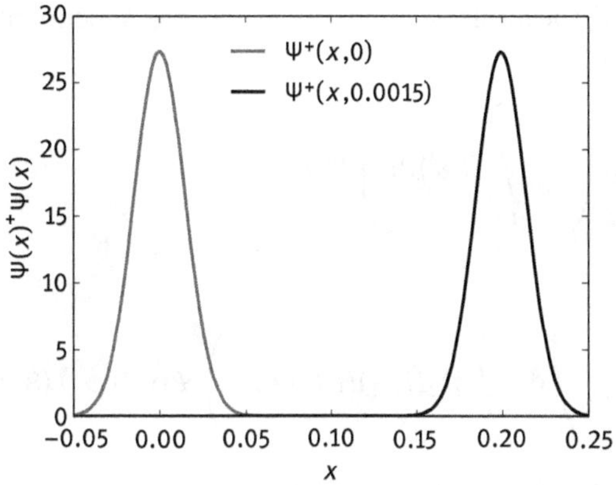

Fig. 5. Same as Figure 4 but for $\Delta p = 0.25 m_0 c$. The whole wave packet moves with velocity close to c without changing its form significantly.

Figure 4. If the momentum distribution is narrow, however, the velocity distribution gets concentrated close to c, and the whole wave packet moves with a velocity close to c without changing its form significantly, see Figure 5.

3 Numerical solution of the Dirac equation

The Dirac equation and the Klein–Gordon equation can be solved exactly only in very few cases. Analytical methods for determining solutions of these equations usually require physical setups with a high degree of symmetry [1, 6, 13, 36, 39, 41]. Also, approximations reach their limits in describing the interaction with high-frequency, few-cycle laser pulses of high intensity and for other relativistic high-energy processes. Thus, numerical approaches via computer simulations are indicated [2, 4, 5, 9, 10, 25, 32].

3.1 General methods for time-dependent quantum mechanics

The basic numerical methods that are used to solve the Dirac equation and the Klein–Gordon equation are often not specific to these equations. In fact, they are based on the (pseudo) Hermitian structure of the time-evolution equation

$$\mathrm{i}\hbar\frac{\partial}{\partial t}\Psi(\mathbf{r},t) = \hat{H}(t)\Psi(\mathbf{r},t). \tag{86}$$

Using Dyson's time-ordering operator $\hat{T}$, the formal solution [31] to the time-dependent equation (86) can be expressed as

$$\Psi(\mathbf{r}, t+\Delta t) = \hat{T} \exp\left(-\frac{\mathrm{i}}{\hbar} \int_{t}^{t+\Delta t} \hat{H}(t')\,\mathrm{d}t'\right) \Psi(\mathbf{r}, t), \tag{87}$$

which reads more explicitly

$$\Psi(\mathbf{r}, t+\Delta t) = \left(1 - \frac{\mathrm{i}}{\hbar} \int_{t_0}^{\Delta t} \mathrm{d}t_1 \hat{H}(t_1) + \frac{\mathrm{i}^2}{\hbar^2} \int_{t_0}^{\Delta t} \mathrm{d}t_1 \int_{t_0}^{t_1} \mathrm{d}t_2 \hat{H}(t_1)\hat{H}(t_2) + \ldots\right) \Psi(\mathbf{r}, t). \tag{88}$$

Usually, it is not possible to cast (87) directly into a numerical scheme. Some approximations have to be introduced. First, we neglect the time-ordering operator, introducing an error on the order of $\mathcal{O}(\Delta t^3)$,

$$\Psi(\mathbf{r}, t+\Delta t) = \exp\left(-\frac{\mathrm{i}}{\hbar} \int_{t}^{t+\Delta t} \hat{H}(t')\,\mathrm{d}t'\right) \Psi(\mathbf{r}, t) + \mathcal{O}\left(\Delta t^3\right). \tag{89}$$

Second, the operator

$$\hat{\mathcal{H}}(t, \Delta t) = \frac{1}{\hbar} \int_{t}^{t+\Delta t} \hat{H}(t')\,\mathrm{d}t' \tag{90}$$

has to be discretized and becomes a matrix $\mathcal{H}(t, \Delta t)$. Thus, time propagation reduces to the calculation of matrix exponentials. In the following, we assume that the discretization preserves the (pseudo) Hermiticity of $\mathcal{H}(t, \Delta t)$.

A standard procedure to compute the exponential of some matrix is to calculate its eigendecomposition. Let $\mathbf{Q}$ be a square matrix whose columns are the normalized eigenvectors of $\mathcal{H}(t, \Delta t)$ in any order. The existence of such a matrix is ensured by $\mathcal{H}(t, \Delta t)$ being (pseudo) Hermitian. Thus,

$$\mathcal{H}(t, \Delta t)\mathbf{Q} = \mathbf{Q}\boldsymbol{\Lambda}, \tag{91}$$

where $\boldsymbol{\Lambda}$ denotes the diagonal matrix of the real eigenvalues $E_i = \boldsymbol{\Lambda}_{i,i}$ of $\mathcal{H}(t, \Delta t)$ such that $\boldsymbol{\Lambda}_{i,i}$ is the eigenvalue associated to the ith column of $\mathbf{Q}$. Thanks to the (pseudo) Hermiticity of $\mathcal{H}(t, \Delta t)$, the inverse of $\mathbf{Q}$ does exist. Consequently, $\mathbf{Q}$ makes $\mathcal{H}(t, \Delta t)$ diagonal

$$\boldsymbol{\Lambda} = \mathbf{Q}^{-1}\mathcal{H}(t, \Delta t)\mathbf{Q}. \tag{92}$$

Knowing the diagonalization of $\mathcal{H}(t, \Delta t)$, the matrix exponential becomes

$$\exp\left(-\mathrm{i}\mathcal{H}(t, \Delta t)\right) = \sum_j \frac{(-\mathrm{i})^j}{j!} \mathcal{H}(t, \Delta t)^j = \sum_j \frac{(-\mathrm{i})^j}{j!} \mathbf{Q}\left(\mathbf{Q}^{-1}\mathcal{H}(t, \Delta t)\mathbf{Q}\right)^j \mathbf{Q}^{-1}$$

$$= \sum_j \frac{(-\mathrm{i})^j}{j!} \mathbf{Q}\mathbf{\Lambda}^j \mathbf{Q}^{-1} = \mathbf{Q}\left(\sum_j \frac{(-\mathrm{i})^j}{j!} \mathbf{\Lambda}^j \right) \mathbf{Q}^{-1} = \mathbf{Q} \exp(-\mathrm{i}\mathbf{\Lambda}) \mathbf{Q}^{-1}. \quad (93)$$

The matrix exponential $\exp(-\mathrm{i}\mathbf{\Lambda})$ is easy to compute. As $\mathbf{\Lambda}$ is diagonal, $\exp(-\mathrm{i}\mathbf{\Lambda})$ is a diagonal matrix with $\exp(-\mathrm{i}E_j)$ on the diagonal. Also, the inverse of $\mathbf{Q}$ is easy to compute. If $\mathcal{H}(t, \Delta t)$ is Hermitian, then

$$\mathbf{Q}^{-1} = \mathbf{Q}^{\dagger}. \quad (94)$$

If $\mathcal{H}(t, \Delta t)$ is $\boldsymbol{\eta}$-pseudo Hermitian and all its eigenvalues are real [24], then

$$\mathbf{Q}^{-1} = \mathbf{Q}^{\dagger}\boldsymbol{\eta}. \quad (95)$$

For the discretization of the Hamiltonian and thus $\mathcal{H}(t, \Delta t)$, many different methods can be applied. Particularly simple schemes are based on finite differences, where the wavefunction is evaluated on a regular rectangular grid of grid spacing Δx. Then, the first- and second-order differential operators can be approximated in second and fourth order as

$$\begin{aligned}
\frac{\mathrm{d}f(x)}{\mathrm{d}x} &= -\frac{f(x-\Delta x)}{2\Delta x} + \frac{f(x+\Delta x)}{2\Delta x} + \mathcal{O}\left(\Delta x^3\right), \\
\frac{\mathrm{d}f(x)}{\mathrm{d}x} &= \frac{f(x-2\Delta x)}{12\Delta x} - \frac{3f(x-\Delta x)}{2\Delta x} + \frac{3f(x+\Delta x)}{2\Delta x} - \frac{f(x+2\Delta x)}{12\Delta x} + \mathcal{O}\left(\Delta x^5\right), \\
\frac{\mathrm{d}^2f(x)}{\mathrm{d}x^2} &= \frac{f(x-\Delta x)}{\Delta x^2} - 2\frac{f(x)}{\Delta x^2} + \frac{f(x+\Delta x)}{\Delta x^2} + \mathcal{O}\left(\Delta x^3\right), \\
\frac{\mathrm{d}^2f(x)}{\mathrm{d}x^2} &= -\frac{f(x-2\Delta x)}{12\Delta x^2} + \frac{4f(x-\Delta x)}{3\Delta x^2} - \frac{5f(x)}{2\Delta x^2} \\
&\quad + \frac{4f(x+\Delta x)}{3\Delta x^2} - \frac{f(x+2\Delta x)}{12\Delta x^2} + \mathcal{O}\left(\Delta x^5\right).
\end{aligned}$$

Listing 1 shows a Python [37] implementation of the numerical solution of the time-dependent Schrödinger equation for the harmonic oscillator with $q\phi(x,t) = \frac{1}{2}m_0\omega^2x^2$ and vanishing vector potential via an exact diagonalization of the Hamiltonian as outlined above. The second-order derivatives are approximated via second-order finite differences.

The diagonalization approach is very general and can be applied to the Schrödinger equation, the Klein–Gordon equation, or the Dirac equation, as well as any other equation with a similar Hamiltonian structure. Employing some standard mathematical software, it can be implemented in just a few lines of code. However, approaches that are based on exact diagonalization suffer from several problems:

- The explicit matrix representation of the Hamiltonian has to be stored.
- The method requires to store all eigenvectors; thus, the required storage scales like $\mathcal{O}(N^2)$, where N is the total number of grid points.
- Practical matrix diagonalization algorithms scale with $\mathcal{O}(N^3)$.

Listing 1. Numerical solution of the time-dependent one-dimensional Schrödinger equation for the harmonic oscillator via exact diagonalization of the Hamiltonian.

```
#!/usr/bin/env python
# -*- coding: utf-8 -*-

from pylab import *

# the harmonic oscillator potential with hbar=m0=omega=1
def V(x):
    return 0.5*x**2

# define computational grid running from -5 to 5 with 512 grid points
N=512
x0, x1=-5., 5.
x=linspace(x0, x1, N)
dx=(x1-x0)/(N-1)  # size of spartial grid spacing
dt=1./64          # temporal step size

# create Hamiltonian matrix and calculate its eigenvectors and eigenvalues
# use finite differences to approximate derivatives to 2nd order
H=( -0.5*(diag(ones(N-1), -1)-diag(2*ones(N))+diag(ones(N-1), +1) )/(dx**2) +
    diag(V(x)) )*dt
Lambda, Q=eigh(H)
Q_inv=Q.conj().transpose()  # inverse of Q

# Gaussian wave packet with mean momentum one as initial condition
Psi=1./(2*pi)**(0.25)*exp(-x**2/4 + 1j*x)

# propagate 512 time steps
for k in range(0, 512):
    clf()
    plot(x, Psi.real, '--', color='r', label=r'$\mathrm{Re}\,\Psi(x)$')
    plot(x, Psi.imag, ':', color='r', label=r'$\mathrm{Im}\,\Psi(x)$')
    plot(x, Psi.real**2+Psi.imag**2, color='#266bbd', label=r'$|\Psi(x)|^2$')
    gca().set_xlim(x0, x1)
    gca().set_ylim(-0.6, 1)
    xlabel(r'$x$')
    legend(loc='upper left')
    draw()
    show()
    pause(0.01)
    Psi=dot(Q_inv, Psi)    # expand into eigen-functions
    Psi*=exp(-1j*Lambda)   # propagate eigen-functions
    Psi=dot(Q, Psi)        # superpose eigen-functions
```

- If the Hamiltonian is time dependent, diagonalization is required at each time step.
- Because of the band gap of $\simeq 2m_0c^2$ of relativistic Hamiltonians, it is difficult to compute all the eigenvalues and eigenvectors of the corresponding Hamiltonian with high accuracy.

Fortunately, there are methods to circumvent these issues. One possible way is to find a representation of the matrix $\mathcal{H}(t, \Delta t)$ in a smaller subspace via the Lanczos algorithm and to perform the diagonalization in this subspace [4, 12, 14, 22, 23, 30, 33, 35]. Other methods are, for example, based on operator splitting, as discussed in Chapter I, which will be adapted for relativistic wave equations in Section 3.4.

3.2 The split operator method

The central idea of the split operator method [2, 5, 7, 25, 32] is to approximate the time-evolution operator

$$\hat{U}(t+\Delta t, t) = \exp\left(-\frac{\mathrm{i}}{\hbar}\int\limits_{t}^{t+\Delta t} \hat{H}(t')\,\mathrm{d}t'\right) \tag{96}$$

by a product of operators that are diagonal in an appropriate space. Let $\hat{O}(t)$ denote some possibly time-dependent operator and define the operator

$$\hat{U}_{\hat{O}}(t_2, t_1, \delta) = \exp\left(-\delta\frac{\mathrm{i}}{\hbar}\int\limits_{t_1}^{t_2} \hat{O}(t')\,\mathrm{d}t'\right), \tag{97}$$

which depends on the times t_1 and t_2 and the auxiliary parameter δ. Expanding $\hat{U}(t+\Delta t, t)$ to the third order in Δt and assuming that the Hamiltonian $\hat{H}(t)$ has the form

$$\hat{H}(t) = \hat{H}_1(t) + \hat{H}_2(t), \tag{98}$$

the time-evolution operator (96) can be factorized [31] into

$$\hat{U}(t+\Delta t, t) = \hat{U}_{\hat{H}_1}\left(t+\Delta t, t, \tfrac{1}{2}\right)\hat{U}_{\hat{H}_2}(t+\Delta t, t, 1)\,\hat{U}_{\hat{H}_1}\left(t+\Delta t, t, \tfrac{1}{2}\right) + \mathcal{O}\left(\Delta t^3\right). \tag{99}$$

Neglecting terms of order $\mathcal{O}(\Delta t^3)$, (99) gives an explicit second-order accurate time-stepping scheme for the propagation of the function $\Psi(\mathbf{r}, t)$

$$\Psi(\mathbf{r}, t+\Delta t) = \hat{U}_{\hat{H}_1}\left(t+\Delta t, t, \tfrac{1}{2}\right)\hat{U}_{\hat{H}_2}(t+\Delta t, t, 1)\,\hat{U}_{\hat{H}_1}\left(t+\Delta t, t, \tfrac{1}{2}\right)\Psi(\mathbf{r}, t) + \mathcal{O}\left(\Delta t^3\right). \tag{100}$$

This scheme translates the difficulty of calculating the action of operator (96) to the calculation of the action of (97) for $\hat{O} \equiv \hat{H}_1$ and $\hat{O} \equiv \hat{H}_2$, respectively. Strang [38] and Galbraith et al. [11] utilized the splitting (100) to calculate the action of (97) in position space by a finite differences scheme. For many Cauchy problems, however, one can find a splitting such that the operator $\hat{U}_{\hat{H}_1}(t+\Delta t, t, \delta)$ is diagonal in real space, and $\hat{U}_{\hat{H}_2}(t+\Delta t, t, \delta)$ is diagonal in Fourier space. Thus, the calculation of these operators becomes feasible in the appropriate space and (100) is then calculated via

$$\Psi(\mathbf{r},t+\Delta t)=\hat{U}_{\hat{H}_1}\left(t+\Delta t,t,\tfrac{1}{2}\right)\mathcal{F}^{-1}\left\{\hat{\tilde{U}}_{\hat{H}_2}(t+\Delta t,t,1)\right.$$
$$\left.\times\mathcal{F}\left\{\hat{U}_{\hat{H}_1}\left(t+\Delta t,t,\tfrac{1}{2}\right)\Psi(\mathbf{r},t)\right\}\right\}+\mathcal{O}\left(\Delta t^3\right) \quad (101)$$

(cf. Section 1.4.3 of Chapter I). The expression $\mathcal{F}\{\cdot\}$ in (101) denotes the Fourier transform of the argument, $\mathcal{F}^{-1}\{\cdot\}$ the inverse.

In an actual implementation of the Fourier split operator method, the function $\Psi(\mathbf{r},t)$ is discretized on a rectangular, regular lattice of N points, and the continuous Fourier transform is approximated by a discrete Fourier transform. The computational complexity of propagating the wavefunction $\Psi(\mathbf{r},t)$ from time t to time $t+\Delta t$ is dominated by the transformation to Fourier space and back. If these transforms are accomplished by the fast (discrete) Fourier transform (FFT), the computation of a single time step of the Fourier split operator method takes $\mathcal{O}(N\log N)$ operations.

In order to understand and implement the Fourier split operator method, it is crucial to establish a correspondence between the "usual" continuous and the computational, discrete Fourier transforms. The continuous Fourier transform $\tilde{h}(\omega)$ of some function $h(t)$ is defined as

$$\tilde{h}(\omega)=\frac{1}{\sqrt{2\pi}}\int_{-\infty}^{\infty}h(t)\mathrm{e}^{-\mathrm{i}\omega t}\,\mathrm{d}t\,. \quad (102)$$

Let us assume that $h(t)$ differs nonnegligibly from zero only in the interval $[t_0,t_0+N\Delta t]$ so that

$$\tilde{h}(\omega)=\frac{1}{\sqrt{2\pi}}\int_{t_0}^{t_0+N\Delta t}h(t)\mathrm{e}^{-\mathrm{i}\omega t}\,\mathrm{d}t\,. \quad (103)$$

This integral may be approximated by the Riemann sum

$$\tilde{h}(\omega)=\frac{\Delta t}{\sqrt{2\pi}}\mathrm{e}^{-\mathrm{i}\omega t_0}\sum_{n=0}^{N-1}h(t_0+n\Delta t)\mathrm{e}^{-\mathrm{i}\omega n\Delta t}\,. \quad (104)$$

The sum (104) is defined for arbitrary arguments ω, but we will evaluate it for discrete arguments $\omega_0,\omega_0+\Delta\omega,\dots,\omega_0+(N-1)\Delta\omega$, where the spacing $\Delta\omega$ is related to Δt via

$$\Delta\omega=\frac{2\pi}{N\Delta t}\,. \quad (105)$$

Introducing $h_n=h(t_0+n\Delta t)$ and $\tilde{h}_m=\tilde{h}(\omega_0+m\Delta\omega)$ with $n=0,1,\dots,N-1$ and $m=0,1,\dots,N-1$, we get

$$\tilde{h}_m=\frac{\Delta t}{\sqrt{2\pi}}\mathrm{e}^{-\mathrm{i}(\omega_0+m\Delta\omega)t_0}\sum_{n=0}^{N-1}h_n\,\mathrm{e}^{-\mathrm{i}\omega_0 n\Delta t}\,\mathrm{e}^{-2\pi\mathrm{i}mn/N}\,. \quad (106)$$

Thus, the continuous Fourier transform of $h(t)$ can be approximated by the discrete Fourier transform of $h_n\,\mathrm{e}^{-\mathrm{i}\omega_0 n\Delta t}$ plus some phase factor $\Delta t\,\mathrm{e}^{-\mathrm{i}(\omega_0+m\Delta\omega)t_0}/\sqrt{2\pi}$. For the inverse Fourier transform, one can find the similar relation

$$h_n = \frac{\Delta\omega}{\sqrt{2\pi}}\,\mathrm{e}^{\mathrm{i}(t_0+n\Delta t)\omega_0}\sum_{m=0}^{N-1}\tilde{h}_m\,\mathrm{e}^{\mathrm{i}t_0 m\Delta\omega}\,\mathrm{e}^{2\pi\mathrm{i}mn/N}. \tag{107}$$

3.3 The Fourier split operator method for the Schrödinger equation

To exemplify the Fourier split operator method, let us consider the time-dependent Schrödinger equation first. Splitting the Hamiltonian of the Schrödinger equation in dipole approximation

$$\mathrm{i}\hbar\frac{\partial\Psi(\mathbf{r},t)}{\partial t} = \left(\frac{1}{2m}\left[-\mathrm{i}\hbar\nabla - q\mathbf{A}(t)\right]^2 + q\phi(\mathbf{r},t)\right)\Psi(\mathbf{r},t) \tag{108}$$

into a potential energy term $\hat{H}_1$ and a kinetic energy term $\hat{H}_2$

$$\hat{H}_1 = q\phi(\mathbf{r},t), \tag{109a}$$

$$\hat{H}_2 = \frac{1}{2m_0}\left[-\mathrm{i}\hbar\nabla - q\mathbf{A}(t)\right]^2 \tag{109b}$$

separates the spatial-dependent parts from spatial derivatives, which makes the operator $\hat{U}_{\hat{H}_1}(t+\Delta t,t,\delta)$ diagonal in position space and $\hat{U}_{\hat{H}_2}(t+\Delta t,t,\delta)$ diagonal in momentum space, respectively. The action of $\hat{U}_{\hat{H}_1}(t+\Delta t,t,\delta)$ on a position-space wavefunction is given by

$$\hat{U}_{\hat{H}_1}(t+\Delta t,t,\delta)\,\Psi(\mathbf{r},t) = \exp\left(-\delta\frac{\mathrm{i}}{\hbar}\int_t^{t+\Delta t} q\phi(\mathbf{r},t')\,\mathrm{d}t'\right)\Psi(\mathbf{r},t), \tag{110}$$

the action of the Fourier space operator $\hat{\tilde{U}}_{\hat{H}_2}(t+\Delta t,t,\delta)$ on a d-dimensional wave function in Fourier space

$$\tilde{\Psi}(\mathbf{p},t) = \mathcal{F}\left\{\Psi(\mathbf{r},t)\right\} = \frac{1}{(2\pi\hbar^2)^{d/2}}\int \Psi(\mathbf{r},t)\exp\left(-\mathrm{i}\mathbf{p}\cdot\mathbf{r}/\hbar\right)\mathrm{d}^d r \tag{111}$$

reads as

$$\hat{\tilde{U}}_{\hat{H}_2}(t+\Delta t,t,\delta)\,\tilde{\Psi}(\mathbf{p},t) = \exp\left(-\delta\frac{\mathrm{i}}{\hbar}\int_t^{t+\Delta t}\frac{1}{2m_0}\left[\mathbf{p}-q\mathbf{A}(t')\right]^2\mathrm{d}t'\right)\tilde{\Psi}(\mathbf{p},t). \tag{112}$$

Listing 2. Numerical solution of the time-dependent Schrödinger equation for the harmonic oscillator via the Fourier split operator method.

```
#!/usr/bin/env python
# -*- coding: utf-8 -*-

from pylab import *

# the harmonic oscillator potential with hbar=m0=omega=1
def V(x):
    return 0.5*x**2

# define computational grid running from -5 to 5 with 256 grid points
N=256
x0, x1=-5., 5.
x=linspace(x0, x1, N)
dx=(x1-x0)/(N-1)  # size of spatial grid spacing
dt=1./64          # temporal step size

# Gaussian wave packet with mean momentum one as initial condition
Psi=1./(2*pi)**(0.25)*exp(-x**2/4 + 1j*x)

# constuct momentum grid
dp=2*pi/(N*dx)
p=linspace(0, (N-1)*dp, N)
p[p>0.5*N*dp]-=N*dp

# construct the two propagators
U_1=exp(-0.5*1j*V(x)*dt)
U_2=exp(-1j*0.5*p**2*dt)

# propagate 512 time steps
for k in range(0, 512):
    clf()
    plot(x, Psi.real, '--', color='r', label=r'$\mathrm{Re}\,\Psi(x)$')
    plot(x, Psi.imag, ':', color='r', label=r'$\mathrm{Im}\,\Psi(x)$')
    plot(x, Psi.real**2+Psi.imag**2, color='#266bbd', label=r'$|\Psi(x)|^2$')
    gca().set_xlim(x0, x1)
    gca().set_ylim(-0.6, 1)
    xlabel(r'$x$')
    legend(loc='upper left')
    draw()
    show()
    pause(0.01)
    Psi*=U_1       # apply U_1 in real space
    Psi=fft(Psi)   # go to Fourier space
    Psi*=U_2       # apply U_2 in Fourier space
    Psi=ifft(Psi)  # go to real space
    Psi*=U_1       # apply U_1 in real space
```

Listing 2 shows a short Python program, which solves the time-dependent one-dimensional Schrödinger equation for a harmonic oscillator via the Fourier split operator method. The wavefunction is represented on a regular grid from x_0 to x_1 with N grid points and spacing $\Delta x = (x_1 - x_0)/(N-1)$. In Fourier (momentum) space, the grid ranges from $-(N/2-1)\Delta p$ to $N/2\Delta p$ (for even N) or from $-(N-1)/2\Delta p$ to $(N-1)/2\Delta p$ (for odd N), respectively, with $\Delta p = 2\pi\hbar/(N\Delta x)$.

Note that it is crucial for the application of the Fourier split operator method that the vector potential $\mathbf{A}(t)$ does not depend on the spatial coordinate $\mathbf{r}$. The expansion of the term $[-\mathrm{i}\hbar\nabla - q\mathbf{A}(\mathbf{r},t)]^2$ in the Hamiltonian of the Schrödinger equation (108) for a particle in an arbitrary vector potential $\mathbf{A}(\mathbf{r},t)$ contains the term $(\mathrm{i}q\hbar/m_0)\mathbf{A}(\mathbf{r},t)\cdot\nabla$, which is spatially dependent and contains spatial derivatives too. Consequently, it is diagonal neither in position space nor in Fourier space, which renders the Fourier split operator method inadequate. This hampering term also appears in the Klein–Gordon equation. In this case, however, the split operator method can be realized entirely in position space because of some specific mathematical properties of the Klein–Gordon Hamiltonian, as explained in detail in [32]. Mixing of momentum and position space is absent for vector potentials in dipole approximation but also for the important case of a vector potential of a linearly polarized plane wave, where the vector potential can be brought, e.g., into the form $\mathbf{A}(\mathbf{r},t) = (A_x(y,t),0,0)$ [16] by choosing the coordinate system appropriately.

3.4 The Fourier split operator method for the Dirac equation

In order to apply the Fourier split operator method to the Dirac equation (43) in $d = 1,2,3$ dimensions, the Hamiltonian is split into an interaction part $\hat{H}_1$ and a free-particle part $\hat{H}_2$,

$$\hat{H}_1 = c\sum_{i=1}^{d}\alpha_i\left[-qA_i(\mathbf{r},t)\right] + q\phi(\mathbf{r},t)\,, \tag{113a}$$

$$\hat{H}_2 = c\sum_{i=1}^{d}\alpha_i\left(-\mathrm{i}\hbar\frac{\partial}{\partial r_i}\right) + m_0c^2\beta\,. \tag{113b}$$

The operator $\hat{U}_{\hat{H}_1}(t+\Delta t,t,\delta)$ may be constructed by splitting $\hat{H}_1$ further into

$$\hat{H}_1 = \hat{H}_{1,1} + \hat{H}_{1,2} \tag{114}$$

with

$$\hat{H}_{1,1} = q\phi(\mathbf{r},t)\,, \tag{115a}$$

$$\hat{H}_{1,2} = c\sum_{i=1}^{d}\alpha_i\left[-qA_i(\mathbf{r},t)\right]\,. \tag{115b}$$

Because $\hat{H}_{1,1}$ and $\hat{H}_{1,2}$ commute, we can factorize the operator $\hat{U}_{\hat{H}_1}(t+\Delta t, t, \delta)$ into

$$\hat{U}_{\hat{H}_1}(t+\Delta t, t, \delta) = \hat{U}_{\hat{H}_{1,1}}(t+\Delta t, t, \delta)\,\hat{U}_{\hat{H}_{1,2}}(t+\Delta t, t, \delta) \tag{116}$$

with the diagonal operator $\hat{U}_{\hat{H}_{1,1}}(t+\Delta t, t, \delta)$ acting in position space,

$$\hat{U}_{\hat{H}_{1,1}}(t+\Delta t, t, \delta)\,\Psi(\mathbf{r}, t) = \exp\left(-\delta\frac{\mathrm{i}}{\hbar}\int_t^{t+\Delta t} q\phi(\mathbf{r}, t')\,\mathrm{d}t'\right)\Psi(\mathbf{r}, t)\,. \tag{117}$$

The operator $\hat{U}_{\hat{H}_{1,2}}(t+\Delta t, t, \delta)$ involves matrix exponentials, which may be calculated by taking into account the Dirac algebra (35). Exponentials of α_i are obtained by summing the Taylor expansion of the exponential explicitly. Introducing some auxiliary complex numbers a_i, we find

$$\begin{aligned}
\exp\left(\mathrm{i}\sum_{i=1}^{d} a_i\alpha_i\right) &= \sum_{k=0}^{\infty}\frac{1}{k!}\left(\mathrm{i}\sum_{i=1}^{d} a_i\alpha_i\right)^k \\
&= \sum_{k=0}^{\infty}\frac{(-1)^k}{(2k)!}\left(\sum_{i=1}^{d} a_i\alpha_i\right)^{2k} + \mathrm{i}\sum_{i=1}^{d} a_i\alpha_i\sum_{k=0}^{\infty}\frac{(-1)^k}{(2k+1)!}\left(\sum_{i=1}^{d} a_i\alpha_i\right)^{2k} \\
&= \sum_{k=0}^{\infty}\frac{(-1)^k}{(2k)!}\left(\sqrt{\sum_{i=1}^{d} a_i^2}\right)^{2k} + \mathrm{i}\sum_{i=1}^{d} a_i\alpha_i\sum_{k=0}^{\infty}\frac{(-1)^k}{(2k+1)!}\left(\sqrt{\sum_{i=1}^{d} a_i^2}\right)^{2k} \\
&= \cos(|a|) + \mathrm{i}\sum_{i=1}^{d}\frac{a_i}{|a|}\alpha_i\sin(|a|)\,,
\end{aligned} \tag{118}$$

where we have defined $|a| = \sqrt{\sum_{i=1}^{d} a_i^2}$ and employed the Dirac algebra (35), which yields the relation

$$\left(\sum_{i=1}^{d} a_i\alpha_i\right)^{2k} = \left(\sum_{i=1}^{d} a_i^2\alpha_i^2 + \sum_{i=1}^{d}\sum_{j=1}^{i-1} a_i a_j(\alpha_i\alpha_j + \alpha_j\alpha_i)\right)^k = \left(\sqrt{\sum_{i=1}^{d} a_i^2}\right)^{2k}.$$

For convenience, let us define

$$\bar{A}_i(\mathbf{r}, t) = \int_t^{t+\Delta t} A_i(\mathbf{r}, t')\,\mathrm{d}t'\,, \qquad \bar{A}(\mathbf{r}, t) = \sqrt{\sum_{i=1}^{d}\bar{A}_i(\mathbf{r}, t)^2}\,, \tag{119}$$

with which we obtain (118)

$$\hat{U}_{\hat{H}_{1,2}}(t+\Delta t,t,\delta)\,\Psi(\mathbf{r},t)$$
$$=\left[\cos\left(-\frac{\delta cq}{\hbar}\bar{A}(\mathbf{r},t)\right)+\mathrm{i}\sum_{i=1}^{d}\frac{\bar{A}_i(\mathbf{r},t)}{\bar{A}(\mathbf{r},t)}\alpha_i\sin\left(-\frac{\delta cq}{\hbar}\bar{A}(\mathbf{r},t)\right)\right]\Psi(\mathbf{r},t). \tag{120}$$

The operator $\hat{U}_{\hat{H}_2}(t_2,t_1,\delta)$ equals the time-evolution operator of the free-particle Dirac Hamiltonian. In Fourier space, it has the form

$$\hat{\tilde{U}}_{\hat{H}_2}(t+\Delta t,t,\delta)=\exp\left[-\delta\Delta t\frac{\mathrm{i}}{\hbar}\left(c\sum_{i=1}^{d}\alpha_i p_i+m_0c^2\beta\right)\right], \tag{121}$$

where p_i denotes the ith component of the momentum vector $\mathbf{p}$. In order to calculate the operator exponential in (121), we have to diagonalize the matrix

$$\hat{\tilde{H}}_2=c\sum_{i=1}^{d}\alpha_i p_i+m_0c^2\beta, \tag{122}$$

which represents the Hamiltonian of the free-particle Dirac equation in momentum space. As shown in Section 3.1, a Hermitian matrix is diagonalized by the matrix of its eigenvectors. The eigenvectors of (122) have been derived earlier, and the matrix of eigenvectors is given in the three-dimensional case with (73) by

$$\mathbf{Q}(\mathbf{p})=\begin{pmatrix}\tilde{\Psi}_1^+(\mathbf{p}) & \tilde{\Psi}_2^+(\mathbf{p}) & \tilde{\Psi}_1^-(\mathbf{p}) & \tilde{\Psi}_2^-(\mathbf{p})\end{pmatrix}, \tag{123}$$

which can also be written as [40]

$$\mathbf{Q}(\mathbf{p})=d_+(\mathbf{p})+d_-(\mathbf{p})\sum_{i=1}^{d}\frac{p_i}{|\mathbf{p}|}\beta\cdot\alpha_i \tag{124}$$

by introducing the scalars

$$d_\pm(\mathbf{p})=\sqrt{\frac{1}{2}\pm\frac{m_0c}{2p_0(\mathbf{p})}}. \tag{125}$$

The matrix $\mathbf{Q}^\dagger(\mathbf{p})\hat{\tilde{H}}_2\mathbf{Q}(\mathbf{p})$ is diagonal,

$$\mathbf{Q}^\dagger(\mathbf{p})\hat{\tilde{H}}_2\mathbf{Q}(\mathbf{p})=E(\mathbf{p})\beta. \tag{126}$$

The Fourier space operator (121) simplifies to

$$\hat{\tilde{U}}_{\hat{H}_2}(t+\Delta t,t,\delta)\,\tilde{\Psi}(\mathbf{p},t)=\mathbf{Q}(\mathbf{p})\exp\left(-\mathrm{i}\delta\Delta t\,E(\mathbf{p})\beta/\hbar\right)\mathbf{Q}^\dagger(\mathbf{p})\tilde{\Psi}(\mathbf{p},t), \tag{127}$$

where

$$\exp\left(-\mathrm{i}\delta\Delta t\,E(\mathbf{p})\beta/\hbar\right) = \begin{pmatrix} \mathrm{e}^{-\mathrm{i}\delta\Delta t\,E(\mathbf{p})/\hbar} & 0 & 0 & 0 \\ 0 & \mathrm{e}^{-\mathrm{i}\delta\Delta t\,E(\mathbf{p})/\hbar} & 0 & 0 \\ 0 & 0 & \mathrm{e}^{\mathrm{i}\delta\Delta t\,E(\mathbf{p})/\hbar} & 0 \\ 0 & 0 & 0 & \mathrm{e}^{\mathrm{i}\delta\Delta t\,E(\mathbf{p})/\hbar} \end{pmatrix} \tag{128}$$

in the case of four-component wavefunctions and

$$\exp\left(-\mathrm{i}\delta\Delta t\,E(\mathbf{p})\beta/\hbar\right) = \begin{pmatrix} \mathrm{e}^{-\mathrm{i}\delta\Delta t\,E(\mathbf{p})/\hbar} & 0 \\ 0 & \mathrm{e}^{\mathrm{i}\delta\Delta t\,E(\mathbf{p})/\hbar} \end{pmatrix} \tag{129}$$

in the case of two-component wavefunctions.

4 Numerical examples

The numerical methods enable us to study various relativistic quantum systems in electromagnetic fields. Let us start with a one-dimensional system, which is technically quite simple but features very interesting physics, e.g., the so-called Klein tunneling [15, 18, 28]. Klein tunneling is the phenomenon that in relativistic quantum mechanics, a steep potential barrier may become transparent if it exceeds a critical height of about $2m_0c^2$, as depicted in Figure 6. The figure shows the initial density of a wave packet (solid line), which approaches from the left the step-potential barrier of the form

$$q\varphi(x) = \frac{V_0}{2}\left(\tanh\frac{x}{w} + 1\right) \tag{130}$$

that has the characteristic width w and vanishes for $x \to -\infty$ and equals V_0 for $x \to +\infty$. After the interaction with the potential, the wave packet has split into a transmitted part and a reflected part (dashed line).

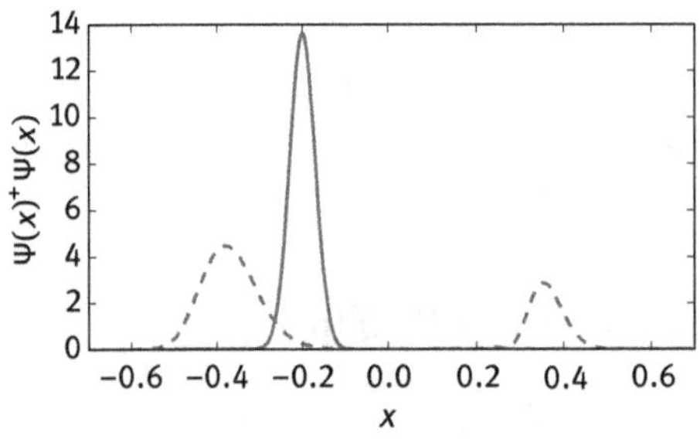

Fig. 6. Klein tunneling of a Dirac wave packet at the step potential (130) with $V_0 = 2m_0c^2 + 10^4$ and $w = 0.3/c$ (all parameters adopted from [21] and given in atomic units).

Listing 3 shows a short Python program, which produces an animation of the scattering dynamics, whose initial and final states are shown in Figure 6. The program utilizes the Fourier split operator method to solve the time-dependent one-dimensional Dirac equation and is very similar to the program in Listing 2 for the Schrödinger equation. However, there are a few noteworthy differences. Because the Dirac wavefunction has two components, it is represented by a $2 \times N$ matrix, where N denotes the number of grid points. The initial wave packet is first constructed in momentum space as a superposition of free-particle solutions with positive energy, as given in (81a). The position-space representation of the initial wave packet is obtained by an inverse Fourier transform, which is implemented via a discrete Fourier transform, taking into account the required phase factors as outlined at the end of Section 3.2. The matrix representation of the interaction propagator $\hat{U}_{\hat{H}_1}$ is diagonal so that it can be implemented via an element-wise multiplication of the wavefunction. The matrix representation of the free propagator in Fourier space $\hat{\hat{U}}_{\hat{H}_2}$, however, is block diagonal, see (127). For the Python implementation, which represents the wavefunction as a $2 \times N$ matrix, it is convenient to represent this block-diagonal matrix as an array of size $2 \times 2 \times N$. Then, the application of $\hat{\hat{U}}_{\hat{H}_2}$ to the wavefunction in Fourier space representation can be carried out via the Python function `einsum`, see [37] for details. Note that the time step for the simulation of Klein tunneling is much smaller than for the harmonic oscillator problem in Section 3.3 because $\Delta t \le \hbar/(m_0 c^2)$ must hold in the simulation of relativistic systems.

Figure 7 shows an application of the Fourier split operator method to the time-dependent Dirac equation in two dimensions. The solid line represents the center-of-mass trajectory of an electron moving in a strong, linearly polarized elec-

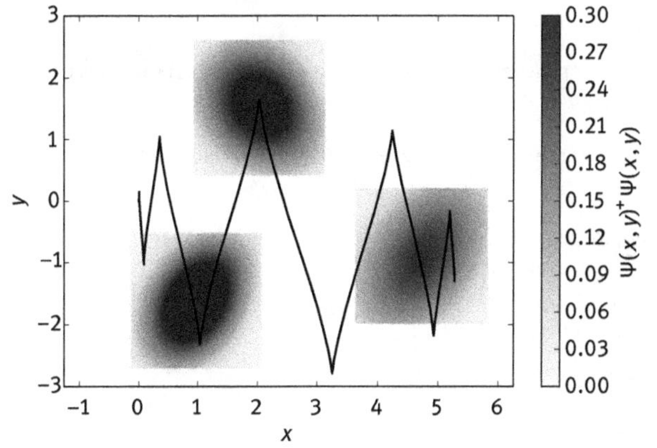

Fig. 7. An electron in a strong electromagnetic pulse with linear polarization and a $\sin^2$ envelope with six cycles. The pulse travels from the left to the right and has a peak electric field strength of 4000 a.u. and a wave length of 20 a.u.

Listing 3. Numerical solution of the time-dependent one-dimensional Dirac equation for a particle scattering at a potential via the Fourier split operator method. Atomic units are employed in the program.

```
#!/usr/bin/env python
# -*- coding: utf-8 -*-

from pylab import *

# smooth step potential
def V(x, V0, w):
    return V0*0.5*(tanh(x/w)+1)

# define computational grid running from -0.7 to 0.7 with 1024 grid points
N=1024
x0, x1=-0.7, 0.7
x=linspace(x0, x1, N)
dx=(x1-x0)/(N-1)  # size of spatial grid spacing
c=137.            # speed of light
dt=1./4/c**2      # temporal step size

# constuct momentum grid
dp=2*pi/(N*dx)
p=linspace(0, (N-1)*dp, N)
p[p>0.5*N*dp]-=N*dp
p=sort(p)

# Gaussian wave packet with mean momentum p_mean composed of
# positive-energy states only as initial condition
x_init=-0.2             # initial position of the wave packet
p_mean=106.4            # mean momentum
Delta_p=0.125*c         # momentum width
p_0=sqrt(p**2 + c**2)   # reduced energy
Psi=zeros((2, N), dtype='complex')
Psi[0, :]=1
Psi[1, :]=p/(c+p_0)
Psi*=sqrt(0.5*(c+p_0)/p_0)  # normalize
g=1/(2*pi*Delta_p**2)**0.25 * exp(-(p-p_mean)**2/4/Delta_p**2) * exp(-1j*p*x_init)
Psi*=g                      # multiply with weight function
# go to real space
Psi=dp/sqrt(2*pi)*exp(1j*x*p[0])*ifft(exp(1j*x[0]*arange(0, N)*dp)*Psi, axis=1)*N

# real space propagator
U_1=exp(-0.5*1j*V(x, 2*c**2+1e4, 0.3/c)*dt)
# construct the free propagator and propagate 512 time steps
p=linspace(0, (N-1)*dp, N)
p[p>0.5*N*dp]-=N*dp
p_0=sqrt(p**2 + c**2)
U_2=zeros((2, 2, N), dtype='complex')
for i in range(0, N):
    Q=array([ [ 1, -p[i]/(c+p_0[i])], [p[i]/(c+p_0[i]), 1] ])
    Q*=sqrt(0.5*(c+p_0[i])/p_0[i])
    U_2[:, :, i]=dot(Q, dot(diag([exp(-1j*dt*c*p_0[i]), exp(+1j*dt*c*p_0[i]) ]), Q.T))

# propagate 512 time steps
for k in range(0, 512):
```

```
clf()
plot(x, Psi[0, :].real**2+Psi[0, :].imag**2+Psi[1, :].real**2+Psi[1, :].imag**2,
     color='#266bbd', label=r'$|\Psi(x)|^2$')
xlim(x0, x1)
ylim(0, 15)
xlabel(r'$x$')
ylabel(r'$\Psi(x)^\dagger\Psi(x)$')
draw()
show()
pause(0.01)
Psi*=U_1                                  # apply U_1 in real space
Psi=fft(Psi, axis=1)                      # go to Fourier space
Psi=einsum('ij...,j...->i...', U_2, Psi)  # apply U_2 in Fourier space
Psi=ifft(Psi, axis=1)                     # go to real space
Psi*=U_1                                  # apply U_1 in real space
```

tromagnetic pulse. The figure also depicts three snapshots of the electron density. The pulse travels from the left to the right. Its electric field component accelerates the electron along the y axis, and the Lorentz force, which results from the electron's fast motion and the strong magnetic field component, accelerates the electron along the propagation direction x. Compared to the one-dimensional simulation of Klein tunneling, the numerical propagation of the two-dimensional Dirac wave packet is much more demanding. Therefore, the simulation program was written in C++ and utilizes graphics cards as accelerators [2]. Furthermore, the computational grid follows the motion of the wave packet to reduce the computational costs.

Figure 8 demonstrates a further application of a numerical solution of the Dirac equation: ionization of hydrogen-like, highly charged ions in strong laser fields. Here, an electron that is initially bound to a soft-core potential is emitted because of a strong electromagnetic wave with wavevector **k**, traveling in y-direction. The electric field component in x-direction tilts the atomic binding potential such that a barrier is formed through which the electron may tunnel. The emitted part of the electron

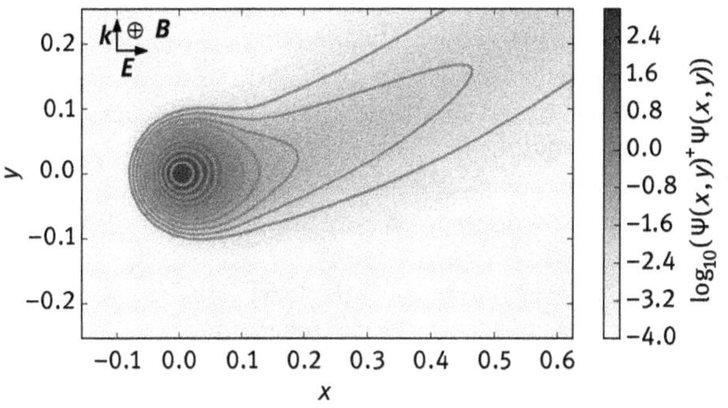

Fig. 8. Numerical simulation of ionization from a soft-core potential via the solution of the Dirac equation. The density plot and the contour lines show the electron density at the moment of maximal laser field strength at the atomic core, see [17] for details.

density is also accelerated into the propagation direction of the electromagnetic field because of the Lorentz force involving the magnetic field component **B**.

Bibliography

[1] V. G. Bagrov and D. Gitman. *Exact Solutions of Relativistic Wave Equations, Mathematics and Its Applications*, vol. 39. Springer, Heidelberg, Berlin, 1990.

[2] H. Bauke and C. H. Keitel. Accelerating the Fourier split operator method via graphics processing units. *Computer Physics Communications*, 182(12):2454–2463, 2011.

[3] H. Bauke, S. Ahrens, C. H. Keitel, and R. Grobe. Relativistic spin operators in various electromagnetic environments. *Physical Review A*, 89(5):052101, 2014.

[4] R. Beerwerth and H. Bauke. Krylov subspace methods for the Dirac equation. *Computer Physics Communications*, 188:189–197, 2015.

[5] J. W. Braun, Q. Su, and R. Grobe. Numerical approach to solve the time-dependent Dirac equation. *Physical Review A*, 59(1):604–612, 1999.

[6] F. Ehlotzky, K. Krajewska, and J. Z. Kamiński. Fundamental processes of quantum electrodynamics in laser fields of relativistic power. *Reports on Progress in Physics*, 72(4):1–32, 2009.

[7] M. D. Feit, J. A. Fleck, and A. Steiger. Solution of the Schrödinger equation by a spectral method. *Journal of Computational Physics*, 47(3):412–433, 1982.

[8] H. Feshbach and F. Villars. Elementary relativistic wave mechanics of spin 0 and spin 1/2 particles. *Reviews of Modern Physics*, 30(1):24–45, 1958.

[9] F. Fillion-Gourdeau, E. Lorin, and A. D. Bandrauk. Numerical solution of the time-dependent Dirac equation in coordinate space without fermion-doubling. *Computer Physics Communications*, 183(7):1403–1415, 2012.

[10] F. Fillion-Gourdeau, E. Lorin, and A. D. Bandrauk. A split-step numerical method for the time-dependent Dirac equation in 3-D axisymmetric geometry. *Journal of Computational Physics*, 272:559–587, 2014.

[11] I. Galbraith, Y. S. Ching, and E. Abraham. Two-dimensional time-dependent quantum-mechanical scattering event. *American Journal of Physics*, 52(1):60–68, 1984.

[12] G. H. Golub and C. F. Van Loan. *Matrix Computations*. Johns Hopkins Studies in the Mathematical Sciences, Johns Hopkins University Press, Baltimore, 1996.

[13] F. Gross. *Relativistic Quantum Mechanics and Field Theory*. Wiley VCH, Weinheim, 2004.

[14] X. Guan, C. J. Noble, O. Zatsarinny, K. Bartschat, and B. I. Schneider. ALTDSE: An Arnoldi-Lanczos program to solve the time-dependent Schrödinger equation. *Computer Physics Communications*, 180(12):2401–2409, 2009.

[15] B. R. Holstein. Klein's paradox. *American Journal of Physics*, 66(6):507, 1998.

[16] S. X. Hu and C. H. Keitel. Dynamics of multiply charged ions in intense laser fields. *Physical Review A*, 63(5):053402, 2001.

[17] M. Klaiber, E. Yakaboylu, H. Bauke, K. Z. Hatsagortsyan, and C. H. Keitel. Under-the-barrier dynamics in laser-induced relativistic tunneling. *Physical Review Letters*, 110(15):153004, 2013.

[18] O. Klein. Die Reflexion von Elektronen an einem Potentialsprung nach der relativistischen Dynamik von Dirac. *Zeitschrift für Physik*, 53(3–4):157–165, 1929.

[19] K. Kowalski and J. Rembieliński. Salpeter equation and probability current in the relativistic Hamiltonian quantum mechanics. *Physical Review A*, 84:012108, 2011.

[20] H. Kragh. Equation with the many fathers. The Klein–Gordon equation in 1926. *American Journal of Physics*, 52(11):1024–1033, 1984.
[21] P. Krekora, Q. Su, and R. Grobe. Klein paradox with spin-resolved electrons and positrons. *Physical Review A*, 72(6):064103, 2005.
[22] C. Lanczos. An iteration method for the solution of the eigenvalue problem of linear differential and integral operators. *Journal of Research of the National Bureau of Standards*, 45(4):255, 1950.
[23] C. Leforestier, R. H. Bisseling, C. Cerjan, M. D. Feit, R. Friesner, A. Guldberg, A. Hammerich, G. Jolicard, W. Karrlein, H.-D. Meyer, N. Lipkin, O. Roncero, and R. Kosloff. A comparison of different propagation schemes for the time dependent Schrödinger equation. *Journal of Computational Physics*, 94(1):59–80, 1991.
[24] Q. Z. Lv, Heiko Bauke, Q. Su, C. H. Keitel, and R. Grobe. Bosonic pair creation and the Schiff-Snyder-Weinberg effect. *Physical Review A*, 93(1):012119, 2016.
[25] G. R. Mocken and C. H. Keitel. FFT-split-operator code for solving the Dirac equation in 2 + 1 dimensions. *Computer Physics Communications*, 178(11):868–882, 2008.
[26] A. Mostafazadeh. Pseudounitary operators and pseudounitary quantum dynamics. *Journal of Mathematical Physics*, 45(3):932–946, 2004.
[27] A. Mostafazadeh. Pseudo-Hermitian representation of quantum mechanics. *International Journal of Geometric Methods in Modern Physics*, 07(07):1191–1306, 2010.
[28] H. Nitta. Motion of a wave packet in the Klein paradox. *American Journal of Physics*, 67(11):966, 1999.
[29] G. Paiano. Wave function in a relativistic quark-diquark model. *Lettere Al Nuovo Cimento Series 2*, 41(3):69–72, 1984.
[30] T. J. Park and J. C. Light. Unitary quantum time evolution by iterative Lanczos reduction. *The Journal of Chemical Physics*, 85(10):5870, 1986.
[31] P. Pechukas and J. C. Light. On the exponential form of time-displacement operators in quantum mechanics. *The Journal of Chemical Physics*, 44(10):3897–3912, 1966.
[32] M. Ruf, H. Bauke, and C. H. Keitel. A real space split operator method for the Klein–Gordon equation. *Journal of Computational Physics*, 228(24):9092–9106, 2009.
[33] Yousef Saad. *Numerical Methods for Large Eigenvalue Problems. Society for Industrial and Applied Mathematics*, Philadelphia, 2nd ed., 2011.
[34] L. I. Schiff, H. Snyder, and J. Weinberg. On the existence of stationary states of the mesotron field. *Physical Review*, 57(4):315–318, 1940.
[35] B. I. Schneider, L. A. Collins, and S. X. Hu. Parallel solver for the time-dependent linear and nonlinear Schrödinger equation. *Physical Review E*, 73(3):036708, 2006.
[36] F. Schwabl. *Advanced Quantum Mechanics*. Springer, Heidelberg, Berlin, 2008.
[37] SciPy. Scientific computing tools for Python. http://www.scipy.org/. Accessed October 1, 2016.
[38] G. Strang. On the construction and comparison of difference schemes. *SIAM Journal on Numerical Analysis*, 5(3):506–517, 1968.
[39] P. Strange. *Relativistic Quantum Mechanics: With Applications in Condensed Matter and Atomic Physics*. Cambridge University Press, Cambridge, 1998.
[40] B. Thaller. *Advanced Visual Quantum Mechanics*. Springer, Heidelberg, Berlin, 2005.
[41] A. Wachter. *Relativistic Quantum Mechanics*. Theoretical and Mathematical Physics. Springer, Heidelberg, Berlin, 2011.

Dieter Bauer

IV Time-dependent density functional theory

Time-dependent density functional theory (TDDFT) is the extension of Nobel-prize winning ground-state density functional theory (DFT) [19] toward time-dependent densities $n(\mathbf{r}, t)$. Most of the literature employing TDDFT remains within the linear response regime where a time-dependent density arises because of a small perturbation of the ground state, e.g., through a gentle "kick" by an external electric field. By Fourier-transforming the dipole response, the elementary excitations of the system can be inferred and compared with experimental absorption or emission spectra. Linear response is not enough for tackling strong-field physics problems though. Fortunately, the generalization of the Hohenberg–Kohn theorem [17] to time-dependent densities – the Runge–Gross theorem [40] – is formally not restricted to small perturbations. It states very generally that different time-dependent external potentials will lead to different time-dependent densities (when starting from the same initial state). Analogously to the time-independent case thus follows that all observables are functionals of the single-particle density $n(\mathbf{r}, t)$, and a wavefunction-free formulation of quantum mechanics should be possible, in principle. Unlike with the many-body wavefunction, the numerical effort to store and propagate a single-particle density may, perhaps, not scale exponentially with the particle number, which makes TDDFT a candidate for overcoming the exponential wall of computational many-body quantum mechanics [19]. The no-free-lunch theorem, on the other hand, tells us that there must be a price to pay for such simplicity. And in fact, there is.

There are excellent reviews of TDDFT, e.g., [43, 44]. The aim of this contribution rather is to illustrate how some of the single-electron TDSE propagation methods introduced in the previous chapters can be augmented for strong-field TDDFT studies employing the time-dependent version of the Kohn–Sham (KS) scheme [20]. We restrict ourselves to a three-dimensional (3D) time-dependent KS (TDKS) solver where the KS orbitals are expanded in spherical harmonics and the KS potential in multipoles. This is the "natural" extension of the TDSE solver in Section 1.5 of Chapter I toward TDDFT, which is also published [5] as a part of the `Qprop` code. A more versatile real-space and real-time TDDFT solver is the publicly available and widely used `octopus` code [1]. The final part of this chapter is devoted to extremely challenging applications for TDDFT where known and practicable exchange-correlation (xc) functionals or the density functionals for the observables of interest do not work.

Dieter Bauer: Institute of Physics, University of Rostock, 18051 Rostock, Germany; email: dieter.bauer@uni-rostock.de

De Gruyter Graduate – Computational Strong-Field Quantum Dynamics, Volume 5, 2017, pp. 111–144.
DOI 10.1515/9783110417265-004

1 A few general remarks on time-dependent many-particle methods

Since this is the first chapter on a time-dependent many-body approach in this book, a few general considerations might be in order. While electronic structure calculations are hard enough, time-dependent problems with strong, external drivers are orders of magnitude harder. In fact, on a TDSE level, only $N = 2$ electrons can be treated in full dimensionality to date. And even if computers will be faster, solving the full N-body TDSE seems the wrong way to go, as it will be increasingly cumbersome to store and analyze the highly dimensional wavefunction. Observables of interest can be obtained from reduced quantities such as reduced density matrices. In that sense, many-body wavefunctions contain too much, redundant information. Unfortunately, the conceptually simple, linear, local, and memory-free N-body TDSE acting on a brobdingnagian wavefunction can only be traded for much more complicated, nonlinear, and possibly nonlocal (in space and time) equations of motion for reduced quantities. The hope then is to find reasonable approximations such that the system of equations of motion remains tractable.

Strategies to tackle the time-dependent N-body problem may be classified according to the "level of quantumness" incorporated:

1. Full quantum field theoretical (including antiparticles, pair production, quantized electromagnetic field, N not fixed – hopeless without severe approximations).
2. Full relativistic quantum treatment (small time and length scales, still unfeasible in full dimensionality for more than one particle, see Chapter III).
3. Full nonrelativistic quantum treatment (only possible for a few particles; scales $\mathcal{O}(G^N)$ with G being a typical number of grid points required to represent the corresponding one-body problem, see Chapter I).
4. In principle exact but in practice approximate time-dependent methods (e.g., multiconfigurational Hartree–Fock (HF), see Chapter V, configuration interaction, see Chapter VI, density matrix–based methods [2, 9, 12, 33], DFT (this chapter), etc., work well or not so well, depending on the problem – tough but doable; need to scale better than $\mathcal{O}(G^N)$; the more mean field–like, i.e., the less correlation, the more particles can be treated).
5. Semiclassical methods such as time-dependent Thomas-Fermi—scaling largely independent of N (only through spatial extension).
6. Classical methods such as molecular dynamics where particle-particle interaction is fully accounted for on a classical level – scales $\mathcal{O}(N^2)$, see Chapter VIII.

7. Classical methods with further simplifying assumptions (particle-in-cell (PIC), see Chapter VIII, tree codes, magnetohydrodynamics, fluid methods – scale better than $\mathcal{O}(N^2)$, e.g., $\mathcal{O}(N)$ (PIC) or with the spatial extension only).

Sometimes it is more important to have correlation properly incorporated rather than quantumness. In such cases, it may be better to choose a classical method with particle-particle interactions taken fully into account rather than a mean-field quantum method.

The word "correlation" is a fancy fuzzy buzz word often used without knowing what it means. In fact, it means different things in different fields and theories. One definition (for Fermions) reads *Correlation is whatever is not included in a HF treatment*. Here, a HF treatment with just a single Slater determinant is meant, not the in-principle-exact multiconfigurational HF. A spin-singlet state can be written as a single Slater determinant and is thus *not* correlated in that sense although it cannot be written as a product of single-particle wavefunctions and is a prime example for an entangled state. Because a single Slater determinant is sufficient to describe noninteracting Fermions, we expect that an interaction potential in the Schrödinger equation in general gives rise to correlation effects. Working out how strong they are and how they manifest themselves is what makes many-body theory interesting. Correlation effects are at the basis of magnetism, superconductivity, chemical bonds, structure formation, life, everything [4].

One should distinguish collective effects from correlation. Collective oscillations or density waves are already included in simple time-dependent HF, Thomas-Fermi, PIC, and fluid treatments, while correlation is not.

Ground-state DFT is a widely applied, extremely successful, and relatively cheap computational method. DFT is based on the Hohenberg–Kohn theorem [17], stating that the single-particle ground-state density $n(\mathbf{r})$ (defined below) determines uniquely whatever one may know about a many-body Fermionic system in a given scalar external potential $v(\mathbf{r})$ and for a given two-body interaction. This is because both the mapping $v(\mathbf{r}) \longrightarrow |\Psi\rangle$ and $|\Psi\rangle \longrightarrow n(\mathbf{r})$ are invertible, $v(\mathbf{r}) \longleftrightarrow |\Psi\rangle \longleftrightarrow n(\mathbf{r})$, so that

$$v(\mathbf{r}) \longleftrightarrow n(\mathbf{r}). \tag{1}$$

Here, $|\Psi\rangle$ is the exact many-body ground state, i.e., the solution of the corresponding Schrödinger equation for the ground state whose calculation one wants to avoid. Thanks to the mapping (1) and the fact that the external potential $v(\mathbf{r})$ determines[1] whatever one may know about the system; knowledge of the single-particle density $n(\mathbf{r})$ is in principle enough to calculate any observable.

1 Through the Hamiltonian in the Schrödinger equation, for a given particle-particle interaction.

2 DFT for effective single-electron potentials

The simplest, not yet time-dependent but important application of DFT is the calculation of effective potentials that can be used in single-active-electron TDSE simulations of strong-field phenomena. Any of the well-established electronic structure DFT codes can be used for that purpose. However, it is particularly elegant and numerically self-consistent to use the same code for both the DFT ground-state calculation and the subsequent real-time propagation. We saw in Chapter I how ground states can be obtained with a TDSE solver by switching from real time to imaginary time. We apply the same methodology in this section.

2.1 KS spin-DFT

While the Hohenberg–Kohn theorem – as a kind of existence theorem – is the basis of ground-state DFT, the KS scheme [20] allows to perform actual calculations in an efficient way. The key idea of KS theory is to apply the Hohenberg–Kohn theorem simultaneously to the original interacting system and to an auxiliary noninteracting system, called the KS system. If the interacting system with its external potential $v(\mathbf{r})$ leads to a unique ground-state density $n(\mathbf{r})$, one may search for a potential $v_{\mathrm{KS}}(\mathbf{r})$ that leads to the same density,

$$n_{\mathrm{KS}}(\mathbf{r}) \stackrel{!}{=} n(\mathbf{r}). \tag{2}$$

We assume that this is possible, ignoring pathological cases where there exists no potential $v(\mathbf{r})$ that generates a given density as its ground-state density ("v-representability problem") or where an "interacting ground-state density" cannot be a ground-state density of the auxiliary noninteracting KS system ("noninteracting v-representability problem").

We consider an application of the KS scheme in atomic physics, i.e., N electrons that interact via

$$v_{ee}(|\mathbf{r}_i - \mathbf{r}_j|) = \frac{1}{|\mathbf{r}_i - \mathbf{r}_j|}, \qquad i,j = 1,2,\ldots,N, \quad i \neq j. \tag{3}$$

In atomic physics, the external potential is that of an ion of charge state Z, i.e., spherically symmetric,

$$v(\mathbf{r}) = v(r) = -\frac{Z}{r}. \tag{4}$$

What about spin? In DFT and KS theory, one distinguishes between spin-neutral and spin-polarized situations, depending on whether one assumes that there are as much $N_\uparrow$ spin-up as $N_\downarrow$ spin-down electrons or not, respectively.

We define the spin density as

$$n_\sigma(\mathbf{r}) = N \sum_{\sigma_2,\ldots,\sigma_N} \int d^3r_2 \cdots d^3r_N \, |\Psi(\mathbf{r}\sigma, \mathbf{r}_2\sigma_2, \ldots, \mathbf{r}_N\sigma_N)|^2. \tag{5}$$

Here, Ψ is the antisymmetric many-body wavefunction (normalized to unity) fulfilling the "interacting Schrödinger equation." The spin-variable $\sigma =\uparrow,\downarrow$ corresponds to magnetic spin quantum number $m_s = \pm 1/2$. The total single-particle density is

$$n(\mathbf{r}) = \sum_\sigma n_\sigma(\mathbf{r}), \qquad \int d^3r\, n(\mathbf{r}) = N. \tag{6}$$

Example: $N = 2$ spin-singlet ground state in atomic helium or helium-like ions. Here, the Schrödinger wavefunction has the form

$$\Psi(\mathbf{r}_1\sigma_1, \mathbf{r}_2\sigma_2) = \varphi(\mathbf{r}_1, \mathbf{r}_2) \frac{1}{\sqrt{2}} (\delta_{\sigma_1\uparrow}\delta_{\sigma_2\downarrow} - \delta_{\sigma_2\uparrow}\delta_{\sigma_1\downarrow}), \qquad \varphi(\mathbf{r}_1, \mathbf{r}_2) = \varphi(\mathbf{r}_2, \mathbf{r}_1).$$

Hence,

$$|\Psi(\mathbf{r}_1\sigma_1, \mathbf{r}_2\sigma_2)|^2 = |\varphi(\mathbf{r}_1, \mathbf{r}_2)|^2 \frac{1}{2} (\delta_{\sigma_1\uparrow}\delta_{\sigma_2\downarrow} + \delta_{\sigma_2\uparrow}\delta_{\sigma_1\downarrow})$$

and

$$n_\sigma(\mathbf{r}) = 2 \sum_{\sigma_2} \int d^3r_2 \, |\varphi(\mathbf{r}, \mathbf{r}_2)|^2 \frac{1}{2} (\delta_{\sigma\uparrow}\delta_{\sigma_2\downarrow} + \delta_{\sigma_2\uparrow}\delta_{\sigma\downarrow}) = (\delta_{\sigma\uparrow} + \delta_{\sigma\downarrow}) \int d^3r_2 \, |\varphi(\mathbf{r}, \mathbf{r}_2)|^2.$$

For symmetry reasons, $n_\uparrow(\mathbf{r}) = n_\downarrow(\mathbf{r})$ so that $n(\mathbf{r}) = 2n_\sigma(\mathbf{r})$. If $\varphi(\mathbf{r}_1, \mathbf{r}_2) = \varphi(\mathbf{r}_1)\varphi(\mathbf{r}_2)$, as in the case of noninteracting systems (such as the KS auxiliary one), we have

$$n_\sigma(\mathbf{r}) = (\delta_{\sigma\uparrow} + \delta_{\sigma\downarrow}) |\varphi(\mathbf{r})|^2$$

and

$$n(\mathbf{r}) = 2|\varphi(\mathbf{r})|^2.$$

KS equation

The exact Fermionic many-body wavefunction for the auxiliary, noninteracting KS system is a single Slater determinant made of single-particle wavefunctions, known as "KS orbitals"

$$\psi_{\alpha\sigma}, \quad \alpha = 1, 2, \ldots N_\sigma, \quad \sigma =\uparrow,\downarrow, \quad N_\uparrow + N_\downarrow = N. \tag{7}$$

Hence, the corresponding time-independent Schrödinger equation separates into the set of KS equations for the KS orbitals $\psi_{\alpha\sigma}(\mathbf{r})$,

$$\left[-\frac{1}{2}\nabla^2 + v_{\mathrm{KS}}^{\sigma}[\{n_\sigma\}](\mathbf{r})\right]\psi_{\alpha\sigma}(\mathbf{r}) = \epsilon_{\alpha\sigma}\psi_{\alpha\sigma}(\mathbf{r}). \tag{8}$$

The spin densities for such a noninteracting system simply are

$$n_\sigma(\mathbf{r}) = \sum_{\alpha=1}^{N_\sigma} |\psi_{\alpha\sigma}(\mathbf{r})|^2. \tag{9}$$

The KS potential v_{KS}^{σ} will, in general, be different for KS particles of opposite spin. However, all KS particles of like spin "see" the same KS Hamiltonian

$$H_{\mathrm{KS}}^{\sigma}[\{n_\sigma\}] = -\frac{1}{2}\nabla^2 + v_{\mathrm{KS}}^{\sigma}[\{n_\sigma\}]. \tag{10}$$

The orbitals of different like-spin KS particles are thus orthogonal, and we can populate them according to the Pauli exclusion principle. The orbitals of KS particles with opposite spin are orthogonal anyway.

The sole purpose of the KS potential $v_{\mathrm{KS}}^{\sigma}[\{n_\sigma\}](\mathbf{r})$ is to reproduce the density of the original, interacting system via the auxiliary KS orbital densities. Hence, one should be careful when interpreting the KS orbitals as if they were single-particle wavefunctions describing physically relevant pseudoparticles. They certainly do not describe the (interacting) electrons in the sense that a Slater determinant built from the KS orbitals is a good approximation of the exact many-body wavefunction. Nevertheless, the KS orbitals and corresponding "KS electrons" are often interpreted in a way as if the "real electrons" behaved similar. We also do this occasionally in the following because otherwise the wording becomes cumbersome. But one should always keep in mind that the KS system is a theoretical construct. The only thing for sure is that the KS single-particle density equals the interacting single-particle density – if v_{KS} was known exactly.

Let us decompose the KS potential into a part that is also present in the original, interacting Schrödinger equation and the rest,

$$v_{\mathrm{KS}}^{\sigma}[\{n_\sigma\}](\mathbf{r}) = v(\mathbf{r}) + v_{\mathrm{Hxc}}^{\sigma}[\{n_\sigma\}](\mathbf{r}). \tag{11}$$

The Hartree-xc potential $v_{\mathrm{Hxc}}^{\sigma}[\{n_\sigma\}](\mathbf{r})$ is further decomposed,

$$v_{\mathrm{Hxc}}^{\sigma}[\{n_\sigma\}](\mathbf{r}) = u[n](\mathbf{r}) + v_{\mathrm{xc}}^{\sigma}[\{n_\sigma\}](\mathbf{r}) \tag{12}$$

with the Hartree potential

$$u[n](\mathbf{r}) = \int d^3r' \, \frac{n(\mathbf{r}')}{|\mathbf{r}-\mathbf{r}'|}, \tag{13}$$

being a functional of the total density $n(\mathbf{r})$ only and the xc-potential functional $v_{\mathrm{xc}}^{\sigma}[\{n_\sigma\}](\mathbf{r})$, which depends on both spin-densities $n_\sigma(\mathbf{r})$ and may be different for $\sigma=\uparrow$ and $\downarrow$. As the Hartree potential describes the potential due to a charge distribution $n(\mathbf{r})$, it is expected to capture the main part of the mean field-like electron-electron interaction. However, we know that because of the required antisymmetry of Fermionic wavefunctions, exchange effects come into play. A part of the exchange potential $v_{\mathrm{x}}^{\sigma}[\{n_\sigma\}](\mathbf{r})$ in

$$v_{\mathrm{xc}}^{\sigma}[\{n_\sigma\}](\mathbf{r}) = v_{\mathrm{x}}^{\sigma}[\{n_\sigma\}](\mathbf{r}) + v_{\mathrm{c}}^{\sigma}[\{n_\sigma\}](\mathbf{r}) \tag{14}$$

is responsible for canceling the self-interaction in $u[n](\mathbf{r})$ (see subsection 2.2.2). Whatever else besides $v(\mathbf{r})$, Hartree and exchange is required to reproduce the interacting density via the KS potential is called "correlation" in the DFT framework.

Contrary to even simplest HF (i.e., with a single Slater determinant), the KS Hamiltonian $\hat{H}_{\mathrm{KS}}$ only contains multiplicative terms, which are easier to handle numerically than the nonlocal Fock integral term in the HF equations. Moreover, even correlation can be included in DFT. So why does anybody bother about HF, which seems to be both harder and worse than DFT? As usual, the devil is in the details and, unfortunately, the no-free-lunch-theorem still holds, as will become clear toward the end of this chapter.

Energy and potential functionals

Consider the total energy functional

$$E = T_s[\{n_\sigma\}] + \int d^3r\, n(\mathbf{r})v(\mathbf{r}) + U[n] + E_{\mathrm{xc}}[\{n_\sigma\}]. \tag{15}$$

The "noninteracting kinetic energy" $T_s[\{n_\sigma\}]$ is evaluated with the KS orbitals,

$$T_s[\{n_\sigma\}] = -\frac{1}{2}\sum_{\sigma}\sum_{\alpha=1}^{N_\sigma}\int d^3r\, \psi_{\alpha\sigma}^{\star}(\mathbf{r})\nabla^2\psi_{\alpha\sigma}(\mathbf{r}). \tag{16}$$

Note that this differs from the exact kinetic $T[\Psi]$ energy evaluated with the exact many-electron wavefunction Ψ. Hence, a part of the correlation originates from the difference between T_s and T. The Hartree energy in (15) reads as

$$U[n] = \frac{1}{2}\int d^3r\, u[n](\mathbf{r})\, n(\mathbf{r}) = \frac{1}{2}\int d^3r \int d^3r'\, \frac{n(\mathbf{r}')n(\mathbf{r})}{|\mathbf{r}-\mathbf{r}'|}. \tag{17}$$

The variational derivative of the Hartree-energy functional gives the Hartree potential,

$$u[n](\mathbf{r}) = \frac{\delta U[n(\mathbf{r})]}{\delta n(\mathbf{r})} \tag{18}$$

with the variational derivative $\frac{\delta F[n(\mathbf{r})]}{\delta n(\mathbf{r})}$ of a functional $F[n(\mathbf{r})]$ defined via

$$\delta F[n(\mathbf{r})] = \int d^3r \, \frac{\delta F[n(\mathbf{r})]}{\delta n(\mathbf{r})} \delta n(\mathbf{r}), \qquad \delta F[n(\mathbf{r})] = F[n(\mathbf{r}) + \delta n(\mathbf{r})] - F[n(\mathbf{r})], \tag{19}$$

understood in the limit $\delta n(\mathbf{r}) \to 0$.

Example: We have for the Hartree part

$$\begin{aligned} \delta U[n] &= \frac{1}{2} \int d^3r \int d^3r' \, \frac{[n(\mathbf{r}') + \delta n(\mathbf{r}')][n(\mathbf{r}) + \delta n(\mathbf{r})] - n(\mathbf{r}')n(\mathbf{r})}{|\mathbf{r} - \mathbf{r}'|} \\ &= \int d^3r \left\{ \int d^3r' \, \frac{n(\mathbf{r}')}{|\mathbf{r} - \mathbf{r}'|} \right\} \delta n(\mathbf{r}), \end{aligned}$$

where we neglected all terms $\mathcal{O}[(\delta n)^2]$ and renamed integration variables, obtaining two times the same term, thus canceling the $1/2$. If the $1/2$ was not there, we would double-count the Hartree interaction. We identify with (19) the expression in the curly brackets as the Hartree potential (13),

$$u[n](\mathbf{r}) = \frac{\delta U[n(\mathbf{r})]}{\delta n(\mathbf{r})} = \int d^3r' \, \frac{n(\mathbf{r}')}{|\mathbf{r} - \mathbf{r}'|}.$$

Local density approximation

There exists a hierarchy of approximations for $E_{\mathrm{xc}}[\{n_\sigma\}]$. At the top end reside "optimized effective potential (OEP) methods" that take exchange and self-interaction correction (SIC) well into account [22]. At the lower end is the widely used local density approximation (LDA) or the local spin density (LSD) approximation. In between LDA and OEP is a zoo of, partially empirically fitted, functionals that take density gradient corrections into account.

The exchange energy functional in LDA reads as

$$E_{\mathrm{X}}^{\mathrm{LDA}}[n] = A_{\mathrm{X}} \int d^3r \, n^{4/3}(\mathbf{r}), \qquad A_{\mathrm{X}} = -\frac{3}{4}\left(\frac{3}{\pi}\right)^{1/3}, \tag{20}$$

which can be written as

$$E_{\mathrm{X}}^{\mathrm{LDA}}[n] = \int d^3r \, n(\mathbf{r}) \varepsilon_{\mathrm{X}}^{\mathrm{LDA}}[n(\mathbf{r})], \qquad \varepsilon_{\mathrm{X}}^{\mathrm{LDA}}[n(\mathbf{r})] = A_{\mathrm{X}} n^{1/3}(\mathbf{r}) \tag{21}$$

with $\varepsilon_{\mathrm{X}}^{\mathrm{LDA}}$ the exchange energy density, i.e., the exchange energy per electron. This exchange energy density of the homogeneous electron gas can be derived by a HF treatment of electrons in a box, charge-neutralized by a smeared-out, so-called jellium background of ions. The correlation energy density $\varepsilon_{\mathrm{C}}^{\mathrm{LDA}}$ of the homogeneous electron gas is not known analytically, only the low- and high-density limits. However, parameterizations exist in the literature (e.g., [14]).

x-only LSD approximation

It is analytically known how the exchange energy of the homogeneous electron gas scales with the spin polarization. Assuming the same for LSD,

$$E_\mathrm{X}^\mathrm{LSD}[\{n_\sigma\}] = \frac{1}{2}\left(E_\mathrm{X}^\mathrm{LDA}[2n_\uparrow] + E_\mathrm{X}^\mathrm{LDA}[2n_\downarrow]\right), \tag{22}$$

leads with (20) to

$$E_\mathrm{X}^\mathrm{LSD}[\{n_\sigma\}] = 2^{1/3}A_\mathrm{X}\int d^3r\left(n_\uparrow^{4/3} + n_\downarrow^{4/3}\right). \tag{23}$$

For the LSD exchange potential, we thus obtain via

$$v_{\mathrm{X},\sigma}^\mathrm{LSD}[\{n_\sigma\}](\mathbf{r}) = \frac{\delta E_\mathrm{X}^\mathrm{LSD}[\{n_\sigma(\mathbf{r})\}]}{\delta n_\sigma(\mathbf{r})} = 2^{1/3}A_\mathrm{X}\frac{4}{3}n_\sigma^{1/3}(\mathbf{r}) = -\left(\frac{6}{\pi}\right)^{1/3} n_\sigma^{1/3}(\mathbf{r}). \tag{24}$$

In the spin-neutral case, $n_\sigma = n/2$, and $v_\mathrm{X}^\mathrm{LDA}[n](\mathbf{r}) = -\left(\frac{3}{\pi}\right)^{1/3} n^{1/3}(\mathbf{r})$ is obtained. We see that the exchange energy and the exchange potential are negative, thus increasing the binding energy. This is partially because the self-energy contained in the Hartree potential needs to be removed. The LDA and LSD exchange potential depend just locally on the density at the position $\mathbf{r}$ in $v_\mathrm{X}(\mathbf{r})$, which is nice and simple from the computational point of view. More advanced OEP potentials involve integrations of combinations of KS orbitals over space.

Central-field approximation

We adopt the so-called "central-field approximation" so that the KS potential is spherically symmetric,

$$v_{\mathrm{KS},l}^\sigma(r) = v_l(r) + u(r) + v_\mathrm{xc}^\sigma(r), \qquad v_l(r) = v(r) + \frac{l(l+1)}{2r^2}. \tag{25}$$

For brevity, we drop all functional dependencies $[n]$ and the like. Note that in the atomic case with $v(r) = -Z/r$, the $v_l(r)$ part is spherically symmetric anyway. However, in general, u and v_xc are not (see below).

In central-field approximation, we can work with radial spin KS orbitals $\phi_{\alpha\sigma lm}(\mathbf{r})$ (cf. section 1.5 of Chapter I),

$$\psi_{\alpha\sigma}(\mathbf{r}) = \frac{1}{r}\sum_{lm}\phi_{\alpha\sigma lm}(r)Y_{lm}(\Omega), \tag{26}$$

and we obtain the set of spherical KS equations

$$\left[-\frac{1}{2}\frac{\partial^2}{\partial r^2} + v_{\mathrm{KS},l}^\sigma(\mathbf{r})\right]\phi_{\alpha\sigma lm}(r) = \epsilon_{\alpha\sigma l}\phi_{\alpha\sigma lm}(r). \tag{27}$$

The total density reads as

$$n(\mathbf{r}) = \sum_{\sigma} \sum_{\alpha=1}^{N_\sigma} \left| \frac{1}{r} \sum_{lm} \phi_{\alpha\sigma lm}(r) Y_{lm}(\Omega) \right|^2 . \tag{28}$$

In general, this density is not spherically symmetric unless all l-subshells are closed (i.e., all states with different ms but otherwise the same indices are populated). This nonsphericality translates to both v_{KS} and $u[n]$, which contradicts our assumption that $v^{\sigma}_{\mathrm{KS},l}$ is spherically symmetric. In order to enforce that $n(\mathbf{r})$ is spherical, we take the average, leading to

$$n(r) = \frac{1}{4\pi} \int d\Omega\, n(\mathbf{r}) = n_{\uparrow}(r) + n_{\downarrow}(r) \tag{29}$$

with

$$n_\sigma(r) = \frac{1}{4\pi r^2} \sum_{\alpha=1}^{N_\sigma} \sum_{lm} |\phi_{\alpha\sigma lm}(r)|^2 . \tag{30}$$

The spherically averaged densities $n_{\uparrow}(r)$, $n_{\downarrow}(r)$ will be used to evaluate the exchange potential.

The multipole expansion of the Hartree potential reads as, using

$$\frac{1}{|\mathbf{r}-\mathbf{r}'|} = \sum_{l=0}^{\infty} \sum_{m=-l}^{l} \frac{4\pi}{2l+1} \frac{r_<^l}{r_>^{l+1}} Y^{\star}_{lm}(\Omega') Y_{lm}(\Omega), \tag{31}$$

$$u(\mathbf{r}) = \sum_{lm} \frac{4\pi Y_{lm}(\Omega)}{2l+1} \sum_{\sigma\alpha l'm'l''m''} \int dr' \frac{r_<^l}{r_>^{l+1}} \phi^{\star}_{\alpha\sigma l'm'}(r') \phi_{\alpha\sigma l''m''}(r') \times \int d\Omega'\, Y^{\star}_{l'm'}(\Omega') Y_{l''m''}(\Omega') Y^{\star}_{lm}(\Omega').$$

Because of angular momentum coupling, we have a $\int d\Omega'$ integral over three spherical harmonics, which can be expressed in terms of Clebsch–Gordan coefficients. However, for the calculation of effective potentials, we are only interested in the monopole term $\sim Y_{00}(\Omega)$. With the spherical average written as

$$u(r) = \frac{1}{4\pi} \int d\Omega\, u(\mathbf{r}) = \frac{1}{\sqrt{4\pi}} \int d\Omega\, u(\mathbf{r}) Y^{\star}_{00}(\Omega) \tag{32}$$

and $Y^{\star}_{00}(\Omega) = 1/\sqrt{4\pi}$, we obtain

$$u(r) = \sum_{\sigma\alpha l'm'} \left\{ \frac{1}{r} \int_0^r dr'\, |\phi_{\alpha\sigma l'm'}(r')|^2 + \int_r^{\infty} dr' \frac{|\phi_{\alpha\sigma l'm'}(r')|^2}{r'} \right\} . \tag{33}$$

We need an efficient way to calculate the radial Hartree potential $u(r)$ for all r on the numerical grid. A straightforward would be to perform the integrals in (33) for each of the discretized r_n values, $r_n = n\Delta r$ from scratch. However, that would require a double loop and thus scale $\mathcal{O}(N_r^2)$, which is not acceptable. In fact, it can be done in $\mathcal{O}(N_r)$. Let

$$g(r) = \int dr' \frac{1}{r_>} f(r'), \tag{34}$$

and the value for r_n be (with, e.g., trapezoidal integration) discretized on a radial grid

$$g(r_n) = g_n = \underline{g}_n + \overline{g}_n \tag{35}$$

with

$$\underline{g}_n = \frac{1}{r_n} \sum_{k=1}^{n} \Delta r f_k, \qquad \overline{g}_n = \sum_{k=n+1}^{N_r} \Delta r \frac{f_k}{r_k}. \tag{36}$$

Then, we can write for the next grid point

$$g_{n+1} = \frac{r_n}{r_{n+1}} \underline{g}_n + \frac{1}{r_{n+1}} \Delta r f_{n+1} + \overline{g}_n - \Delta r \frac{f_{n+1}}{r_{n+1}}. \tag{37}$$

Hence, it is enough to calculate, e.g., g_1, which is $\mathcal{O}(N_r)$, and all other g_n follow, also in $\mathcal{O}(N_r)$.

2.2 Actual implementation

We want to amend our TDSE solver from Section 1.5 of Chapter I by the extra terms in the potential, i.e., the Hartree potential $u[n]$ and the exchange potential $v_{\mathrm{x},\sigma}^{\mathrm{LSD}}[\{n_\sigma\}]$. Moreover, we are interested in the total energy $E[\{n_\sigma\}]$ (15) and the orbital energies $\epsilon_{\alpha\sigma l}$ (27).[2] The spherical DFT KS solver then works as follows:

1. **Set up the configuration.** As a first example, we consider a spin-neutral system, namely, the neon atom for which $Z = 10$. We expect a configuration $1\mathrm{s}^2\ 2\mathrm{s}^2\ 2\mathrm{p}^6$, $N = 10 = N_\uparrow + N_\downarrow$, $N_\uparrow = N_\downarrow = 5$ so that we could initialize as shown in Table 1. That would require five spin-up radial KS orbitals and five spin-down. If we make use of the fact that the radial KS orbitals with different ms but otherwise same quantum numbers and indices will be identical, we could get along with three KS orbitals per spin. Moreover, in the case of Ne, we know that the spin-up radial orbitals are the same as the spin-down ones because the KS-Hamiltonian is the same for $\sigma = \uparrow$ and $\sigma = \downarrow$. Hence, we could even get along with three radial KS orbitals in total. In the calculation of the density, one then has to work with degeneracy factors d_i

2 Note that, as usual in theories with nonlinear Hamiltonians, the sum of the orbital energies does not equal the total energy.

Tab. 1. Possible initialization of the KS solver for imaginary-time propagation in the case of Ne. In the very right column, we use the usual spectroscopic notation nl_m with s, p, d, f, ... corresponding to $l = 0, 1, 2, 3, \ldots$.

Ne	α	Spin up	Spin down	
	1	$l = 0, m = 0$	$l = 0, m = 0$	1s
	2	$l = 0, m = 0$	$l = 0, m = 0$	2s
	3	$l = 1, m = 1$	$l = 1, m = 1$	$2p_1$
	4	$l = 1, m = -1$	$l = 1, m = -1$	$2p_{-1}$
	5	$l = 1, m = 0$	$l = 1, m = 0$	$2p_0$

Tab. 2. Possible initialization for Ne^+ (or F). Same as in Table 1 but with one electron less.

Ne^+ or F	α	Spin up	Spin down	
	1	$l = 0, m = 0$	$l = 0, m = 0$	1s
	2	$l = 0, m = 0$	$l = 0, m = 0$	2s
	3	$l = 1, m = 1$	$l = 1, m = 1$	$2p_1$
	4	$l = 1, m = -1$	$l = 1, m = -1$	$2p_{-1}$
	5	$l = 1, m = 0$	Unoccupied	$2p_0$

that take the degeneracy properly into account, e.g.,

$$n(r) = \frac{1}{4\pi r^2} \sum_{i=1}^{3} d_i |\phi_i(r)|^2, \qquad d_1 = 2,\ d_2 = 2,\ d_3 = 6,$$

where $i = 1$ corresponds to the 1s states ($\alpha = 1$, any spin), $i = 2$ to the 2s ($\alpha = 2$, any spin), and $i = 3$ to the 2p states ($\alpha = 3, 4, 5$, any spin).

Now consider a nine-electron system, e.g., Ne^+ with still $v(r) = -10/r$ or F where $v(r) = -9/r$. Which KS electron should be removed? Because of spin neutrality, it should not matter whether we remove the "outermost" spin-up or spin-down KS particle.[3] One option is shown in Table 2. The radial spin-up KS orbitals $\alpha = 3, 4, 5$ will still be the same. However, because $\hat{H}_{KS}^{\uparrow} \neq \hat{H}_{KS}^{\downarrow}$, they will be different now from the spin-down orbitals with $\alpha = 3, 4$ (themselves identical).

In removing further electrons (creating Ne^{2+} or O, Ne^{3+} or N, etc.), there are more possibilities, e.g., one may remove alike spins or opposite ones. By trying all possibilities and comparing the total energy, one should find out the energetically favorable configuration. In fact, LSD is sufficient to "confirm" Hund's rules. One

3 But what about removing "half a KS electron" from both spin types? This can be investigated by introducing occupation numbers $f_i \in [0, 1]$.

could even create holes, removing an s-electron before the p-electrons. Although this formally is outside the realm of ground-state DFT, it works remarkably well in practice as far as energy differences between different configurations are concerned (so-called ΔSCF methods, SCF for “self-consistent field”).

2. **Initialize** the radial KS orbitals. **Orthonormalize** those radial KS orbitals that have same σ, l, and m.
3. **Imaginary-time propagation** (until total energy E and orbital energies $\epsilon_{\alpha\sigma l}$ converged):
   (a) calculate the spin densities n_σ according (30),
   (b) calculate the exchange potential v_X (24),
   (c) calculate the Hartree potential u (33),
   (d) propagate radial KS orbitals in imaginary time with respective $\hat{H}^\sigma_\mathrm{KS}$ (10),
   (e) orthonormalize.
   How the orthonormalization is achieved has been addressed already in Section 1.3 of Chapter I. KS orbitals that are different either in σ, l, or m are already orthogonal. They only need to be normalized after the imaginary propagation step. Then consider the first KS orbital $\tilde{\phi}_{\alpha'\sigma lm}$ that has σ, l, and m in common with a previous (already normalized) KS orbital $\phi_{\alpha\sigma lm}$, $\alpha' > \alpha$ (it is always possible to order the KS orbitals in that way). Then, project out,

$$\tilde{\phi}'_{\alpha'\sigma lm}(r) = \tilde{\phi}_{\alpha'\sigma lm}(r) - \phi_{\alpha\sigma lm}(r) \int dr\, \phi^*_{\alpha\sigma lm}(r)\tilde{\phi}_{\alpha'\sigma lm}(r),$$

   and normalize the result,

$$\phi_{\alpha'\sigma lm}(r) = \tilde{\phi}'_{\alpha'\sigma lm}(r) \Bigg/ \sqrt{\int dr \left|\tilde{\phi}'_{\alpha'\sigma lm}(r)\right|^2}.$$

   If there is a third KS orbital $\tilde{\phi}_{\alpha''\sigma lm}$ also with the same σ, l, and m, one has to project out of it both $\phi_{\alpha\sigma lm}$ and $\phi_{\alpha'\sigma lm}$, and so on. It is easy to write a routine that does the complete orthonormalization automatically for an arbitrary number of KS orbitals $\phi_{\alpha\sigma lm}(r)$.
   (f) calculate entities of interest (total energy E, orbital energies, $\epsilon_{\alpha\sigma}$ etc.)
4. **Output.** Dump KS orbitals, spin densities, potentials, etc. to files.

2.2.1 Some results from this implementation

For the Ne atom, we obtain the orbital energies (in atomic units) given in Table 3. For the total energy, we obtain

$$E_\mathrm{Ne} = -127.48. \tag{38}$$

Tab. 3. KS orbital energies $\epsilon_{\alpha\sigma l}$ in x-only LSD.

Ne	α	Spin up	Spin down		$\epsilon_{\alpha\sigma l}$
	1	$l = 0, m = 0$	$l = 0, m = 0$	1s	−30.23
	2	$l = 0, m = 0$	$l = 0, m = 0$	2s	−1.27
	3,4,5	$l = 1, m = -1, 0, 1$	$l = 1, m = -1, 0, 1$	2p	−0.44

The experimental ionization potential for Ne (i.e., the energy required to remove the "outermost" electron) is

$$I_p = 0.793. \tag{39}$$

We see that

$$\epsilon_{3\sigma 1m} = \epsilon_{2\text{p}} = -0.44 \neq -I_p.$$

In the HF and DFT jargon, one says that the so-called Koopmans' and Janak's "theorem," respectively,[4] are violated. However, there is a more rigorous way to calculate the ionization energy in ground-state DFT:

$$I_p^{\text{DFT}} = E_{\text{Ne}^+} - E_{\text{Ne}} = 0.77, \tag{40}$$

where E_{Ne^+} is the ground-state energy obtained from a x-only LSD-KS calculation for Ne^+. This value is quite close to the experimental value (39), with a relative error of only 3%. The take-home message here is: be careful with overinterpreting KS level energies for fictitious KS particles; sticking to rigorously meaningful quantities like total ground-state energies is safer.

As an example for a spin-polarized system, let us consider the potassium atom. We expect a configuration $(1\text{s})^2(2\text{s})^2(2\text{p})^6(3\text{s})^2(3\text{p})^6(4\text{s})^1$, and if we did not know that we would be able to figure it out with our DFT code by trying other configurations. Potassium, with its 19 electrons, is the first atom where an – at first sight – strange anomaly occurs: the 4s shell is filled before the 3d. This is because the 4s electron has a higher probability close to the core and thus "sees" a less-screened potential than the 3d. Although that is age-old textbook knowledge, it is interesting to check whether x-only LSD-KS-DFT is able to capture this effect.

Exploiting degeneracy, we still need eleven orbitals to accommodate all KS electrons. The configuration and the orbital energies are given in Table 4. Note that despite the spin polarization, the orbital energies are almost equal because the extra

4 According to the Koopmans theorem, the (absolute value of the) outermost HF orbital energy should equal the ionization energy. This is only true if one "freezes" the "other" electrons. If they are allowed to "relax," there is no reason to expect that $\epsilon_{\text{valence}} = -I_p$ is exactly fulfilled. In DFT, Janak's theorem states that the outermost KS orbital energy should equal $-I_p$ because relaxation is already taken into account [27].

Tab. 4. KS configuration and orbital energies for the K atom (x-only LSD calculation on a radial grid with $\Delta r = 0.01$, $N_r = 5000$).

K	α			$\epsilon_{\alpha\uparrow l}$	$\epsilon_{\alpha\downarrow l}$
	1	$l = 0, m = 0$	1s	−128.30	−128.31
	2	$l = 0, m = 0$	2s	−12.77	−12.77
	3	$l = 1, m = -1, 0, 1$	2p	−10.21	−10.21
	4	$l = 0, m = 0$	3s	−1.23	−1.23
	5	$l = 1, m = -1, 0, 1$	3p	−0.642	−0.640
	6	$l = 0, m = 0$	4s	−0.0797	Unoccupied

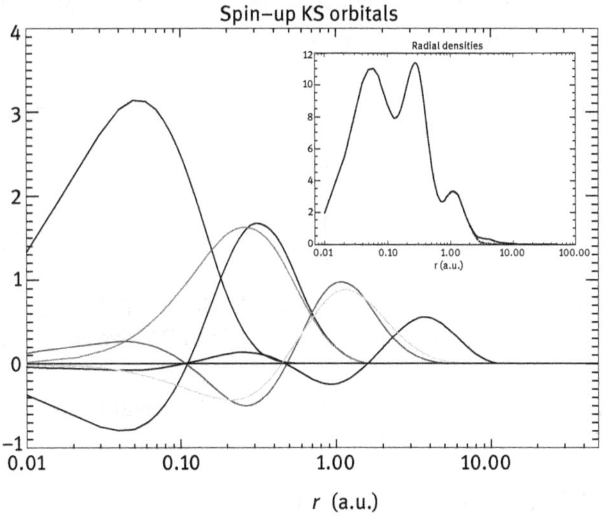

Fig. 1. The radial KS orbitals for K in x-only LSD. It is left as an exercise to identify which is which (count nodes and look at the slope as $r \to 0$). The inset shows the radial spin densities (spin-up drawn solid, spin-down broken).

outermost KS electron does not have a strong effect on the more tightly bound inner ones. The spin-up orbitals are shown in Figure 1. The total energy we obtain is

$$E_{\mathrm{K}} = -596.75. \tag{41}$$

If we populate the 3d orbital instead of the 4s, we obtain a higher value for the total energy E_{K}. In principle, one should try all configurations possible for $N = 19$ and then conclude that the one with the lowest total energy is the one realized in nature. But once again, the KS orbitals are just constructs that lead to the same electron density as the real system. They are not single-electron states. Nevertheless, we have an *aufbau principle* based on their population according to Pauli's exclusion principle that explains the periodic table of elements (and many more things). Janak's theorem is violated again for K treated in x-only LSD approximation, as $\epsilon_{4s} \neq I_p = 0.160$, while the removal energy calculated via $E_{K^+} - E_K = -596.60 - (-596.75) = 0.15$ is only 0.27 eV away from the exact, experimental value.

2.2.2 Perdew–Zunger SIC

We observe in the Ne and K examples above that the ionization energy in x-only LSD is underestimated. One reason for that is the self-energy error of LDA and LSD and many other xc-potentials from the lower end of the $E_{xc}[\{n_\sigma\}]$ hierarchy mentioned above.

Consider the limit of a single electron, i.e., the hydrogen atom. In this limit, the KS equation should reduce to the Schrödinger equation

$$\epsilon_{1s}\phi_{1s}(r) = \left(-\frac{1}{2}\frac{\partial^2}{\partial r^2} - \frac{1}{r}\right)\phi_{1s}(r) \tag{42}$$

so that

$$v_{KS}(r) = -\frac{1}{r} + u + v_{XC} \overset{!}{=} -\frac{1}{r} \tag{43}$$

and thus $u + v_{XC} = 0$, that is, the xc-potential has to cancel the Hartree potential for $N = 1$. Unfortunately, the LDA does not do this. The lack of "SIC" in LDA gives rise to a too low ionization potential.

An obvious idea to remove self-interaction is to subtract u and v_{XC} evaluated with the respective orbital's density from the KS potential,

$$\left[H_\sigma^{KS} + H_{\alpha\sigma}^{PZ}\right]\psi_{\alpha\sigma}(\mathbf{r}) = \epsilon_{\alpha\sigma}\psi_{\alpha\sigma}(\mathbf{r}) \tag{44}$$

with the usual

$$H_\sigma^{KS} = -\frac{1}{2}\nabla^2 + v(\mathbf{r}) + v_{Hxc}^\sigma[\{n_\sigma\}](\mathbf{r}) \tag{45}$$

and the Perdew–Zunger (PZ) SIC [27]

$$H_{\alpha\sigma}^{PZ} = -\, v_{Hxc}^\sigma\left[|\psi_{\alpha\sigma}|^2\right](\mathbf{r}). \tag{46}$$

This approach yields the correct Schrödinger equation in the limit of one particle. In the two-electron spin-singlet case, it reduces to HF because there is effectively only one spatial KS orbital,

$$\psi_{1s\uparrow}(\mathbf{r}) = \psi_{1s\downarrow}(\mathbf{r}) =: \psi_{1s}(\mathbf{r}), \qquad n_\sigma(\mathbf{r}) = |\psi_{1s}(\mathbf{r})|^2, \qquad n(\mathbf{r}) = 2|\psi_{1s}(\mathbf{r})|^2,$$

and the PZ SIC KS equation becomes

$$\epsilon_{1s}\psi_{1s}(\mathbf{r}) = \left[-\frac{1}{2}\nabla^2 - \frac{2}{r} + \int d^3r' \frac{n(\mathbf{r}') - |\psi_{1s}(\mathbf{r}')|^2}{|\mathbf{r}-\mathbf{r}'|} - \left(\frac{6}{\pi}\right)^{1/3}\left[n_\sigma^{1/3}(\mathbf{r}) - |\psi_{1s}(\mathbf{r})|^{2/3}\right]\right]\psi_{1s}(\mathbf{r})$$

$$= \left[-\frac{1}{2}\nabla^2 - \frac{2}{r} + \int d^3r' \frac{|\psi_{1s}(\mathbf{r}')|^2}{|\mathbf{r}-\mathbf{r}'|}\right]\psi_{1s}(\mathbf{r}).$$

In this case, the PZ-LSD-KS Hamiltonian $H_\sigma^{\text{KS}} + H_{\alpha\sigma}^{\text{PZ}}$ yields the so-called exact x-only Hamiltonian, which is equivalent to HF. Its interpretation is intuitive: one of the electrons only sees "the other" electron (which sits in the same orbital). Hence, only half the Hartree interaction should appear in the Hamiltonian.

For more than one electron per spin orientation, the PZ-SIC applied to LSD is not exact anymore. While it always works to subtract the orbital density in the Hartree part, the functional form of the x-potential prevents this, simply because $(a+b)^{1/3} - b^{1/3} \neq a^{1/3}$.

One formal issue with PZ-SIC is that it is an *ad hoc* procedure that lacks a formal foundation. In fact, it is not KS-DFT anymore. The idea of the KS scheme was to introduce an auxiliary system of noninteracting particles. Such noninteracting, indistinguishable particles should be all governed by the same Hamiltonian. However, the Hamiltonian $H_\sigma^{\text{KS}} + H_{\alpha\sigma}^{\text{PZ}}$ is orbital-index dependent, meaning that different KS particles see different KS potentials (even if their spin-projections σ are the same). As a result, the KS orbitals are not orthogonal anymore. Assuming a pragmatic view point, that does not sound too disturbing as long as the method works, i.e., the energies improve compared to the non-SIC results, and the numerical extra cost (calculating N different $H_{\alpha\sigma}^{\text{PZ}}$ instead of just two different H_σ^{KS}) is worth the effort. One may either orthogonalize the KS orbitals or simply accept a small nonorthogonality. PZ-SIC results for the KS orbital energies for the K atom are presented in Table 5. The ionization potential $I_p = 0.160$ is now closer to $|\epsilon_{4s}| = 0.154$, i.e., Janak's theorem is better

Tab. 5. As Table 4 but with PZ SIC.

K	α			$\epsilon_{\alpha\uparrow l}$	$\epsilon_{\alpha\downarrow l}$
	1	$l=0, m=0$	1s	−133.19	−133.19
	2	$l=0, m=0$	2s	−13.67	−13.67
	3	$l=1, m=-1,0,1$	2p	−11.31	−11.31
	4	$l=0, m=0$	3s	−1.50	−1.50
	5	$l=1, m=-1,0,1$	3p	−0.893	−0.891
	6	$l=0, m=0$	4s	−0.154	Unoccupied

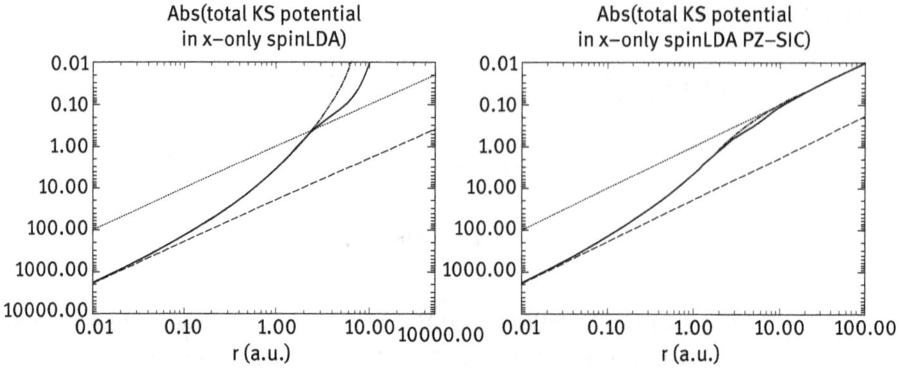

Fig. 2. Double-logarithmic plots of the absolute value of the total KS potentials (spin-up drawn bold solid, spin-down bold broken) in x-only LSD (left) and with PZ-SIC (right, for $\alpha = 1$) for the K atom. The $r \to 0$-limit Z/r is drawn dashed, the $r \to \infty$-limit $1/r$ dotted.

fulfilled. From the energy difference, we obtain for the ionization energy

$$E_{\mathrm{K,PZ^+}} - E_{\mathrm{K,PZ}} = -599.92 - (-600.08) = 0.16, \tag{47}$$

which is in excellent agreement with the experimental value. Another great improvement with PZ-SIC when applied to LSD is that the correct asymptotic behavior of the KS potential is ensured. Any KS potential for neutral atoms should roll-off $\sim -1/r$ as $r \to \infty$. Plain LSD KS potentials do not meet this requirement while the PZ-SIC corrected ones do, as shown in Figure 2.

3 Time-dependent calculations

So far in this chapter, we used a TDKS solver in imaginary-time mode to solve the set of time-independent KS equations (8). In adiabatic approximation, (8) becomes [43]

$$i\frac{\partial}{\partial t}\psi_{\alpha\sigma}(\mathbf{r},t) = \left[-\frac{1}{2}\nabla^2 + v(\mathbf{r},t) + u[n(t)](\mathbf{r}) + v_{\mathrm{xc}}^{\sigma}[\{n_\sigma(t)\}](\mathbf{r})\right]\psi_{\alpha\sigma}(\mathbf{r},t). \tag{48}$$

The external potential $v(\mathbf{r}, t)$ now may include, besides a binding potential, a driver such as $\mathbf{r}\cdot\mathbf{E}(t)$ or $\mathbf{p}\cdot\mathbf{A}(t)$. How to implement such drivers in a single-active-electron TDSE solver has been extensively discussed in Chapter I, and we can apply the same strategies for a TDKS solver. The Hartree and xc-potentials become time dependent via the time-dependent densities

$$n_\sigma(\mathbf{r},t) = \sum_{\alpha=1}^{N_\sigma} |\psi_{\alpha\sigma}(\mathbf{r},t)|^2, \quad n(\mathbf{r},t) = \sum_{\sigma} n_\sigma(\mathbf{r},t). \tag{49}$$

The approximate effective, time-dependent potential $u[n(t)](\mathbf{r}) + v_{xc}^{\sigma}[\{n_\sigma(t)\}](\mathbf{r})$ is called "adiabatic" because it consists of density functionals evaluated with the time-dependent density. In particular, the potential is assumed to be local in time, i.e., without memory effects.

3.1 Time-dependent KS solver with spherical harmonics and multipole expansion

For the imaginary-time propagation according to the algorithm outlined in Section 2.2, we adopted the central-field approximation so that the monopole terms in the expansions of $n(\mathbf{r})$ and $u(\mathbf{r})$ suffice. A linearly polarized laser field leads to a dipole coupling (see Section 1.5.2 of Chapter I) and thus breaks the spherical symmetry. We therefore need to take higher terms in the multipole expansions of $v_{KS}^{\sigma}[\{n_\sigma\}]$ into account.

Instead of the TDSE (105) in Chapter I, we now have the TDKS equation

$$
\begin{aligned}
i\frac{\partial}{\partial t}\phi_{\alpha\sigma lm} = {} & \left(-\frac{1}{2}\frac{\partial^2}{\partial r^2} + v_{\mathrm{eff}\,l}(r) + v_{\mathrm{Hxc}}^{\sigma 0}(r,t) + p_{lm} v_{\mathrm{Hxc}}^{\sigma 2}(r,t)\right)\phi_{\alpha\sigma lm} \\
& - iA(t)\Big(c_{l-1,m}\frac{\partial}{\partial r}\phi_{\alpha\sigma,l-1,m} + c_{lm}\frac{\partial}{\partial r}\phi_{\alpha\sigma,l+1,m} \\
& \qquad - \frac{1}{r} l c_{l-1,m}\phi_{\alpha\sigma,l-1,m} + \frac{1}{r}(l+1)c_{lm}\phi_{\alpha\sigma,l+1,m}\Big) \\
& + \left(rE(t) + v_{\mathrm{Hxc}}^{\sigma 1}(r,t)\right)\left(c_{l-1,m}\phi_{\alpha\sigma,l-1,m} + c_{lm}\phi_{\alpha\sigma,l+1,m}\right) \\
& + v_{\mathrm{Hxc}}^{\sigma 2}(r,t)\left(q_{l-2,m}\phi_{\alpha\sigma,l-2,m} + q_{lm}\phi_{\alpha\sigma,l+2,m}\right),
\end{aligned}
\tag{50}
$$

where

$$v_{\mathrm{eff}\,l}(r) = v(r) + \frac{l(l+1)}{2r^2}, \tag{51}$$

$$v_{\mathrm{Hxc}}^{\sigma}(\mathbf{r},t) = v_{\mathrm{Hxc}}^{\sigma 0}(r,t) + v_{\mathrm{Hxc}}^{\sigma 1}(r,t)\cos\theta + v_{\mathrm{Hxc}}^{\sigma 2}(r,t)\frac{1}{2}(3\cos^2\theta - 1) + \cdots, \tag{52}$$

$$c_{lm} = \sqrt{\frac{(l+1)^2 - m^2}{(2l+1)(2l+3)}}, \tag{53}$$

$$p_{lm} = \frac{l(l+1) - 3m^2}{(2l-1)(2l+3)}, \tag{54}$$

$$q_{lm} = \frac{3}{2(2l+3)}\sqrt{\frac{[(l+1)^2 - m^2][(l+2)^2 - m^2]}{(2l+1)(2l+5)}}. \tag{55}$$

Here, we allowed for both a vector potential $\mathbf{A}(t) = A(t)\mathbf{e}_z$ and an electric field $\mathbf{E}(t) = E(t)\mathbf{e}_z$. In pure velocity or length gauge, only one of the two is present (see discussion in Section 1.5.5 of Chapter I). We terminated the multipole expansion of

$v^{\sigma}_{\mathrm{Hxc}}(\mathbf{r},t)$ in (52) after the quadrupole. More terms may be taken into account. However, they lead to more and more couplings between KS orbital components of different l. The quadrupole term, for instance, leads to the term $\sim p_{lm}$ diagonal in l and to couplings of l with $l \pm 2$. As a consequence, the time-evolution operator requires additional splittings, as discussed below. Note that while σ and m remain good quantum numbers during real-time propagation also for linearly polarized drivers, many l components of the KS orbitals α will in general be populated during the interaction with the external field.

The expansion (52) contains the spin-independent Hartree part

$$u(\mathbf{r},t) = u_0(r,t) + u_1(r,t)\cos\theta + u_2(r,t)\frac{1}{2}(3\cos^2\theta - 1) + \cdots . \tag{56}$$

With the auxiliary quantities

$$\Lambda_\sigma(r,t) = \sum_{\alpha=1}^{N_\sigma}\sum_l |\phi_{\alpha\sigma lm}|^2, \tag{57}$$

$$\Theta_\sigma(r,t) = \sum_{\alpha=1}^{N_\sigma}\sum_l \left(c_{l-1,m}\phi^*_{\alpha\sigma,l-1,m} + c_{lm}\phi^*_{\alpha\sigma,l+1,m}\right)\phi_{\alpha\sigma lm}, \tag{58}$$

$$\Xi_\sigma(r,t) = \sum_{\alpha=1}^{N_\sigma}\sum_l \left(p_{lm}\phi^*_{\alpha\sigma lm} + q_{lm}\phi^*_{\alpha\sigma,l+2,m} + q_{l-2,m}\phi^*_{\alpha\sigma,l-2,m}\right)\phi_{\alpha\sigma lm}, \tag{59}$$

the monopole, dipole, and quadrupole can be written as

$$u_0(r,t) = \int \mathrm{d}r'\,\frac{1}{r_>}\sum_\sigma \Lambda_\sigma(r',t), \tag{60}$$

$$u_1(r,t) = \int \mathrm{d}r'\,\frac{r_<}{r_>^2}\sum_\sigma \Theta_\sigma(r',t), \tag{61}$$

$$u_2(r,t) = \int \mathrm{d}r'\,\frac{r_<^2}{r_>^3}\sum_\sigma \Xi_\sigma(r',t), \tag{62}$$

respectively. In a similar manner, the xc-potential v^{σ}_{xc} can be expanded. In the x-only LSD case, for instance, one needs a multipole expansion of $n_\sigma^{1/3}$. This can be achieved by first expanding the spin densities

$$n_\sigma(\mathbf{r},t) = \frac{1}{r^2}\left(n_{\sigma 0}(r,t) + n_{\sigma 1}(r,t)\cos\theta + n_{\sigma 2}(r,t)\frac{1}{2}(3\cos^2\theta - 1) + \cdots\right), \tag{63}$$

where one finds that the $n_{\sigma i}(r,t)$, $i = 0,1,2$ can be expressed in terms of $\Lambda_\sigma(r,t)$, $\Theta_\sigma(r,t)$, and $\Xi_\sigma(r,t)$. Then, by making an analogous ansatz for $n_\sigma^{1/3}(\mathbf{r},t)$,

$$n_\sigma^{1/3}(\mathbf{r},t) = \frac{1}{r^{2/3}}\left(\eta_{\sigma 0}(r,t) + \eta_{\sigma 1}(r,t)\cos\theta + \eta_{\sigma 2}(r,t)\frac{1}{2}(3\cos^2\theta - 1) + \cdots\right), \tag{64}$$

one can work out the $\eta_{\sigma i}(r,t)$ in terms of the $n_{\sigma i}(r,t)$ by taking the cube of $n_\sigma^{1/3}(\mathbf{r},t)$ from (64) and setting it equal to the expansion of $n_\sigma(\mathbf{r},t)$ in (63).

3.1.1 Propagation

The TDSE propagator (119) in Chapter I needs to be adapted for propagating the TDKS equation (50):

$$\begin{aligned}\mathbf{U}(\Delta t) = & \prod_{l=N_l-3}^{0} \exp\left(-i\frac{\Delta t}{2}\mathbf{H}_{\text{ang}}^{lm(3)}\right)\\ & \times \left\{\prod_{l=N_l-2}^{0} \exp\left(-i\frac{\Delta t}{2}\mathbf{H}_{\text{ang}}^{lm(1,2)}\right)\exp\left(-i\frac{\Delta t}{2}\mathbf{H}_{\text{mix}}^{lm}\right)\right\}\exp(-i\Delta t\mathbf{H}_{\text{at}})\\ & \times \left\{\prod_{l=0}^{N_l-2} \exp\left(-i\frac{\Delta t}{2}\mathbf{H}_{\text{mix}}^{lm}\right)\exp\left(-i\frac{\Delta t}{2}\mathbf{H}_{\text{ang}}^{lm(1,2)}\right)\right\}\\ & \times \prod_{0}^{l=N_l-3} \exp\left(-i\frac{\Delta t}{2}\mathbf{H}_{\text{ang}}^{lm(3)}\right) + \mathcal{O}(\Delta t^3)\end{aligned} \tag{65}$$

with

$$\mathbf{H}_{\text{ang}}^{lm(1,2)} = -i\frac{A(t)}{r}\mathbf{T}^{lm} + \left(rE(t) + v_{\text{Hxc}}^{\sigma 1}(r,t)\right)\mathbf{L}^{lm}, \tag{66}$$

where the 2×2 matrices $\mathbf{L}^{lm}$ and $\mathbf{T}^{lm}$ are given in (114) of Chapter I and act in $l, l+1$-subspace, and

$$\mathbf{H}_{\text{ang}}^{lm(3)} = v_{\text{Hxc}}^{\sigma 2}(r,t)\mathbf{P}^{lm} \tag{67}$$

with another 2×2 matrix

$$\mathbf{P}^{lm} = \begin{pmatrix} 0 & q_{lm} \\ q_{lm} & 0 \end{pmatrix} \tag{68}$$

acting in $l, l+2$-subspace. The new exponential factors with $\mathbf{H}_{\text{ang}}^{lm(1,2)}$ and $\mathbf{H}_{\text{ang}}^{lm(3)}$ can be further treated like the $\mathbf{H}_{\text{ang}}^{lm}$ in Chapter I. The propagation of N KS orbitals according (65) is thus essentially like propagating N wavefunctions with the TDSE solver in Section 1.5.3 of Chapter I. The main difference is that the $v_{\text{Hxc}}^{\sigma i}(r,t)$ need to be calculated each time step.

Further, it is advisable to introduce a predictor-corrector step. Because of the nonlinear Hamiltonian depending on the KS orbitals, observables tend to drift away in a free, real-time propagation. A predictor-corrector step prevents this:[5]

1. Perform for all KS orbitals a propagation step, $\Phi(\Delta t) = \mathbf{U}(\Delta t)\Phi(0)$ using $v_{\text{Hxc}}^{\sigma}(0)$.

5 Observables may still oscillate around a mean value, but these oscillations can be made very small by a well-converged ground state to start with and a sufficiently small time step.

2. Use $\Phi(\Delta t)$ to calculate $v^{\sigma}_{\mathrm{Hxc}}(\Delta t)$.
3. Perform again for all KS orbitals a propagation step $\Phi(\Delta t) = \mathbf{U}(\Delta t)\Phi(0)$ but now using the averaged potential $\left(v^{\sigma}_{\mathrm{Hxc}}(0) + v^{\sigma}_{\mathrm{Hxc}}(\Delta t)\right)/2$.

3.1.2 Example: C_{60} jellium model

TDDFT is particularly well suited for systems with delocalized electrons like metal clusters or fullerenes. Prominent features in such systems are collective modes (Mie plasmons) that also govern their optical properties. Moreover, in so-called jellium models, the treatment of systems with delocalized electrons may be further simplified by smearing out the ionic background. The binding potentials of metal clusters and C_{60} then are smooth and spherically symmetric so that a TDKS solver employing a spherical-harmonics expansion of the KS orbitals is particularly efficient.

Consider the following jellium potential for C_{60} [3, 30, 31, 37]:

$$v(r) = \begin{cases} -r_s^{-3}3(R_o^2 - R_i^2)/2 & \text{for} \quad r \le R_i \\ -r_s^{-3}(3R_o^2/2 - [r^2/2 + R_i^3/r]) - v_0 & \text{for} \quad R_i < r < R_o, \\ -r_s^{-3}(R_o^3 - R_i^3)/r & \text{for} \quad r \ge R_o \end{cases} \tag{69}$$

where $R_i = 5.3$, $R_o = 8.1$, $r_s^{-3} = N/(R_o^3 - R_i^3)$, $N = 250$, and $v_0 = 0.68$. We seek a spin-neutral configuration, which is only achieved with 250 electrons instead of the 240 “real” valence electrons in C_{60}. As $N_\uparrow = N_\downarrow = 125$, we need to solve the TDKS equation only for, say, $\sigma = \uparrow$, knowing that $n = 2n_\uparrow$. The x-only LDA potential in terms of the total density is $v_x(\mathbf{r}) = -[3n(\mathbf{r})/\pi]^{1/3}$. One obtains 200 so-called σ and 50 π electrons without node and with one node in the radial wavefunction, respectively. The node for the π-electrons is located close to the C_{60} radius $R = (R_i + R_o)/2$. The free parameter $v_0 = 0.68$ is used to adjust the KS energy of the highest occupied molecular orbital (HOMO) to the ionization potential of “real” C_{60}, $-\epsilon_{\mathrm{HOMO}} = I_p \simeq 0.28$. The ground-state configuration, obtained via imaginary-time propagation and in central-field approximation, is illustrated in Figure 3(a). It was obtained by building up the C_{60} shell-wise. Keeping in mind that we will switch-on a linearly polarized laser soon, we need to treat KS orbitals with (initial) quantum number l but different $|m|$ differently.[6] So, how many KS orbitals are required to accommodate the $N = 250$ electrons with laser on? It turns out that with the above external binding potential (69) and x-LDA, the energetically favorable configuration is as listed in Table 6, i.e., 70 KS orbitals need to be propagated in time.

Typically, a linear-response calculation follows the ground-state determination in order to see whether the excitations of the implemented system are in reasonable

6 Orbitals starting with the same initial l but $m = \pm|m|$ behave the same in a linearly polarized laser field. This degeneracy should be exploited.

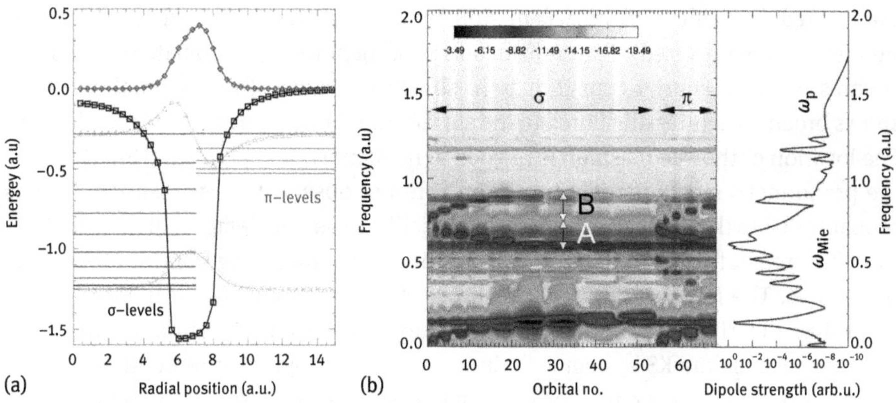

Fig. 3. Panel (a) shows the net KS potential (black, squares), total density (gray, diamonds), wavefunctions of the lowest KS orbital, and the highest occupied orbital (light-gray crosses and triangles, respectively). The σ- and π-levels are indicated. Density and wavefunctions are scaled to fit into the plot (from [37]). Panel (b), right, shows the dipole response of the C_{60} model system. One observes narrow lines (single-particle transitions) on top of two broad structures, the surface or Mie plasmon around $\omega_{\mathrm{Mie}} \simeq 0.7$ and the "volume plasmon" around $\omega_{\mathrm{p}} \simeq 1.4$. The left panel in (b) shows the individual, logarithmically scaled KS orbital dipole strengths, enumerated as in Table 6. Branch A consists of transitions of the type $\sigma l \to \pi(l-1)$ and $\pi l \to \sigma(l+1)$, branch B *vice versa*, i.e., $\sigma l \to \pi(l+1)$ and $\pi l \to \sigma(l-1)$. The "volume plasmon" corresponds to transitions between σ and (initially unpopulated) δ states.

Tab. 6. The 70 KS orbitals required to cover the 250-electron C_{60} jellium model.

l	$\lvert m \rvert$	σ orbital #	π orbital #
0	0	0	55
1	0,1	1,2	56,57
2	0,1,2	3,4,5	58–60
3	0–3	6–9	61–64
4	0–4	10–14	65–69
5	0–5	15–20	
⋮	($l+1$ orbitals)	⋮	
9	0–9	45–54	

agreement with the real, physical system. The result of such a calculation is shown in Figure 3(b). The system was perturbed by a small, step-like vector potential $A(t) = A_0\Theta(t)$, corresponding to a δ-kick in the electric field.[7] After recording the total and the individual KS orbital dipoles[8] for a sufficiently long time (depending on

7 As long as it is small enough, the actual value of A_0 only affects the strength of the response but not the shape of the spectrum, as it should be in the linear response regime.

8 Evaluated according (107) of Chapter II.

the desired frequency resolution) in sufficiently small time steps (determining the frequency range covered), Fourier transforms of them were performed. The modulus squares of these Fourier transforms are shown in Figure 3(b). The total spectrum shows broad structures attributed to the surface (Mie) plasmon and the bulk plasmon. The location of the Mie plasmon is in good agreement with experiment [16]. However, the pronounced single-particle transitions are not observed experimentally. This is probably due to the too high symmetry of our jellium system. Removal of degeneracies would fragment further and thus wash out these lines stemming from single-particle transitions. The left part of panel (b) shows the individual dipole response of each KS orbital. Clearly, this is not an observable not only because it cannot be measured but also because the KS system is fictitious and just designed to reproduce the total density. However, out of curiosity, we may nevertheless analyze which KS transitions make up the total dipole response. The Mie surface plasmon is commonly understood as a dipole oscillation of the electrons against the ionic background. It is thus not surprising that in terms of KS levels, it consists of transitions between σ and π states with l changing by one. The appearance of a bulk plasmon at ω_p is a bit surprising because a breathing mode should not be excited by a δ kick in z direction for which the electric dipole selection rule $l \to l \pm 1$ holds. In fact, in terms of KS states, this transition originates from dipole-allowed transitions between σ and (initially unpopulated) δ states (i.e., with two nodes in the radial KS orbital).

As an example for a strong-field TDDFT calculation beyond linear response, we consider harmonic generation in C_{60} in an eight-cycle trapezoidal laser pulse with up and down ramping over two cycles. The wavelength is 2280 nm ($\omega = 0.02$); the peak electric field strength $\hat{E} = 0.03$ (3.2×10^{13} Wcm^{-2}). We apply a time-frequency analysis as outlined in Section 3 of Chapter II. The result is shown in Figure 4. The emission times of certain harmonics match the return times of electrons of corresponding energy (white lines). An enhanced emission around the Mie plasmon frequency is observed. The time-frequency analysis reveals that this emission is clearly linked to the electron return times. It may therefore be interpreted as recollision-induced plasmon excitation [37].

3.2 Low-dimensional benchmark studies

Approximate N-particle methods should be benchmarked by as much exact results as available. In the few-body limit, $N = 1$ does not pose a problem for most modern quantum many-particle techniques.[9] However, already $N = 2$ is not as simple as it seems. Helium-like ions, for instance, exhibit phenomena that rely on strong electron correlation, examples being single-photon double ionization, autoionization, and

9 Although we saw in Section 2.2.2 that self-interaction is an issue here for DFT.

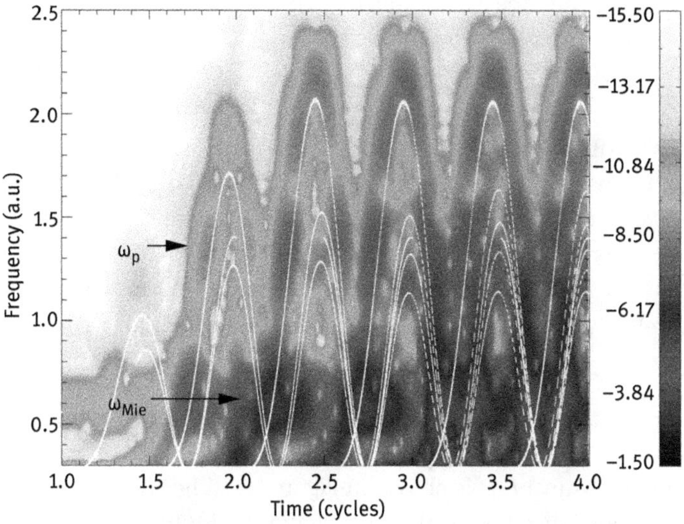

Fig. 4. Time-frequency analysis of harmonic generation in a C_{60} jellium model. The dipole strength is logarithmically scaled and color-coded over 14 orders of magnitude. The laser parameters were $\lambda = 2280$ nm, $\hat{E} = 0.03$, (2,4,2)-trapezoidal pulse. The white lines are expected electron return energies [37]. Copyright © 2008 American Physical Society (APS).

nonsequential double ionization. Helium in a strong laser field in full dimensionality is computationally very expensive [41, 48]. For benchmarking TDDFT or other time-dependent many-electron methods, one may consider a low-dimensional model helium atom whose TDSE can be solved exactly without too much effort. Such a test is at least useful for falsification: if an approximate approach does not work in one dimension (1D), there is no reason to believe that it works in 3D. The converse is also true but involves optimism: if an approximate approach works in 1D, there is reason to believe that it may work in 3D as well.[10]

In computational strong-field physics, low-dimensional models of He [29] (or H^- [13]) in a laser field are extensively used, as all interesting correlation effects can be studied with them on a qualitative level. Consider the TDSE

$$i\frac{\partial}{\partial t}\psi(x_1,x_2,t) = \left(\sum_{i=1}^{2}\hat{h}_i(x_i,t) + v_{\mathrm{ee}}(|x_1-x_2|,t)\right)\psi(x_1,x_2,t) \tag{70}$$

with the one-body part

$$\hat{h}_i(x_i,t) = -\frac{1}{2}\frac{\partial^2}{\partial x_i^2} - \frac{Z}{\sqrt{x_i^2+a^2}} + \hat{v}^{(i)}_{\mathrm{laser}}(t), \tag{71}$$

10 Provided the numerical demand does not grow too rapidly with dimension.

including the laser in $\hat{v}^{(i)}_{\text{laser}}(t)$, and the electron-electron interaction

$$v_{ee}(|x_1 - x_2|, t) = \frac{1}{\sqrt{(x_1 - x_2)^2 + b^2}}. \tag{72}$$

The length parameters a and b smooth the Coulomb interactions. One may think of them as mimicking finite impact parameters in 3D. Note that the model was not designed to describe two electrons of a physical 1D electronic system but to emulate 3D strong-field electron dynamics along the polarization direction.

The interaction with the laser may be taken as $v^{(i)}_{\text{laser}}(t) = x_i E(t)$ in length gauge or $\hat{v}^{(i)}_{\text{laser}}(t) = -iA(t)\partial_{x_i}$ in velocity gauge. A snapshot of the effective potential

$$v_{\text{eff}}(x_1, x_2, t) = -\frac{Z}{\sqrt{x_i^2 + a^2}} - \frac{Z}{\sqrt{x_i^2 + a^2}} + \frac{1}{\sqrt{(x_1 - x_2)^2 + b^2}} + (x_1 + x_2)E(t)$$

for $a = b = 1$, $Z = 2$, and a positive $E(t)$ is shown in Figure 5(a). The channels along the axes correspond to single ionization if $|x_1|$ is small and $|x_2|$ is large or *vice versa*. Probability density in the region close to the origin where both $|x_1|$ and $|x_2|$ are small represents neutral He. The four quadrants where both $|x_1|$ and $|x_2|$ are $\gg 1$ correspond to He^{2+}. The flow of probability density in the tilted potential of Figure 5(a) will mainly follow the He^+ channels along the axes. However, for sufficiently high electric field amplitude, the slope can be strong enough such that probability density enters the double-ionization He^{2+} regions. The potential hill along $x_1 = x_2$ is due to electron-electron repulsion. Figure 5(b) shows the ground-state probability density in $v_{\text{eff}}(x_1, x_2)$ without electric field. The Hamiltonian in the TDSE (70) does not affect

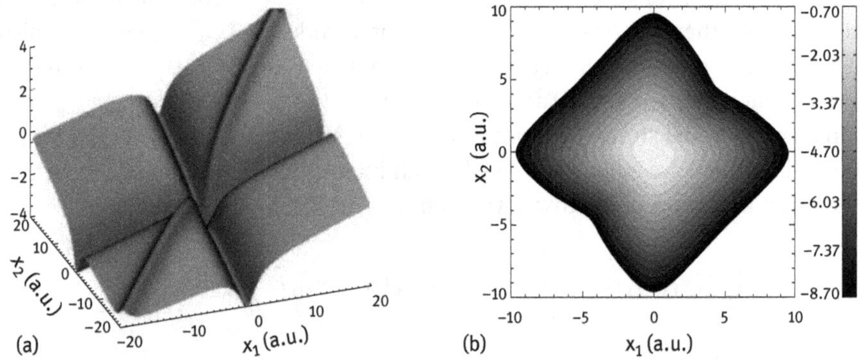

Fig. 5. (a) Effective model He potential, tilted by an electric field. (b) Logarithmically scaled contour plot of the ground-state probability density in $v_{\text{eff}}(x_1, x_2)$ for $E(t) \equiv 0$, $a = b = 1$, and $Z = 2$ over eight orders of magnitude. The ground state was obtained through imaginary-time propagation using operator splitting and the Crank–Nicolson propagator (see Chapter I). The ground-state energy is $E_0 = -2.238$.

spin degrees of freedom so that we consider only the spatial part of the wavefunction $\psi(x_1, x_2, t)$. Nonetheless, spin dictates the symmetry of the spatial wavefunction. Assuming a spin-singlet configuration,[11] i.e.,

$$\Psi(x_1\sigma_1, x_2\sigma_2, t) = \psi(x_1, x_2, t)\frac{1}{\sqrt{2}}(\delta_{\sigma_1\uparrow}\delta_{\sigma_2\downarrow} - \delta_{\sigma_2\uparrow}\delta_{\sigma_1\downarrow}), \tag{73}$$

antisymmetry under particle exchange of $\Psi(x_1\sigma_1, x_2\sigma_2, t)$ implies symmetry of the spatial wavefunction,

$$\psi(x_1, x_2, t) = \psi(x_2, x_1, t). \tag{74}$$

Note that only for two electrons, such a simple separation in spin and spatial degrees of freedom is possible. An analogous model for three electrons (Li) has been studied in [32, 39].

The TDSE (70) can be solved using the methods discussed in detail in Chapter I. In that way, the exact benchmark results for a correlated model system are obtained and can be compared to the corresponding results from the TDKS equation. In so-called exact x-only TDDFT for the spin-singlet configuration, the TDKS equation is equivalent to time-dependent HF. There is only one spatial KS orbital that fulfills

$$i\frac{\partial}{\partial t}\phi(x,t) = \left(\hat{h}(t) + \frac{1}{2}u[n(x,t)]\right)\phi(x,t), \tag{75}$$

where single-particle Hamiltonian and Hartree potential are

$$\hat{h}(t) = -\frac{1}{2}\frac{\partial^2}{\partial x^2} - \frac{Z}{\sqrt{x^2 + a^2}} + \hat{v}_{\text{laser}}(t), \tag{76}$$

$$u[n(x,t)] = \int dx' \frac{n(x',t)}{\sqrt{(x-x')^2 + b^2}}, \qquad n(x,t) = 2|\phi(x,t)|^2, \tag{77}$$

respectively. The 1D nonlinear TDKS equation (75) can be solved by the same method as the (effectively two dimensions) TDSE (70). The exact exchange cancels half of the Hartree potential, taking self-interaction correctly into account.[12] In the spherical 3D case of Section 3.1, the Hartree potential was expanded in multipoles. Here, in 1D, the Hartree potential may be calculated brute force, i.e., performing the integral for all fixed x over x'. With N_x grid points, this is an $\mathcal{O}(N_x^2)$ operation, which is bearable. A smarter way makes use of the fast Fourier transform (FFT). As u has the form of a convolution, $u[n(x,t)] = \int dx'\, n(x',t)v_{\text{ee}}(x-x')$, one may calculate it as the back-FFTed product of FFTed n and v_{ee}. This scales only $\mathcal{O}(N_x \log N_x)$. The other option often employed in 3D Cartesian DFT solvers, namely, calculating the Hartree potential as

11 See also the example in Section 2.1.

12 Each KS particle sees "the other" KS particle of opposite spin, sitting spatially in the same KS orbital though.

the solution of the Poisson equation by some relaxation method, would be applicable here if we knew the analogue of a Poisson equation, which $u[n(x,t)]$ obeys.

The KS ground-state energy

$$E_0^{\mathrm{KS}}[n_0] = 2\,\langle\phi|\hat{h}_0|\phi\rangle + \int dx\,|\phi(x)|^2 \int dx'\,\frac{|\phi(x')|^2}{\sqrt{(x-x')^2+b^2}} \tag{78}$$

of the system with $a = b = 1$ and $Z = 2$ is $E_0^{\mathrm{KS}} = -2.224$. Here, $\hat{h}_0$ is $\hat{h}(t=0)$, i.e., with the laser off. Compared to the exact value $E_0 = -2.238$ from the TDSE, E_0^{KS} is only 0.6% off. The KS or HF spin-singlet ground-state wavefunction is given by the Slater determinant

$$\Psi_0^{\mathrm{HF}}(x_1\sigma_1, x_2\sigma_2) = \phi(x_1)\phi(x_2)\frac{1}{\sqrt{2}}(\delta_{\sigma_1\uparrow}\delta_{\sigma_2\downarrow} - \delta_{\sigma_2\uparrow}\delta_{\sigma_1\downarrow})$$

and does not show the indentation feature along the diagonal $x_1 = x_2$ caused by the electron-electron repulsion in Figure 5(b).

3.2.1 Exact xc-potential

Constructing the exact xc-potential from the exact TDSE wavefunction is not feasible in practical applications of TDDFT where the TDSE solution is not available. After all, the whole idea of TDDFT is to avoid a TDSE solution. In benchmark calculations of the kind discussed here, the situation is different: the TDSE result is available, and investigating the exact xc-potential is not only satisfying one's curiosity[13] but may also help construct better approximative xc-potentials. However, it may well be that, while watching the exact xc-potential at work, one comes to the conclusion that a useful analytical expression for an xc-functional capturing a certain phenomenon of interest will most likely never be found. It might then be better to switch to another many-body method where correlation can be added in a more systematic manner than in TDDFT, e.g., multiconfigurational time-dependent Hartree–Fock (MCTDHF) (Chapter V) or time-dependent configuration interaction singles (Chapter VI).

In general, constructing the exact xc-potential, given the exact density from the reference TDSE solution, is possible but nontrivial, and computationally cumbersome [26, 36, 38]. It is straightforward only in the case of a single spatial KS orbital, as in the TDKS equation (75) for the He spin-singlet configuration [10, 23]. Writing (75) as

$$i\frac{\partial}{\partial t}\phi(x,t) = \left(-\frac{1}{2}\frac{\partial^2}{\partial x^2} + v_{\mathrm{KS}}(x,t)\right)\phi(x,t), \tag{79}$$

13 "How does the exact xc-potential miraculously do that?"

we can solve for the total effective potential

$$v_{\mathrm{KS}}(x,t)=\frac{i\partial_t\phi(x,t)+\frac{1}{2}\partial_x^2\phi(x,t)}{\phi(x,t)}. \tag{80}$$

Writing

$$\phi(x,t)=\sqrt{\frac{n(x,t)}{2}}\,\exp[iS(x,t)] \tag{81}$$

and employing continuity

$$\frac{\partial}{\partial t}n(x,t)+\frac{\partial}{\partial x}j(x,t)=0 \tag{82}$$

with

$$j(x,t)=-i[\phi^*(x,t)\partial_x\phi(x,t)-\phi(x,t)\partial_x\phi^*(x,t)]=n(x,t)\partial_x S(x,t), \tag{83}$$

we obtain

$$v_{\mathrm{KS}}(x,t)=\frac{1}{2}\frac{\partial_x^2\sqrt{n(x,t)}}{\sqrt{n(x,t)}}-\partial_t S(x,t)-\frac{1}{2}[\partial_x S(x,t)]^2. \tag{84}$$

With the TDSE solution $\psi(x_1,x_2,t)$ at hand, we can calculate the exact density and current density,

$$n(x,t)=2\int dx'\,|\psi(x,x')|^2, \tag{85}$$

$$j(x,t)=-i\int dx'\,[\psi^*(x,x',t)\partial_x\psi(x,x',t)-\psi(x,x',t)\partial_x\psi^*(x,x',t)], \tag{86}$$

and the phase, up to a time-dependent constant, by integration of

$$\partial_x S(x,t)=\frac{j(x,t)}{n(x,t)}. \tag{87}$$

Equation (84) is one option to calculate the exact $v_{\mathrm{KS}}(x,t)$. After subtracting the external potential and the Hartree potential, the exact $v_{\mathrm{xc}}(x,t)$ is obtained. Another option [23] works via the inversion of the split-operator time evolution operator. Let us suppose we have determined $\phi(x,t)$ and $\phi(x,t+\Delta t)$ according (81) with the corresponding n and S from the TDSE using (85), (86), and (87). Then, since

$$\phi(x,t+\Delta t)\simeq\exp(-i\hat{T}\Delta t/2)\exp[-iv_{\mathrm{KS}}(x,t+\Delta t/2)\Delta t]\exp(-i\hat{T}\Delta t/2)\phi(x,t),$$

where $\hat{T}=-\frac{1}{2}\partial_x^2$, we find

$$\exp[-iv_{\mathrm{KS}}(x,t+\Delta t/2)\Delta t]\simeq\frac{\exp(i\hat{T}\Delta t/2)\phi(x,t+\Delta t)}{\exp(-i\hat{T}\Delta t/2)\phi(x,t)},$$

i.e., the KS potential for each x can be extracted from the corresponding complex phase of the right-hand side,

$$v_{\mathrm{KS}}(x,t+\Delta t/2) \simeq \frac{i}{\Delta t} \arg\left[\frac{\exp(i\hat{T}\Delta t/2)\phi(x,t+\Delta t)}{\exp(-i\hat{T}\Delta t/2)\phi(x,t)}\right]. \tag{88}$$

Care has to be taken to keep the numerically extracted phase continuous.

Exact KS potentials have been analyzed in, e.g., [11, 18, 23, 24]. An interesting aspect, for instance, is to quantify the nonadiabaticity of the exact xc-potential, i.e., whether memory effects are important or not [42]. Note that (84) or (88) does not tell directly whether $v_{\mathrm{xc}}(x,t) = v_{\mathrm{KS}}(x,t) - v(x,t) - u[n(x,t)]$ can be well approximated in a time-local, adiabatic manner $v_{\mathrm{xc}}[n(x,t)]$.

A typical feature of the exact KS potential in the nonlinear regime relevant to strong-field applications is the development of dynamical steps. These steps are related to the derivative discontinuity in static DFT [28]. If the first electron has left during an ionization process, the second gets a higher ionization potential. However, this shift should not be a smooth function of the fractional occupation number.

3.3 Where TDDFT fails in practice

Low-dimensional benchmark studies helped reveal numerous deficiencies of TDDFT in practice [25]. Although slightly depressing, such studies yield valuable insight and give directions toward potential improvements. The following list of failures is from a strong-field person's perspective and, unfortunately, not exhaustive.

- **Resonant transitions:** As the KS potential depends on the time-dependent density, its levels will depend on the density as well. Consider a resonant laser, driving Rabi oscillations. Although the energy gap to the lowest unpopulated state in the ground-state KS potential is energetically close to the corresponding transition in the interacting, exact system, there is no population inversion in the KS system. In fact, the exact xc-potential generates the true population-inverted density as a ground-state density in another KS potential. Known and practicable approximate xc-potentials do not do that. Funnily enough, the KS exact x-only dipole expectation value nevertheless shows Rabi-like beating, but for a different reason, and the density after half a Rabi cycle is not the correct excited-state density but close to the ground-state density [35].
- **Autoionization:** The linear response spectra of both real 3D helium and the model system (70) show peaks due to doubly excited states while the corresponding KS system in exact x-only approximation (or with other time-local xc-potentials) does not [35]. As a consequence, no Fano profiles in the photoabsorption or emission spectra are reproduced [6, 8]. It is interesting to calculate

the exact xc-potential from the TDSE solution to figure out how it virtuosically manages autoionization [18]: it does so by developing a step-like potential barrier through which the KS orbital tunnels out. Outside this barrier, the KS orbital represents the escaping electron, inside the "other" electron that remains bound.

- **Nonsequential double ionization:** It is probably the largest dynamical correlation effect in laser-atom interaction: an electron is emitted, oscillates back to the parent ion due to the laser field, and kicks out another electron (see, e.g., [34]). How is this to be described by just one KS orbital? In fact, TDDFT fails not only because the xc-potential is not good enough [23]. Even if the exact density was generated by a TDKS equation using the exact xc-potential, the density functionals to calculate the ionization probabilities for, e.g., He^+ and He^{2+}, are unknown [46], let alone a functional to calculate correlated photoelectron spectra [7, 47].
- **Harmonic generation:** It is seemingly harmless for TDDFT because the essential ingredient to calculate the spectrum is just the dipole expectation value, which is an explicit functional of the density. However, exact high-harmonic generation spectra for He show features that are not well covered by TDDFT with standard xc-potentials, such as resonant enhancements, two plateaus with cutoffs[14] according to neutral He and He^+ [8], or additional plateaus due to nonsequential double recombination [21].
- **Photoelectron spectra:** In the absence of a better alternative, KS orbitals may be interpreted as single-particle wavefunctions and used in that way to calculate photoelectron spectra employing the methods for TDSEs outlined in Chapter II. Correlated photoelectron momentum spectra cannot be reproduced in that way [7, 47]. But even total spectra might be wrong. Imagine a high-frequency laser pulse that significantly depletes the ground state. As the population of the KS ground-state level diminishes in time, the photoelectron peak moves in energy from the initial position $E^{\mathrm{He}} = -I_p^{\mathrm{He}} + \omega$ down to $E^{\mathrm{He}^+} = -I_p^{\mathrm{He}^+} + \omega$, which is unphysical. The exact xc-potential's job is rather demanding: it has to generate peaks at E^{He^+} and at E^{He} but also in between because of single-photon double ionization, where the electrons share the photon energy. Moreover, one outgoing electron can give energy to the other, leaving an excited He^+ ion behind. The corresponding photoelectron peak is then at an energy $< E^{\mathrm{He}}$. Although it is fun to investigate how the exact xc-potential constructed from the TDSE does the job, it is doubtful that one can write down a general-enough xc-potential that is useful in practice, i.e., without having the information from the TDSE solution.
  Resorting just to the single-particle density, one may try to extract total photoelectron spectra by field-free postpropagation of the density after the laser pulse, measuring the speed of density wave packets [45]. This somewhat cumbersome

14 See Section 3 of Chapter II.

procedure may overcome the problem of the density functional for the total photoelectron spectrum. Yet, the problem of a sufficiently accurate xc-potential remains.

Methods capable of overcoming the failures of TDDFT in the above list are available. MCTDHF with a sufficient number of Slater determinants works (see Chapter V). One may also use the stationary or TDKS orbitals in configuration interaction–like approaches (cf. Chapter VI). For two electrons, recently developed time-dependent renormalized natural orbital theory has been successfully tested against all the issues in the above list [6–8, 15, 33]. However, complying with the no-free-lunch theorem, all these methods are *much* more demanding than TDDFT. There is urgent need for more efficient time-dependent many-body methods beyond linear response and with predictive power, which capture satisfactorily all the interesting dynamical correlation effects in strong-field laser-matter interaction.

Bibliography

[1] X. Andrade, D. Strubbe, U. De Giovannini, A. H. Larsen, M. J. T. Oliveira, J. Alberdi-Rodriguez, A. Varas, I. Theophilou, N. Helbig, M. J. Verstraete, L. Stella, F. Nogueira, A. Aspuru-Guzik, A. Castro, M. A. L. Marques, and A. Rubio. Real-space grids and the Octopus code as tools for the development of new simulation approaches for electronic systems. *Physical Chemistry Chemical Physics*, 17:31371–31396, 2015.

[2] H. Appel. *Time-Dependent Quantum Many-Body Systems: Linear Response, Electronic Transport, and Reduced Density Matrices*. PhD thesis, Freie Universität Berlin, 2007.

[3] D. Bauer, F. Ceccherini, A. Macchi, and F. Cornolti. C_{60} in intense femtosecond laser pulses: Nonlinear dipole response and ionization. *Physical Review A*, 64:063203, 2001.

[4] D. Bauer, T. Brabec, H. Fehske, S. Lochbrunner, K.-H. Meiwes-Broer, and R. Redmer. Focus on correlation effects in radiation fields. *New Journal of Physics*, 15(6):065015, 2013.

[5] D. Bauer and P. Koval. Qprop: A Schrödinger-solver for intense laser–atom interaction. *Computer Physics Communications*, 174(5):396–421, 2006. www.qprop.de.

[6] M. Brics and D. Bauer. Time-dependent renormalized natural orbital theory applied to the two-electron spin-singlet case: Ground state, linear response, and autoionization. *Physical Review A*, 88:052514, 2013.

[7] M. Brics, J. Rapp, and D. Bauer. Nonsequential double ionization with time-dependent renormalized-natural-orbital theory. *Physical Review A*, 90:053418, 2014.

[8] M. Brics, J. Rapp, and D. Bauer. Strong-field absorption and emission of radiation in two-electron systems calculated with time-dependent natural orbitals. *Physical Review A*, 93: 013404, 2016.

[9] A. J. Daley, C. Kollath, U. Schollwöck, and G. Vidal. Time-dependent density-matrix renormalization-group using adaptive effective hilbert spaces. *Journal of Statistical Mechanics: Theory and Experiment*, 2004(04):P04005, 2004.

[10] I. D'Amico and G. Vignale. Exact exchange-correlation potential for a time-dependent two-electron system. *Physical Review B*, 59:7876–7887, 1999.

[11] P. Elliott, J. I. Fuks, A. Rubio, and N. T. Maitra. Universal dynamical steps in the exact time-dependent exchange-correlation potential. *Physical Review Letters*, 109:266404, 2012.

[12] K. J. H. Giesbertz. *Time-Dependent One-Body Reduced Density Matrix Functional Theory*. PhD thesis, Free University Amsterdam, 2010.

[13] R. Grobe and J. H. Eberly. Photoelectron spectra for a two-electron system in a strong laser field. *Physical Review Letters*, 68:2905–2908, 1992.

[14] O. Gunnarsson and B. I. Lundqvist. Exchange and correlation in atoms, molecules, and solids by the spin-density-functional formalism. *Physical Review B*, 13:4274–4298, 1976.

[15] A. Hanusch, J. Rapp, M. Brics, and D. Bauer. Time-dependent renormalized-natural-orbital theory applied to laser-driven H_2^+. *Physical Review A*, 93:043414, 2016.

[16] I. V. Hertel, H. Steger, J. de Vries, B. Weisser, C. Menzel, B. Kamke, and W. Kamke. Giant plasmon excitation in free C_{60} and C_{70} molecules studied by photoionization. *Physical Review Letters*, 68:784–787, 1992.

[17] P. Hohenberg and W. Kohn. Inhomogeneous electron gas. *Physical Review*, 136:B864–B871, 1964.

[18] V. Kapoor. Autoionization in time-dependent density-functional theory. *Physical Review A*, 93: 063408, 2016.

[19] W. Kohn. Nobel lecture: Electronic structure of matter—wave functions and density functionals. *Reviews of Modern Physics*, 71:1253–1266, 1999.

[20] W. Kohn and L. J. Sham. Self-consistent equations including exchange and correlation effects. *Physical Review*, 140:A1133–A1138, 1965.

[21] P. Koval, F. Wilken, D. Bauer, and C. H. Keitel. Nonsequential double recombination in intense laser fields. *Physical Review Letters*, 98:043904, 2007.

[22] S. Kümmel and L. Kronik. Orbital-dependent density functionals: Theory and applications. *Reviews of Modern Physics*, 80:3–60, 2008.

[23] M. Lein and S. Kümmel. Exact time-dependent exchange-correlation potentials for strong-field electron dynamics. *Physical Review Letters*, 94:143003, 2005.

[24] K. Luo, J. I. Fuks, E. D. Sandoval, P. Elliott, and N. T. Maitra. Kinetic and interaction components of the exact time-dependent correlation potential. *The Journal of Chemical Physics*, 140(18): 18A515, 2014.

[25] N. T. Maitra. Perspective: Fundamental aspects of time-dependent density functional theory. *The Journal of Chemical Physics*, 144(22):220901, 2016.

[26] S. E. B. Nielsen, M. Ruggenthaler, and R. van Leeuwen. Many-body quantum dynamics from the density. *EPL (Europhysics Letters)*, 101(3):33001, 2013.

[27] J. P. Perdew and A. Zunger. Self-interaction correction to density-functional approximations for many-electron systems. *Physical Review B*, 23:5048–5079, 1981.

[28] J. P. Perdew, R. G. Parr, M. Levy, and J. L. Balduz. Density-functional theory for fractional particle number: Derivative discontinuities of the energy. *Physical Review Letters*, 49: 1691–1694, 1982.

[29] M. S. Pindzola, D. C. Griffin, and C. Bottcher. Validity of time-dependent Hartree–Fock theory for the multiphoton ionization of atoms. *Physical Review Letters*, 66:2305–2307, 1991.

[30] M. J. Puska and R. M. Nieminen. Photoabsorption of atoms inside C_{60}. *Physical Review A*, 47: 1181–1186, 1993.

[31] M. J. Puska and R. M. Nieminen. Erratum: Photoabsorption of atoms inside C_{60}. *Physical Review A*, 49:629–629, 1994.

[32] J. Rapp and D. Bauer. Effects of inner electrons on atomic strong-field-ionization dynamics. *Physical Review A*, 89:033401, 2014.

[33] J. Rapp, M. Brics, and D. Bauer. Equations of motion for natural orbitals of strongly driven two-electron systems. *Physical Review A*, 90:012518, 2014.

[34] A. Rudenko, K. Zrost, B. Feuerstein, V. L. B. de Jesus, C. D. Schröter, R. Moshammer, and J. Ullrich. Correlated multielectron dynamics in ultrafast laser pulse interactions with atoms. *Physical Review Letters*, 93:253001, 2004.
[35] M. Ruggenthaler and D. Bauer. Rabi oscillations and few-level approximations in time-dependent density functional theory. *Physical Review Letters*, 102:233001, 2009.
[36] M. Ruggenthaler and R. van Leeuwen. Global fixed-point proof of time-dependent density-functional theory. *EPL (Europhysics Letters)*, 95(1):13001, 2011.
[37] M. Ruggenthaler, S. V. Popruzhenko, and D. Bauer. Recollision-induced plasmon excitation in strong laser fields. *Physical Review A*, 78:033413, 2008.
[38] M. Ruggenthaler, K. J. H. Giesbertz, M. Penz, and R. van Leeuwen. Density-potential mappings in quantum dynamics. *Physical Review A*, 85:052504, 2012.
[39] C. Ruiz, L. Plaja, and L. Roso. Lithium ionization by a strong laser field. *Physical Review Letters*, 94:063002, 2005.
[40] E. Runge and E. K. U. Gross. Density-functional theory for time-dependent systems. *Physical Review Letters*, 52:997–1000, 1984.
[41] E. S. Smyth, J. S. Parker, and K. T. Taylor. Numerical integration of the time-dependent Schrödinger equation for laser-driven helium. *Computer Physics Communications*, 114(1):1–14, 1998.
[42] M. Thiele, E. K. U. Gross, and S. Kümmel. Adiabatic approximation in nonperturbative time-dependent density-functional theory. *Physical Review Letters*, 100:153004, 2008.
[43] C. A. Ullrich. *Time-Dependent Density-Functional Theory*. Oxford University Press, 2011.
[44] C. A. Ullrich and Z.-H. Yang. A brief compendium of time-dependent density functional theory. *Brazilian Journal of Physics*, 44(1):154–188, 2013. arXiv:1305.1388.
[45] V. Véniard, R. Taïeb, and A. Maquet. Photoionization of atoms using time-dependent density functional theory. *Laser Physics*, 13:465, 2003.
[46] F. Wilken and D. Bauer. Adiabatic approximation of the correlation function in the density-functional treatment of ionization processes. *Physical Review Letters*, 97:203001, 2006.
[47] F. Wilken and D. Bauer. Momentum distributions in time-dependent density-functional theory: Product-phase approximation for nonsequential double ionization in strong laser fields. *Physical Review A*, 76:023409, 2007.
[48] A. Zielinski, V. P. Majety, and A. Scrinzi. Double photoelectron momentum spectra of helium at infrared wavelength. *Physical Review A*, 93:023406, 2016.

Chris R. McDonald, Gisela Pöplau, and Thomas Brabec

V The multiconfiguration time-dependent Hartree–Fock method

Computing the time-dependent wavefunction for a multielectron system subject to a nonperturbative field is a formidable task that can require significant amounts of computational effort and resources. The time-dependent Hartree–Fock (TDHF) method provides an efficient way to calculate the wavefunction. However, this method represents the f-electron wavefunction as a single Slater determinant constructed from f time-dependent, single-particle spin-orbitals. These spin-orbitals move in time and at each time provide the best representation of the f-electron wavefunction within the space of single Slater determinants. As such, the interaction of the electrons is only treated in an average way and provides a poor representation of the correlated wavefunction. At the other extreme lies the time-dependent configuration interaction (TDCI) method (see, e.g., [1, 8, 11, 12] and Chapter VI). Here, the wavefunction is expanded over many Slater determinants using fixed spin-orbitals; the time dependence is carried in the expansion coefficients. In TDCI, electron correlation can be accounted for in a systematic way by increasing the number of Slater determinants used to represent the electronic wavefunction. In the limit of a complete set, the TDCI wavefunction will converge to the true wavefunction. However, as the number of configuration needed grows exponentially, for systems beyond two electrons, computational capacities can be quickly overwhelmed.

The multiconfiguration time-dependent Hartree–Fock (MCTDHF) method seeks to balance the speed of TDHF and the accuracy of TDCI by representing the wavefunction using time-dependent Slater determinants (similar to TDHF) and time-dependent expansion coefficients (similar to TDCI). The Slater determinants and the expansion coefficients are optimized by a variational principle. As with TDCI, MCTDHF provides a way for systematically accounting for two-body interactions. Since the spin-orbitals are optimized at each time, the MCTDHF approach has the advantage in that it requires much fewer Slater determinants to represent the electronic wavefunction than TDCI.

In this chapter, we will introduce the MCTDHF method and discuss how this method can be implemented.

Chris R. McDonald: Department of Physics, University of Ottawa, Ottawa, ON K1N 6N5, Canada; email: cmcdo059@uottawa.ca
Gisela Pöplau: Institute of General Electrical Engineering, Rostock University, 18051 Rostock, Germany; email: gisela.pöplau@uni-rostock.de
Thomas Brabec: Department of Physics, University of Ottawa, Ottawa, ON K1N 6N5, Canada; email: Thomas.Brabec@uottawa.ca

De Gruyter Graduate – Computational Strong-Field Quantum Dynamics, Volume 5, 2017, pp. 145–167.
DOI 10.1515/9783110417265-005

1 Multiconfiguration time-dependent Hartree–Fock

Here, we will give an overview of the MCTDHF method; suggestions for further reading are given at the end of this chapter. The MCTDHF method provides an approximate solution to the time-dependent Schrödinger equation for f interacting electrons in the presence of a strong perturbation, such as a laser electric field. The Hamiltonian for such a system, in dipole approximation, is given by

$$H = \sum_{i=1}^{f} \left\{ \frac{1}{2} \left[\frac{1}{i} \nabla + \boldsymbol{A}(t) \right]^2 + U(\mathbf{r}_i) \right\} + \sum_{i<j}^{n} V(\mathbf{r}_i - \mathbf{r}_j), \tag{1}$$

where $\boldsymbol{A}(t)$ is the time-dependent vector potential, $U(\mathbf{r})$ is the binding potential, and $V(\mathbf{r} - \mathbf{r}')$ is the interaction between the particles. Typically, V is the standard Coulomb interaction in atoms and molecules; however, for quantum gases, it can also take the form of a Dirac δ function or some other short-range potential. The Hamiltonian above is presented in atomic units; unless otherwise indicated, this is the system of units that will be used throughout this chapter.

MCTDHF represents the f-electron wavefunction $\Psi(\mathbf{q}_1, \dots, \mathbf{q}_f; t)$ using a set of n time-dependent spin-orbitals $\varphi_i(\mathbf{q}; t)$ and a set of time-dependent linear expansion coefficients $A_{j_1 \dots j_f}(t)$ as

$$\Psi(\mathbf{q}_1, \dots, \mathbf{q}_f; t) = \sum_{j_1=1}^{n} \cdots \sum_{j_f=1}^{n} A_{j_1 \dots j_f}(t) \varphi_{j_1}(\mathbf{q}_1; t) \cdots \varphi_{j_f}(\mathbf{q}_f; t), \tag{2}$$

where $\mathbf{q}_i$ represents both the spatial and spin coordinates of the ith electron. The antisymmetry of the wavefunction is preserved by requiring the expansion coefficients to be antisymmetric under exchange of any two indices; that is,

$$A_{j_1 \dots j_k \dots j_l \dots j_f} = -A_{j_1 \dots j_l \dots j_k \dots j_f}. \tag{3}$$

This antisymmetry requirement reduces the number of independent expansion coefficients to $\binom{n}{f}$. The ansatz in (2) is invariant under transformations of the form

$$A_{j_1 \dots j_f} \to \tilde{A}_{j_1 \dots j_f} = \sum_{k_1=1}^{n} \cdots \sum_{k_f=1}^{n} B^{-1}_{j_1 k_1} \cdots B^{-1}_{j_1 k_f} A_{k_1 \dots k_f}, \tag{4}$$

$$\varphi_j \to \tilde{\varphi}_j = \sum_{k=1}^{n} B_{jk} \varphi_k, \tag{5}$$

where B is any invertible $n \times n$ matrix [4]. In order to restrict B and ensure orthonormality of the spin-orbital functions, the following constraints are used:

$$\langle \varphi_j | \varphi_k \rangle = \delta_{jk}, \tag{6}$$

$$i\langle\varphi_j|\dot{\varphi}_k\rangle = g_{jk}, \tag{7}$$

where $\boldsymbol{g}$ is an arbitrary (possibly time dependent), single-particle, Hermitian operator with matrix elements $g_{ij} = \langle\varphi_i|g(t)|\varphi_j\rangle$.

Equations of motion for the As and φs are determined from the Dirac–Frenkel variational principle [9]

$$\langle\delta\Psi|i\frac{\partial}{\partial t} - H|\Psi\rangle = 0. \tag{8}$$

For Ψ in the approximation manifold $\mathcal{M}$—the set of all wavefunctions constructed as in (2)—the tangent space of Ψ, denoted by $T_\Psi\mathcal{M}$, is the set of all allowed variations. The variational principle states that, at each time, the residual $(i\partial_t - H)|\Psi\rangle$ is orthogonal to $T_\Psi\mathcal{M}$. This ensures that, within the ansatz, $i\dot{\Psi}$ is closest to its true value. The variational principle also ensures that the norm and energy are conserved when H is explicitly time independent and $\Psi \in T_\Psi\mathcal{M}$ [3]. Performing the variation leads to

$$i\dot{A}_J = \sum_L H_{JL}A_L - \sum_{i=1}^{f}\sum_{k=1}^{n} g_{j_ik}A_{j_1\ldots j_{i-1}kj_{i+1}\ldots j_f}, \tag{9}$$

$$i\dot{\boldsymbol{\varphi}} = (1-P)\boldsymbol{\rho}^{-1}\langle H\rangle\boldsymbol{\varphi} + P\boldsymbol{g}\boldsymbol{\varphi}, \tag{10}$$

where J stands for the set of j_is and similarly for L, $H_{JL} = \langle\Phi_J|H|\Phi_L\rangle$ with $|\Phi_J\rangle = |\varphi_{j_1}\cdots\varphi_{j_f}\rangle$, and $\boldsymbol{\varphi}$ is the column vector containing the spin-orbital functions. The density matrix $\boldsymbol{\rho}$ is given by

$$\rho_{kl} = \sum_{j_2\ldots j_f} A^*_{kj_2\ldots j_f}A_{lj_2\ldots j_f}, \tag{11}$$

and the mean-field operator $\langle H\rangle$ is given by

$$\langle H\rangle_{kl} = \left\langle \Psi^{(k)} \middle| H \middle| \Psi^{(l)} \right\rangle, \tag{12}$$

with

$$\Psi^{(l)} = \frac{\delta\Psi}{\delta\varphi_l} \tag{13}$$

the single-particle-hole function. In terms of spin-orbitals and expansion coefficients, the single-particle-hole function can be written as

$$|\Psi^{(l)}\rangle = \sum_{j_2\ldots j_f} A_{lj_2\ldots j_f}|\varphi_{j_2}\cdots\varphi_{j_f}\rangle. \tag{14}$$

The projector P onto the space of spin-orbitals is given by

$$P = \sum_k |\varphi_k\rangle\langle\varphi_k|. \tag{15}$$

Before moving forward, it is interesting to note the effect that the choice of the constraint operator $\mathbf{g}$ will have on (9) and (10). Here, we will discuss two possible choices. For the sake of this discussion, we write the Hamiltonian as $H = H^{(1)} + H^{(2)}$ where $H^{(1)}$ and $H^{(2)}$ represent the one-body and two-body terms, respectively. The simplest choice for $\mathbf{g}$ is $\mathbf{g} = 0$, which gives the equations of motion

$$i\dot{A}_J = \sum_L H_{JL} A_L, \tag{16}$$

$$i\dot{\boldsymbol{\varphi}} = (1 - P)\boldsymbol{\rho}^{-1}\langle H\rangle\boldsymbol{\varphi}. \tag{17}$$

This constraint is used in most of our applications of MCTDHF. However, another possibility is $\mathbf{g} = H^{(1)}$. In this case, the equations of motion become

$$i\dot{A}_J = \sum_L H^{(2)}_{JL} A_L, \tag{18}$$

$$i\dot{\boldsymbol{\varphi}} = (1 - P)\boldsymbol{\rho}^{-1}\langle H\rangle^{(2)}\boldsymbol{\varphi} + H^{(1)}\mathbf{1}\boldsymbol{\varphi}. \tag{19}$$

Comparing (16) and (18), it can be seen that in (16), the time evolution of the A_Js depends on the full Hamiltonian, whereas in (18), it depends solely on the interaction terms. The motion of the A_Js that is lost in (18) is transferred to the spin-orbitals in (19). Here, the motion of the spin-orbitals is less restricted than it is in (17) because the projector only acts on the interaction part of the equation. Even in the case of a complete basis set, where the $1 - P$ term vanishes, the spin-orbitals in (19) will continue to evolve in time.

To illustrate how MCTDHF accounts for electron correlation, consider the case with $f = 2$ and $n = 3$. The MCTDHF wavefunction can be written as

$$\Psi(t) = A_{12}\text{Det}[\varphi_1(t)\varphi_2(t)] + A_{13}\text{Det}[\varphi_1(t)\varphi_3(t)] + A_{23}\text{Det}[\varphi_2(t)\varphi_3(t)], \tag{20}$$

where $\text{Det}[\varphi_i(t)\varphi_j(t)] = \varphi_i(t)\varphi_j(t) - \varphi_j(t)\varphi_i(t)$ and the coordinates $\mathbf{q}_i$ have been dropped for simplicity. The first term on the right-hand side in (20) is the Hartree–Fock part of the wavefunction; the second and third add correlation. For the $n = f$ case, MCTDHF reduces to the TDHF method. For $n > f$, MCTDHF is better able to treat the electron-electron interaction because, as with TDCI, it uses a larger number of configurations. However, MCTDHF uses a spin-orbital basis set that is optimized at each time. This allows MCTDHF to be better able to represent the wavefunction using a much smaller basis set than is required by TDCI. The outcome of this is that MCTDHF is able to handle time-dependent problems for larger numbers of electrons than can be treated by TDCI.

2 Implementing the MCTDHF method

In our implementation of MCTDHF, we represent the spin-orbitals on a tensor product grid. We will first demonstrate the use of MCTDHF on a uniform grid and will then generalize it to a nonuniform grid at the end of this chapter. Other schemes exist for representing the spin-orbitals, each having their own merits, but these are beyond the scope of this chapter. For more information on different implementation of MCTDHF, see the suggested readings at the end of the chapter.

2.1 Uniform grids

A general way to implement MCTDHF is to represent the spin-orbitals on a uniform Cartesian tensor product grid. While it is possible to compute derivatives $f^{(n)}(\mathbf{r})$ on such a grid using finite difference formulas or other methods, it is beneficial to use a fast Fourier transform (FFT) algorithm [6] and the property of Fourier transforms,

$$f^{(n)}(\mathbf{r}) = \mathcal{F}^{-1}\left\{(i\mathbf{k})^n \mathcal{F}\left[f(\mathbf{r})\right]\right\}, \tag{21}$$

where $\mathcal{F}$ denotes the Fourier operator and $\mathbf{k}$ the momentum coordinates. The workload to compute the derivatives using the FFT method requires more computational effort than the finite differences. However, the improved accuracy using the FFT method over finite differences is well worth the effort since this part of the computation is negligible compared to the computational workload for the two-body terms.

Integrals to be evaluated numerically come in two forms: (i) one-body and (ii) two-body integrals. For one-body integrals of the form

$$\mathbf{O}_{\mathrm{ij}} = \int \varphi_{\mathrm{i}}^{\star}(\mathbf{r})\mathbf{O}(\mathbf{r})\varphi_{\mathrm{j}}(\mathbf{r})\,d\mathbf{r}, \tag{22}$$

where $\mathbf{O}$ is a one-body operator, the trapezoidal method [7] is sufficient providing that the grid spacing is not chosen too large. The two-body integrals take the form

$$\mathbf{V}_{\mathrm{ijkl}} = \int \varphi_{\mathrm{i}}^{\star}(\mathbf{r})\left(\int \varphi_{\mathrm{j}}^{\star}(\mathbf{r}')\mathbf{V}(\mathbf{r}-\mathbf{r}')\varphi_{\mathrm{l}}(\mathbf{r}')d\mathbf{r}'\right)\varphi_{\mathrm{k}}(\mathbf{r})\,d\mathbf{r}, \tag{23}$$

where $\mathbf{V}(\mathbf{r}-\mathbf{r}')$ is a two-body operator depending on difference coordinates. The integrals can be evaluated efficiently by first, performing integration over $\mathbf{r}'$ using the convolution theorem and the FFT algorithm. What remains is a one-body integral that can be calculated using the trapezoidal rule or a higher-order numerical integration scheme.

2.2 Computation of the mean-field operator

In MCTDHF, much of the computational time is required to calculate the mean-field operator $\langle H \rangle$. The elements $\langle H \rangle_{kl}$ can be written as

$$\langle H \rangle_{kl} = \sum_{i=1}^{f} \langle \Psi^{(k)} | H_1(\mathbf{q}_i; t) | \Psi^{(l)} \rangle + \sum_{j>i}^{f} \langle \Psi^{(k)} | H_2(\mathbf{q}_i, \mathbf{q}_j) | \Psi^{(l)} \rangle , \tag{24}$$

where we have broken the Hamiltonian H up into one-body and two-body parts. Inserting the definition of the single-particle-hole function gives

$$\begin{aligned} \langle H \rangle_{kl} = & \sum_{i=1}^{f} \sum_{j_2 \ldots j_f} \sum_{l_2 \ldots l_f} A^{\star}_{kj_2 \ldots j_f} A_{ll_2 \ldots l_f} \langle \varphi_{j_2} \cdots \varphi_{j_f} | H_1(\mathbf{q}_i; t) | \varphi_{l_2} \cdots \varphi_{l_f} \rangle \\ & + \sum_{j>i}^{f} \sum_{j_2 \ldots j_f} \sum_{l_2 \ldots l_f} A^{\star}_{kj_2 \ldots j_f} A_{ll_2 \ldots l_f} \langle \varphi_{j_2} \cdots \varphi_{j_f} | H_2(\mathbf{q}_i, \mathbf{q}_j) | \varphi_{l_2} \cdots \varphi_{l_f} \rangle . \end{aligned} \tag{25}$$

We first examine the one-body part of (25), which we label $\langle H \rangle^{(1)}_{kl}$. For $i = 1$, we have

$$\begin{aligned} \langle H \rangle^{(1)}_{kl}(\mathbf{q}_1; t) &= \sum_{j_2 \ldots j_f} \sum_{l_2 \ldots l_f} A^{\star}_{kj_2 \ldots j_f} A_{ll_2 \ldots l_f} \langle \varphi_{j_2} \cdots \varphi_{j_f} | H_1(\mathbf{q}_1; t) | \varphi_{l_2} \cdots \varphi_{l_f} \rangle \\ &= \sum_{j_2 \ldots j_f} A^{\star}_{kj_2 \ldots j_f} A_{lj_2 \ldots j_f} H_1(\mathbf{q}_1; t) \\ &= \rho_{kl} H_1(\mathbf{q}_1; t). \end{aligned} \tag{26}$$

When $i \geq 2$, we can rearrange the multi-indices to get $(f-1)$ equal terms giving

$$\begin{aligned} & \sum_{i=2}^{f} \sum_{j_2 \ldots j_f} \sum_{l_2 \ldots l_f} A^{\star}_{kj_2 \ldots j_f} A_{ll_2 \ldots l_f} \langle \varphi_{j_2} \cdots \varphi_{j_f} | H_1(\mathbf{q}_i; t) | \varphi_{l_2} \cdots \varphi_{l_f} \rangle \\ & \quad = (f-1) \sum_{j_2, l_2} \sum_{j_3 \ldots j_f} A^{\star}_{kj_2 j_3 \ldots j_f} A_{ll_2 j_3 \ldots j_f} \langle \varphi_{j_2} | H_1(\mathbf{q}_2; t) | \varphi_{l_2} \rangle . \end{aligned} \tag{27}$$

Similarly for the two-body parts, which we label $\langle H \rangle^{(2)}_{kl}$, we separate for $i = 1$, $j = 2$ to get

$$\begin{aligned} \langle H \rangle^{(1)}_{kl}(\mathbf{q}_1; t) &= \sum_{j_2 \ldots j_f} \sum_{l_2 \ldots l_f} A^{\star}_{kj_2 \ldots j_f} A_{ll_2 \ldots l_f} \langle \varphi_{j_2} \cdots \varphi_{j_f} | H_2(\mathbf{q}_1, \mathbf{q}_2) | \varphi_{l_2} \cdots \varphi_{l_f} \rangle \\ &= (f-1) \sum_{j_2, l_2} \sum_{j_3 \ldots j_f} A^{\star}_{kj_2 j_3 \ldots j_f} A_{ll_2 j_3 \ldots j_f} \langle \varphi_{j_2} | H_2(\mathbf{q}_1, \mathbf{q}_2) | \varphi_{l_2} \rangle , \end{aligned} \tag{28}$$

and for $i \geq 2, j > i$,

$$\sum_{i=2}^{f} \sum_{j=i+1}^{f} \sum_{j_2 \ldots j_f} \sum_{l_2 \ldots l_f} A^*_{kj_2 \ldots j_f} A_{ll_2 \ldots l_f} \langle \varphi_{j_2} \cdots \varphi_{j_f} | H_2(\mathbf{q}_i, \mathbf{q}_j) | \varphi_{l_2} \cdots \varphi_{l_f} \rangle$$
$$= \frac{(f-1)(f-2)}{2} \sum_{j_2, l_2} \sum_{j_3, l_3} \sum_{j_4 \ldots j_f} A^*_{kj_2 j_3 j_4 \ldots j_f} A_{ll_2 l_3 j_4 \ldots j_f}$$
$$\times \langle \varphi_{j_2} \varphi_{j_3} | H_2(\mathbf{q}_1, \mathbf{q}_2) | \varphi_{l_2} \varphi_{l_3} \rangle \,. \tag{29}$$

The terms in (27) and (29) are not complex functions but merely complex numbers. When the operator $1 - P$, which projects onto the space orthogonal to the φ_is, is applied, these terms will disappear. The components of the mean-field operator can now be written as

$$\langle H \rangle_{kl} = \rho_{kl} H_1(\mathbf{q}_1; t) \tag{30}$$
$$+ (f-1) \sum_{j_2, l_2} \sum_{j_3 \ldots j_f} A^*_{kj_2 j_3 \ldots j_f} A_{ll_2 j_3 \ldots j_f} \langle \varphi_{j_2} | H_2(\mathbf{q}_1, \mathbf{q}_2) | \varphi_{l_2} \rangle \,.$$

Finally, the second term in (30) has many summations that, for more than two electrons, can require significant computing time. The amount of these summations can be greatly reduced by rearranging the indices $j_3, \ldots, j_f$ to yield $(f-2)!$ identical terms. The mean-field elements can then be computed as

$$\langle H \rangle_{kl} = \rho_{kl} H_1(\mathbf{q}_1; t) \tag{31}$$
$$+ (f-1)! \sum_{j_2, l_2} \sum_{j_3 < \ldots < j_f} A^*_{kj_2 j_3 \ldots j_f} A_{ll_2 j_3 \ldots j_f} \langle \varphi_{j_2} | H_2(\mathbf{q}_1, \mathbf{q}_2) | \varphi_{l_2} \rangle \,.$$

2.3 Restricted vs unrestricted

The TDHF wavefunction can be represented in a restricted form or an unrestricted form. In the unrestricted form, the spin-orbital with spin state $|s_1\rangle$ will have a different spatial function than the spin-orbital with spin state $|s_2\rangle$. In this sense, each spin state has its own subspace of spatial functions. For unrestricted TDHF, each of these subspaces contains a single function. In this case, the wavefunction is

$$\Psi_u(\mathbf{q}_1, \mathbf{q}_2) = \varphi_1(\mathbf{r}_1)\varphi_2(\mathbf{r}_2) |s_1 s_2\rangle - \varphi_2(\mathbf{r}_1)\varphi_1(\mathbf{r}_2) |s_2 s_1\rangle \,. \tag{32}$$

When $s_1 = s_2$, this form is appropriate because φ_1 cannot be equal to φ_2 or the Pauli's exclusion principle will be violated. When solving the TDHF equations, two spatial functions must be propagated in time. When H is spin-independent and $s_1 \neq s_2$—as in a singlet state—there is no differentiating between the two spatial functions; this leads to restricted TDHF. In the restricted case, the spatial parts of the spin-orbitals are the same,

$$\Psi_r(\mathbf{q}_1, \mathbf{q}_2) = \varphi_1(\mathbf{r}_1)\varphi_1(\mathbf{r}_2)(|s_1 s_2\rangle - |s_2 s_1\rangle). \tag{33}$$

Now only a single one-body spatial function needs to be propagated to advance the wavefunction in time.

This idea can be extended to MCTDHF. For a system with spin $S = 0$, as is common in atomic and molecular systems, the subspace of spatial functions for electrons with spin $|\uparrow\rangle$ will be equal to the subspace of spatial functions with spin $|\downarrow\rangle$. If n spin-orbitals are used to represent the MCTDHF wavefunction, then only $n/2$ spatial functions will need to be propagated in time. Furthermore, the work to compute all one- and two-body operators will be reduced because the matrix representation of these operators becomes block diagonal because of the orthogonality of the two subspaces. For our calculations, we will use this restricted MCTDHF scheme.

2.4 Time integration

The most straightforward way to solve (16) and (17) [or (18) and (19)] is the *variable mean-field* (VMF) approach where the mean-field operator $\langle H\rangle$, the density matrix $\boldsymbol{\rho}$, and the matrix elements H_{JL} are calculated at each step. For systems in a strong laser field, high-frequency oscillations can develop, imposing a small time step on our integrator. The result of this is that $\langle H\rangle$, $\boldsymbol{\rho}$, and the H_{JL}s must be calculated many times. These quantities are computationally work intensive and thus make the VMF scheme not ideal.

A much more efficient integration scheme—known as the *constant mean-field* (CMF) approach—was presented in reference [2]. A brief sketch of the second-order CMF scheme is given here; for full details, we refer the reader to [2]. The CMF scheme is based on the notion that the matrix elements H_{JL} and the product of the inverse density with the mean field matrix $\boldsymbol{\rho}^{-1}\langle H\rangle$ evolve much slower than the MCTDHF coefficients and the spin-orbital functions. This allows for H_{JL}, $\boldsymbol{\rho}^{-1}$, and $\langle H\rangle$ to be calculated on a coarser time mesh.

To advance the wavefunction from time $t = 0$ to time $t = \tau$, the coefficients $\boldsymbol{A}(0)$ and spin-orbitals $\boldsymbol{\varphi}(0)$ are used to calculate $\langle H\rangle(0)$, $\boldsymbol{\rho}^{-1}(0)$, and $\mathcal{H}(0)$ where $\mathcal{H}$ denotes the matrix elements H_{JL}. These quantities are then used to propagate $\boldsymbol{A}$ and $\boldsymbol{\varphi}$ from $t = 0$ to $t = \tau/2$. While many smaller time steps may be taken between $t = 0$ and $t = \tau/2$, the quantities $\langle H\rangle(0)$, $\boldsymbol{\rho}^{-1}(0)$, and $\mathcal{H}(0)$ remain constant until $t = \tau/2$ where the quantities $\boldsymbol{A}(\tau/2)$ and $\tilde{\boldsymbol{\varphi}}(\tau/2)$ are obtained. From $\boldsymbol{A}(\tau/2)$ and $\tilde{\boldsymbol{\varphi}}(\tau/2)$, the quantities $\langle H\rangle(\tau/2)$ and $\boldsymbol{\rho}^{-1}(\tau/2)$ are obtained and used to propagate the spin-orbitals from $t = 0$ to $t = \tau/2$ yielding $\boldsymbol{\varphi}(\tau/2)$. If the difference between $\boldsymbol{\varphi}(\tau/2)$ and $\tilde{\boldsymbol{\varphi}}(\tau/2)$ is above the set error threshold, the step is rejected and propagation begins again at $t = 0$ with a smaller time step. If the difference is below the error threshold, $\boldsymbol{\varphi}(\tau)$ is calculated by propagating $\boldsymbol{\varphi}(\tau/2)$ to $t = \tau$ using $\langle H\rangle(\tau/2)$ and $\boldsymbol{\rho}^{-1}(\tau/2)$. The next step is to obtain an error estimate for the $\boldsymbol{A}$ coefficients. To achieve this, $\mathcal{H}(\tau)$ is calculated from $\boldsymbol{\varphi}(\tau)$ and then $\boldsymbol{A}(\tau/2)$ is propagated back to $t = 0$, yielding $\tilde{\boldsymbol{A}}(0)$. Similarly to $\boldsymbol{\varphi}$, if the difference between $\boldsymbol{A}(0)$ and $\tilde{\boldsymbol{A}}(0)$ is above the set error threshold, the step is rejected and propagation begins at $t = 0$ with a smaller time step. Otherwise, $\mathcal{H}(\tau)$ is used to calculate $\boldsymbol{A}(\tau)$, completing one full CMF step.

We have found the CMF scheme presented above to be a reliable and efficient method to solve the MCTDHF equations of motion. For intermediate steps between $t = 0$ and $t = \tau/2$, and $t = \tau/2$ and $t = \tau$, we use a high-order embedded Runge-Kutta method [7] to advance the wavefunction in time.

2.5 Computing the ground state

Before we investigate systems subjected to nonperturbative fields, we need to find the MCTDHF ground state of the system; this is accomplished by using imaginary time propagation, as introduced in Section 1.3 of Chapter I. Here, t is set to $t = -i\tau$ making the propagator $e^{-H\tau}$. Applied to a wavefunction, this operator yields

$$|\Psi(\tau)\rangle = \alpha_0 e^{-E_0\tau}|\Psi_0\rangle + \alpha_1 e^{-E_1\tau}|\Psi_1\rangle + \alpha_2 e^{-E_2\tau}|\Psi_2\rangle + \ldots, \tag{34}$$

where E_i and Ψ_i are the eigenvalues and eigenfunctions of the full Hamiltonian H, and the α_is are the probability amplitudes. Since E_0 is the lowest eigenvalue, as $\tau \to \infty$, we have $\Psi \to \Psi_0$. The wavefunction will lose norm during imaginary-time propagation so that it is necessary to renormalize after each step; this can be achieved during the Gram-Schmidt orthonormalization procedure (see, e.g., [7] or Section 1.3.2 of Chapter I).

3 Applications of MCTDHF

In this section, we will apply MCTDHF to problems involving correlated electrons, using spin-orbitals represented on a uniform grid.

3.1 Calculation of highly correlated ground states

Because MCTDHF uses an optimized set of spin-orbitals and expansion coefficients, it can be used to compute the ground states of systems with a high degree of correlation where the (TD)HF method would not give an accurate result. To demonstrate convergence of the ground state, we consider four electrons confined to a harmonic oscillator potential in one dimension (1D). The Hamiltonian in atomic units for this system is given by

$$H = \sum_{i=1}^{4} \left(-\frac{1}{2}\nabla_{x_i}^2 + \frac{\Omega^2}{2} x_i^2 \right) + \sum_{i<j}^{4} \frac{1}{\sqrt{(x_i - x_j)^2 + a^2}}, \tag{35}$$

where Ω is the harmonic oscillator frequency Ω and a is the shielding parameter. We calculate the ground state with $\Omega = 0.25$ and $a = 0.25$—a system for which there is a high degree of correlation—for $n = 4$, 8, 12, 16, 24, and 32 spin-orbitals. We use a lattice range of ±25 with 512 grid points.

Table 1 shows the ground-state energies as a function of the number of spin-orbitals used; the ground-state energies are given up to six decimal places. It

Tab. 1. Calculated ground-state energies for four electrons in a 1D harmonic oscillator potential with $\Omega = 0.25$ and electron-electron shielding of $a = 0.25$.

n	Number of configurations	E_0 (a.u.)
4	1	4.4661
8	70	3.8915
12	495	3.8127
16	1820	3.7878
24	10626	3.7835
32	35960	3.7826
48	194580	3.7823

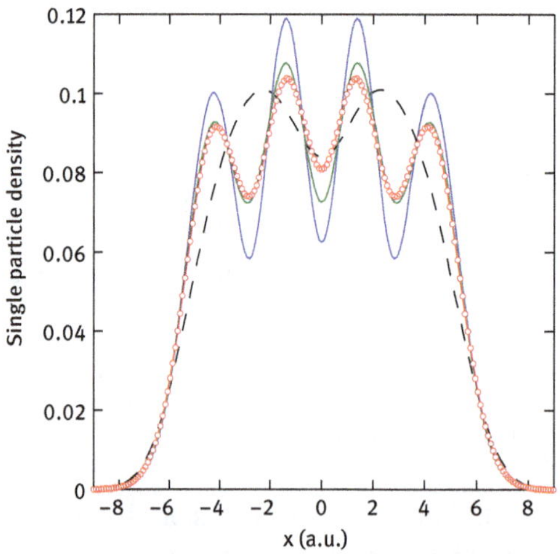

Fig. 1. Single-particle density for $n = 8$ (blue line), $n = 12$ (green line), $n = 24$ (red line), and $n = 32$ (circles) spin-orbitals. The TDHF result is also plotted for comparison (black dashed line).

is clear from Table 1 that for highly correlated systems, many configurations will be needed to reach convergence. Figure 1 shows the single-particle density of the ground state for $n = 4, 8, 12, 24$, and 32 spin-orbitals. When $n = 4$, we get the TDHF result. It can be seen that the TDHF result (black dashed line) does not even produce the correct characteristics of the single-particle density. For the TDHF result, there are only two peaks present, whereas the result with $n = 8$ (blue line) already shows the proper characteristics. The single-particle density between $n = 24$ (red line) and $n = 32$ (circles) is almost indistinguishable.

3.2 Nonsequential double ionization

When atoms and molecules interact with a strong laser field, two electrons can be emitted by a the process of nonsequential double ionization (NSDI) [17]. NSDI is a correlated process by which the first electron is liberated by the laser field through tunnel ionization. This electron is then accelerated by the field and can return and collide with parent ion. Through energy exchange with the remaining bound electrons and assistance from the laser field, one or more of the bound electrons can be released.

The wavefunction for NSDI presents an extreme computational challenge. Unlike a process such as HHG where there is only a single continuum electron, NSDI requires an accurate calculation of the two-body continuum portion of the wavefunction. The use of TDHF for such a problem is out of the question since it does not even properly represent the ground state. In addition, while TDCI can in principle treat NSDI, due to the fixed basis, the number of configurations required would be extremely prohibitive for such a calculation. Because of its optimized spin-orbital basis, MCTDHF allows for the basis size to remain manageable while still adequately representing the two-electron part of the wavefunction.

To demonstrate MCTDHF's ability to address NSDI, we consider a $2 \times 1\mathrm{D}$ model diatomic molecule with nuclei positioned at $x = \pm R/2$.[1] The binding potential for such a system is given by

$$U(x) = -\frac{1}{\sqrt{(x - R/2)^2 + a^2}} - \frac{1}{\sqrt{(x + R/2)^2 + a^2}}, \tag{36}$$

where a is the Coulomb softening parameter. The Hamiltonian for this system subjected to a strong laser field in dipole approximations is

$$H = \sum_{i=1}^{2} \left[T_i + U(x_i) + x_i F(t)\right] + V(x_1, x_2). \tag{37}$$

1 This low-dimensional model is analogous to the one used for He in Section 3.2 of Chapter IV.

T is the kinetic energy operator, xF represents the laser electron dipole interaction, and V is the electron-electron interaction

$$V(x_1, x_2) = 1/\sqrt{(x_2 - x_1)^2 + a^2}. \tag{38}$$

The softening parameter is chosen to make the one- and two-electron binding energies of the model system match the binding energies typical for diatomic molecules with a σ^2 configuration, such as N_2 and H_2. For our calculations, we choose $a = 1.12$, which reproduces the N_2 ionization potential of 15.8 eV and the two-electron binding energy of 47.1 eV at the N_2 equilibrium internuclear distance of $R = 2.08$.

The total simulation interval is $\pm x_b, \pm y_b$ with $x_b = y_b = 400$ and 1024 grid points. This grid spacing is sufficient to converge the simulation; this has been verified by a comparison to calculations with a smaller grid spacing. The use of fairly coarse grids is possible because the Coulomb potentials are softened and because a high numerical accuracy is achieved by the FFT-based calculation of the kinetic energy operator and of the electron-electron interaction.

The laser electric field is $F(t) = -\dot{A}(t)$ with the vector potential $A(t) = F_0/\omega \cos^2(\pi t/T) \sin(\omega t)$ for $-T/2 \leq t \leq T/2$ and $A(t) = 0$ otherwise. The pulse duration

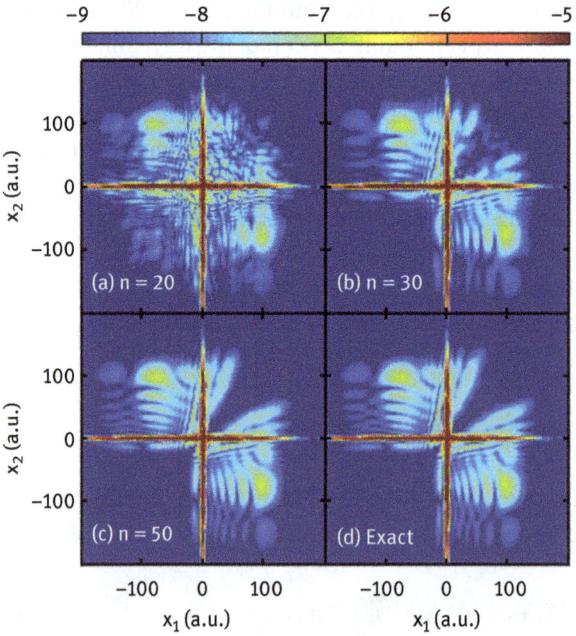

Fig. 2. Probability density of a two-electron model molecule in an intense laser field with 1D per electron (2 × 1D). Panels (a–d) show $|\Psi|^2$ after the laser pulse at $t = 220$; (a–c) MCTDH results with $n = 20, 30, 50$ spin-orbitals, respectively; (d) exact numerical integration of the Schrödinger equation.

is $T = 2T_0$, where $T_0 = 2\pi/\omega = 110.23$ ($T_0 = 2.6$ fs, $\lambda = 0.8\,\mu$m) is the optical cycle. The simulation interval extends from $-T_0 \le t \le t_f = 140$, and $F_0 = 0.075$ ($I = 2\times10^{14}$W/cm^2).

The two-dimensional (2D) Schrödinger equation for this system can be integrated exactly using a split operator method (see Chapter I). Figure 2 shows a comparison of MCTDHF solution for $n = 20, 30$, and 50 spin-orbitals with the wavefunction obtained from a split-step integration of the 2D Schrödinger equation. For $n = 50$, the exact result is well reproduced.

3.3 High-harmonic generation

MCTDHF can also be used to treat problems involving the process of HHG [5]. HHG takes place when a gas of atoms or molecules is exposed to an intense laser field, yielding harmonics of the field's fundamental frequency. This process can be described by the three-step model where (i) an electron is emitted from the highest occupied orbital, (ii) the electron is accelerated in the field, and (iii) the electron recombines with the parent ion and emits a harmonic photon in the process.

To demonstrate the convergence of MCTDHF for HHG, we consider a diatomic molecule similar to Section 3.2. However, here, we use a 2×3D system with nuclei positioned at $(\pm R/2, 0, 0)$ and shielding parameter $a = 0.1$. A simulation interval of ± 30 is used along each coordinate direction. A complex absorbing potential of the form $W(x) = -\eta|x - x_c|^3$ is used for $|x| \ge x_c$ with $x_c = 16$ and $\eta = 0.0005$. This potential suppresses the co-called long trajectories while keeping the short ones, which tend to dominate experimental spectra. A six-cycle pulse of intensity $I = 10^{14}$ W/cm^2 and a fundamental frequency of $\omega = 0.057$ (800 nm) was applied. The vector potential had the form $A(t) = F_0/\omega \sin^2(\omega t/12)\sin(\omega t)$. The harmonic spectrum was calculated by taking the Fourier transform of the time-dependent dipole $x(t) = \langle \Psi(t)| x |\Psi(t)\rangle$ (cf. Section 3 of Chapter II).

Figure 3 shows the calculated harmonic spectrum for 4, 8, and 10 spin-orbitals. The spectrum is well converged for 10 spin-orbitals. At first, this may seem surprising, given the number of spin-orbitals required for convergence in the NSDI calculations in Section 3.2. However, HHG is predominantly due to a single continuum electron, whereas for NSDI, we had two correlated continuum electrons. Furthermore, the HHG spectrum relies on the calculation of an expectation value of a one-body operator; these tend to converge with much fewer orbitals than their two-body counterparts.

High-harmonic spectra for multielectron systems often resemble spectra for single-electron systems. In order to investigate multielectron effects, special diagnostic tools such as ionic eigenstate-resolved wavefunctions [14] are required but this is beyond the scope of this chapter. It should be noted that multielectron effects often appear as a dip or irregularity in the harmonic spectra. In Figure 3, the 15th harmonic is a possible candidate for displaying multielectron effects.

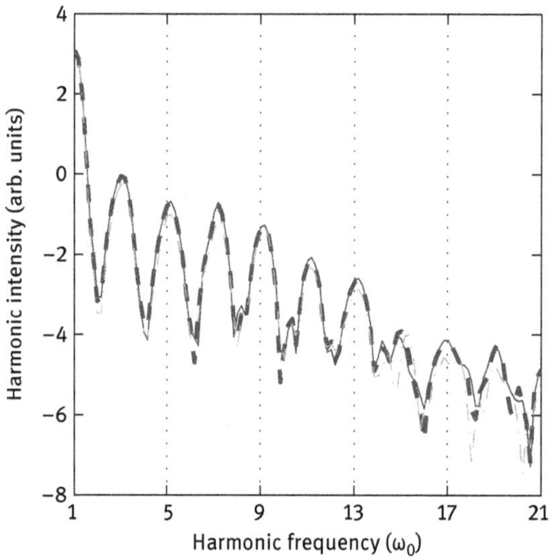

Fig. 3. High-harmonics spectra for a 2 × 3*D* diatomic molecule calculated using MCTDHF with $n = 4$ (dashed-dotted), $n = 8$ (solid), and $n = 10$ (dashed) spin-orbitals.

4 Extending MCTDHF to nonuniform grids

The ultimate goal of MCTDHF is to perform calculations on atoms and molecules with up to 10 electrons moving in three dimensions (3D) in the presence of a strong laser field. On a uniform grid, this may require solving $\simeq 10^7$–10^8 coupled, nonlinear, ordinary differential equations for the spin-orbital part of the wavefunction. Reducing this number will allow the calculations to proceed much faster.

On a uniform grid, the smallest grid spacing Δx required by the system is used on every region of the grid. This forces each part of the spin-orbital to be resolved equally well over the whole grid, even if this resolution is not required in certain regions (near the edges, for example). As such, more grid points will be used than are required. One possible way to overcome this challenge is to use a nonuniform tensor product grid. This will allow for the grid to be tailored to have a fine resolution in regions where such resolution is required and to have a coarse resolution where a fine resolution is not required. Other coordinate choices—spherical, cylindrical, or prolate spheroidal—may be more beneficial to problems with specific geometries. Here, we will focus our discussion on Cartesian tensor product grids as these are a general way to treat most problems. In order to perform MCTDHF calculations on a nonuniform grid, accurate methods for differentiation and integration will be required. In addition, since the FFT algorithm is not applicable to nonuniform grids, a new method of calculating the two-body terms is needed.

4.1 Differentiation on a nonuniform grid

In Section 2.1, differentiation in MCTDHF could be achieved through the use of an FFT. When using a nonuniform grid, the FFT is no longer ideal for obtaining an accurate derivative. An FFT method for nonuniform grids does exist. However, we have found its numerical accuracy insufficient to maintain the stability of our calculations when repeatedly performing pairs of forward and inverse Fourier transforms. In order to treat derivatives on the nonuniform grid, we look to finite difference methods. While there are well-known formulas for the first and second derivatives on uniform grids, there is a paucity of options for nonuniform grids. Here, a method is presented for computing formulas for the first and second derivatives based on the given grid. The points at the boundary of the grid will need to be handled by setting appropriate boundary conditions.

Let x_j be the points along a single grid line and $f_j = f(x_j)$ where $j = 1, \dots, N$. In what follows, we take N_d, i.e., the number of points used to approximate the derivative at x_j, to be an odd integer and approximate $f^{(n)}(x_j)$ using M points on either side of x_j; see Figure 4. We are searching for a formula for the nth derivative at x_j of the form

$$f^{(n)}(x_j) \simeq \sum_{k=1}^{N_d} a_{jk}^{(n)} f(x_{j+k-M-1}), \tag{39}$$

where $M = (N_\mathrm{d} - 1)/2$ and the $a_{jk}^{(n)}$ are the coefficients for the nth derivative formula at x_j. Introducing the notation $h_{j+k} = x_{j+k} - x_j$ and $f_{j+k} = f(x_{j+k})$, it is possible to write the

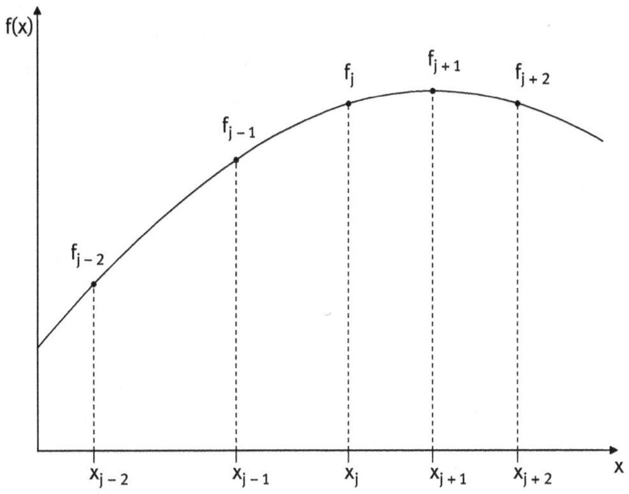

Fig. 4. Schematic of differentiation on a nonuniform grid.

system of equations

$$f(x_{j+k}) \simeq f(x_j) + h_{j+k} f'(x_j) + \frac{h_{j+k}^2}{2!} f''(x_j) + \ldots + \frac{h_{j+k}^{N_d-1}}{(N_d-1)!} f^{(N_d-1)}(x_j). \tag{40}$$

Thus, the set of equations centered at x_j is of the form $\mathbf{f}_j = \mathbf{A}_j \mathbf{F}_j$ where

$$\mathbf{f}_j = \begin{pmatrix} f(x_{j-M}) \\ \vdots \\ f(x_j) \\ \vdots \\ f(x_{j+M}) \end{pmatrix}, \qquad \mathbf{F}_j = \begin{pmatrix} f(x_j) \\ f'(x_j) \\ \vdots \\ f^{(N_d-1)}(x_j), \end{pmatrix} \tag{41}$$

and the nonsingular matrix $\mathbf{A}_j$ is given by

$$\mathbf{A}_j = \begin{pmatrix} 1 & h_{j-M} & \frac{h_{j-M}^2}{2!} & \cdots & \frac{h_{j-M}^{N_d-1}}{(N_d-1)!} \\ 1 & h_{j-M+1} & \frac{h_{j-M+1}^2}{2!} & \cdots & \frac{h_{j-M+1}^{N_d-1}}{(N_d-1)!} \\ \vdots & \vdots & \vdots & \ddots & \vdots \\ 1 & h_{j+M} & \frac{h_{j+M}^2}{2!} & \cdots & \frac{h_{j+M}^{N_d-1}}{(N_d-1)!} \end{pmatrix}. \tag{42}$$

Inverting $\mathbf{A}_j$ then gives the solution $\mathbf{F}_j = \mathbf{A}_j^{-1}\mathbf{f}_j$. From this, it is clear that the ith row of $\mathbf{A}_j^{-1}$ gives the coefficients $a_{jk}^{(i-1)}$. That is, the dot product of the first row with $\mathbf{f}_j$ just reproduces $f(x_j)$, the dot product of the second row with $\mathbf{f}_j$ gives $f'(x_j)$, the dot product of the third row with $\mathbf{f}_j$ gives $f''(x_j)$, and so on. In fact, for N_d points, this method gives an approximate expression for the first $N_d - 1$ derivatives. However, the quality of the approximation decreases with increasing order of the derivative.

4.2 Integration on nonuniform grids

On a uniform grid, integration may be handled using the well-known trapezoidal rule. For a grid spacing h, the error then is $\mathcal{O}(h^3)$. However, on a nonuniform grid, the error is increased to $\mathcal{O}(h^2)$ and thus is not sufficient for accurate computation. As with the case of differentiation, consider the N grid points x_j, where $j = 1, \ldots, N$, along a single (1D) grid line and the values of a function $f(x_j)$ at these points. Let N_p be the even number of points used to calculate the integral for the interval $[x_j, x_{j+1}]$. We wish to find an expression for the integral of the form

$$\int_{x_j}^{x_{j+1}} f(x)dx \simeq \sum_{k=1}^{N_p} b_{jk} f(x_{j+k-M}), \qquad (43)$$

where $M = N_p/2$. Since N_p is even, there are $M - 1$ points on each side of the interval $[x_j, x_{j+1}]$, see Figure 5. To determine the b_{jk}s, we will apply the method of undetermined coefficients using a polynomial basis. Thus, it is important that the integration interval be $[x_j, x_{j+1}]$ and not $[x_{j-M+1}, x_{j+M}]$ for the following reason. The method of undetermined coefficients will yield a polynomial interpolation over the interval $[x_{j-M+1}, x_{j+M}]$; however, the interpolating polynomial will not be accurate near the boundaries of the interval. For larger N, the interpolation in this range will get worse—known as the Runge effect. However, the polynomial will be accurate at the center of the interval $[x_{j-M+1}, x_{j+M}]$ and will allow for an accurate approximation of the integral over the range $[x_j, x_{j+1}]$.

We begin by defining $h_{j+k} = x_{j+k} - x_j$ and note that (43) will be exact for polynomials up to order $N - 1$, that is,

$$\int_{x_j}^{x_{j+1}} (x - x_j)^n dx = \frac{1}{n+1}(x_{j+1} - x_j) = \frac{1}{n+1} h_{j+1}^{n+1}, \qquad (44)$$

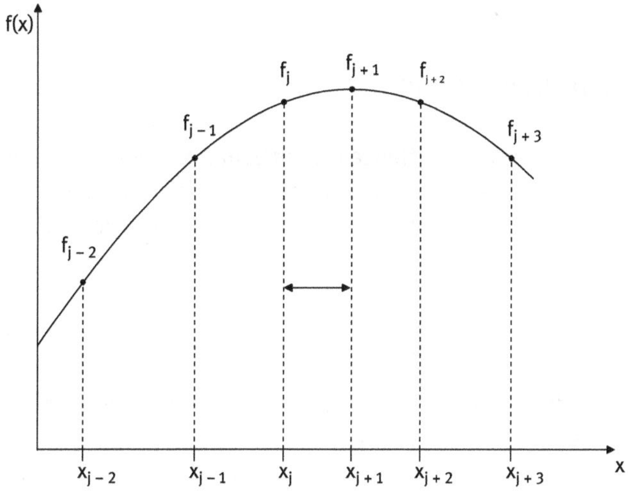

Fig. 5. Schematic for integration on a nonuniform grid.

for $n = 0, \ldots, N-1$. Using (43) allows us to write the set of equations as

$$b_{j(-M+1)}h^n_{j-M+1} + b_{j(-M+2)}h^n_{j-M+2} + \ldots + b_{j(M-1)}h^n_{j+M-1} + b_{jM}h^n_{j+M} = \frac{1}{n+1}h^{n+1}_{j+1}, \tag{45}$$

where the polynomial basis is centered about x_j for numerical considerations. This system is in the form $\mathbf{A}_j\mathbf{b}_j = \mathbf{h}_j$ where

$$\mathbf{b}_j = \begin{pmatrix} b_{j(-M+1)} \\ b_{j(-M+2)} \\ \vdots \\ b_{jM} \end{pmatrix}, \qquad \mathbf{h}_j = \begin{pmatrix} h_{j+1} \\ \frac{1}{2}h^2_{j+1} \\ \vdots \\ \frac{1}{N}h^N_{j+1} \end{pmatrix}, \tag{46}$$

and $\mathbf{A}_j$ is the nonsingular matrix given by

$$\mathbf{A}_j = \begin{pmatrix} 1 & 1 & \cdots & 1 \\ h_{j-M+1} & h_{j-M+2} & \cdots & h_{j+M} \\ h^2_{j-M+1} & h^2_{j-M+2} & \cdots & h^2_{j+M} \\ \vdots & \vdots & \ddots & \vdots \\ h^{N-1}_{j-M+1} & h^{N-1}_{j-M+2} & \cdots & h^{N-1}_{j+M} \end{pmatrix}. \tag{47}$$

The coefficients required in (43) can then be determined by $\mathbf{b}_j = \mathbf{A}_j^{-1}\mathbf{h}_j$. When all grid points are uniformly spaced, the above method will yield the well-known Newton–Coates formulas for integration [7].

4.3 Treatment of the two-body terms

The FFT is not suitable for nonuniform grids. Hence the convolution theorem cannot be applied when solving two-body integrals. It is not feasible to directly integrate (48) because, for N grid points, N one-body integrations would need to be performed. Fortunately, for Coulomb interaction, the integrals have the form

$$\phi(\mathbf{r}) = \int \frac{\rho(\mathbf{r}')}{|\mathbf{r} - \mathbf{r}'|} d\mathbf{r}'. \tag{48}$$

These integrals can still be evaluated in an efficient manner without the need to directly perform the integration. The Poisson equation

$$\nabla^2 \Phi(\mathbf{r}) = -\rho(\mathbf{r}) \tag{49}$$

has the solution

$$\Phi(\mathbf{r}) = \frac{1}{4\pi} \int \frac{\rho(\mathbf{r}')}{|\mathbf{r} - \mathbf{r}'|} d\mathbf{r}', \tag{50}$$

which, apart from a constant prefactor, is of the form (48). To solve Poisson's equation, we use the MOEVE package [10]. MOEVE is a multigrid algorithm adapted for nonuniform tensor product grids. The speed of MOEVE is comparable to that of FFTW [6].

4.4 Ground state of small sodium clusters

As a test of MCTDHF using nonuniform grids, we will calculate the ground states of two small sodium clusters in 3D by performing imaginary-time propagation. A single sodium atom has 11 electrons, which already pushes the limits of MCTDHF. When the system contains more than a single Na atom, it is not feasible to model every electron. In order to overcome this hurdle, we use a pseudopotential to account for the core electrons and only model those in the valence shell. Sodium has the electron configuration $1s^2 2s^2 2p^6 3s^1$, or $[\mathrm{Ne}]3s^1$, and thus only the single electron in the $3s$ shell will be active. The remaining ten electrons will be accounted for by the pseudopotential. The pseudopotential used for these calculations is the Topp–Hopfield potential [16] given by

$$U(r) = \begin{cases} 0.1790\cos(1.224r) - 0.179 & \text{for } r \le 3 \\ -1/r & \text{for } r > 3 \end{cases}, \tag{51}$$

where $r = \sqrt{x^2 + y^2 + z^2}$ is the radial coordinate.

We will calculate the ground state of the sodium clusters Na_4-D_{4h} and Na_6-C_{5v} using the structures given in [13]. The geometry of Na_4-D_{4h} is planar; the sodium atoms form the corners of square with sides of length 3.355Å. The geometry of the Na_6-C_{5v} cluster is a pentagonal pyramid. The sides of the pentagon are of length 3.610Å, and the height of the pyramid is 2.913Å. For our calculations, the Na_4-D_{4h} cluster lies in the x-y plane with the center of the square at the origin. For the Na_6-C_{5v} cluster, the pentagonal base of the pyramid lies in the x-y plane, and the peak of the pyramid is centered at the origin and extends up into the z-axis. In our calculations, the Na_4-D_{4h} and Na_6-C_{5v} clusters contain four and six active electrons, respectively.

For our calculations, we use a nonuniform grid with $N_x = N_y = N_z = 120$ with a minimum grid spacing of $\Delta x_{\min} = 0.38$ around the origin. The x-axis is grown at a rate of 2% for $|x| \le 10$; for $|x| > 10$, the grid is grown at a rate of 5% to a maximum step size $\Delta x_{\max} = 2$. The total range on the x-axis is then ± 106. The y and z axes are chosen in the same manner. In order to achieve that same range $N_x = N_y = N_z = 557$, grid points would be needed on a uniform grid with a spacing equal to the minimum spacing of the nonuniform grid. This makes the number of points on the uniform grid $N_u = 557^3$, as opposed to $N_{nu} = 120^3$ for the nonuniform grid, i.e., $N_u \simeq 100 N_{nu}$. For testing and comparison, we also performed our calculations on a uniform grid.

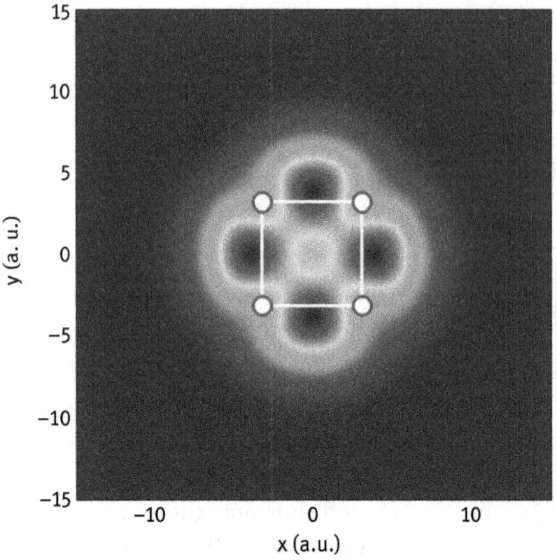

Fig. 6. Ground-state density of the Na_4-D_{4h} cluster. The dots show the position of the Na atoms.

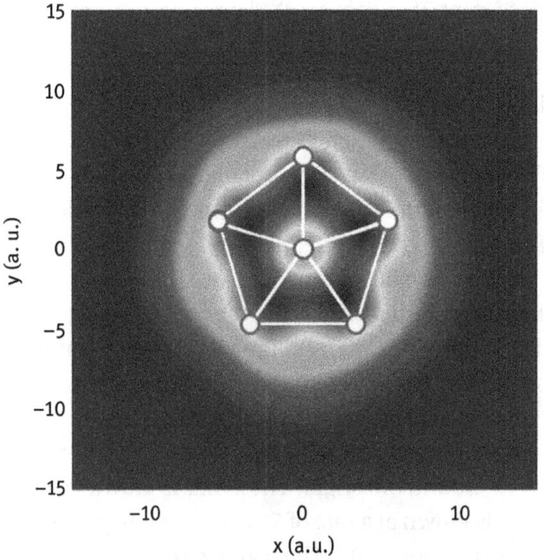

Fig. 7. Ground-state density of the Na_6-C_{5v} cluster. The dots show a birds eye view of the position of the Na atoms.

For the Na_4-D_{4h} cluster, we performed the MCTDHF imaginary-time calculation of the ground state using 12 spin-orbitals. The calculated ground-state energy for this system is found to be $E_0(Na_4) = -1.692$.

For the Na_6-C_{5v} cluster, the MCTDHF calculation is performed using 16 spin-orbitals, and a ground-state energy of $E_0(Na_6) = -3.316$ is calculated. The calculations on the uniform grid are performed with the same number of spin-orbitals as their nonuniform grid counterparts. The ground-state energies found by these calculations are in agreement.

Figure 6 shows the ground-state density

$$\rho(x, y) = \int |\Psi_0(x, y, z)|^2 \, dz \tag{52}$$

for the Na_4-D_{4h} cluster where Ψ_0 is the ground-state wavefunction, and integration is performed over the z-coordinate. The cluster configuration is superposed in the plot. Figure 7 shows the same for the Na_6-C_{5v} cluster. A bird's eye view of the pentagonal pyramid structure is superposed.

5 Conclusion

Even with the computing power available today, the solution of the time-dependent Schrödinger equation for correlated few-electron systems exposed to a strong perturbation remains a formidable task. The MCTDHF method provides a viable approach for making progress toward the simulation of few-electron dynamics in atoms and molecules exposed to strong laser fields. In particular, it performs well for systems where only a single electron is promoted to the continuum, and the motion of the remaining electrons consists of bound state dynamics. The real challenge lies in solving problems with two or more continuum electrons, such as NSDI in 3D. These problems require many configurations to converge the two-body continuum portion of the wavefunction and are unlikely to be solved by MCTDHF alone. That being said, MCTDHF still provides a promising way to investigate phenomena such as the orbital rearrangement of molecules after ionization, the complex ionization processes that occur in noble gas atoms, and the nonlinear effects observed in XFEL experiments on neon atoms [15], to name a few.

Further Reading

- M. H. Beck, A. Jäckle, G. A. Worth, and H.-D. Meyer. The multiconfiguration time-dependent Hartree (MCTDH) method: A highly efficient algorithm for propagating wave packets. *Physics Reports*, 342:1, 2000.
- J. Zanghellini, M. Kitzler, C. Fabian, T. Brabec, and A. Scrinzi. An MCTDHF approach to multielectron dynamics in laser fields. *Laser Physics*, 13:1064, 2003.

– J. Zanghellini, M. Kitzler, T. Brabec, and A. Scrinzi. Testing the multi-configuration time-dependent Hartree–Fock method. *Journal of Physics B: Atomic, Molecular and Optical Physics*, 37:763, 2004.
– J. Caillat, J. Zanghellini, M. Kitzler, O. Koch, W. Kreuzer, and A. Scrinzi. Correlated multielectron systems in strong laser fields: A multiconfiguration time-dependent Hartree–Fock approach. *Physical Review A*, 71:012712, 2005.
– C. F. Destefani, C. McDonald, R. M. Abolfath, P. Hawrylak, and T. Brabec. Ab initio approach to the optimization of qubit manipulation. *Physical Review B*, 78:165331, 2008.
– G. Jordan and A. Scrinzi. Strongly driven few-fermion systems—MCTDHF. In H. D. Meyer, F. Gatti, and G. Worth (eds.), *Multidimensional Quantum Dynamics: MCTDH Theory and Applications*. Wiley-VCH Verlag GmbH & Co., Weinheim, Germany, pp. 161–183, 2009.
– C. F. Destefani, C. R. McDonald, S. Sukiasyan, and T. Brabec. Plasmon dynamics in strongly driven finite few-electron quantum systems: The role of the surface. *Physical Review B*, 81:045314, 2010.
– D. Hochstuhl, S. Bauch, and M. Bonitz. Multiconfigurational time-dependent Hartree–Fock calculations for photoionization of one-dimensional Helium. *Journal of Physics: Conference Series*, 220:012019, 2010.
– D. J. Haxton, K. V. Lawler, and C. W. McCurdy. Multiconfiguration time-dependent Hartree–Fock treatment of electronic and nuclear dynamics in diatomic molecules. *Physical Review A*, 83:063416, 2011.
– D. J. Haxton, K. V. Lawler, and C. W. McCurdy. Single photoionization of Be and HF using the multiconfiguration time-dependent Hartree–Fock method. *Physical Review A*, 86:013406, 2012.
– C. R. McDonald, G. Orlando, J. W. Abraham, D. Hochstuhl, M. Bonitz, and T. Brabec. Theory of the quantum breathing mode in harmonic traps and its use as a diagnostic tool. *Physical Review Letters*, 111:256801, 2013.
– D. Hochstuhl, C. Hinz, and M. Bonitz. Time-dependent multiconfiguration methods for the numerical simulation of photoionization processes of many-electron atoms. *European Physical Journal: Special Topics*, 223:177–336, 2014.
– D. J. Haxton and C. W. McCurdy. Two methods for restricted configuration spaces within the multiconfiguration time-dependent Hartree–Fock method. *Physical Review A*, 91:012509, 2015.

Bibliography

[1] S. Bauch, L. K. Sørenson, and L. B. Madsen. Time-dependent generalized-active-space configuration-interaction approach to photoionization dynamics of atoms and molecules. *Physical Review A*, 90:062508, 2014.
[2] M. H. Beck and H. D. Meyer. An efficient and robust integration scheme for the equations of motion of the multiconfiguration time-dependent Hartree (MCTDH) method. *Zeitschrift fuer Physik D*, 42:113, 1997.
[3] M. H. Beck, A. Jäckle, G. A. Worth, and H.-D. Meyer. The multiconfiguration time-dependent hartree (MCTDH) method: A highly efficient algorithm for propagating wavepackets. *Physics Reports*, 324(1):1–105, 2000.
[4] J. Caillat, J. Zanghellini, M. Kitzler, O. Koch, W. Kreuzer, and A. Scrinzi. Correlated multielectron systems in strong laser fields: A multiconfiguration time-dependent Hartree–Fock approach. *Physical Review A*, 71:012712, 2005.
[5] P. B. Corkum. Plasma perspective on strong field multiphoton ionization. *Physical Review Letters*, 71:1994–1997, 1993.

[6] M. Frigo and S. G. Johnson. FFTW. http://www.fftw.org/. Accessed February 10, 2017.

[7] D. Kincaid and W. Cheney. *Numerical Analysis: Mathematics of Scientific Computing*. Brooks/Cole, Pacific Grove, CA, 3rd ed., 2002.

[8] T. Klamroth. Laser-driven electron transfer through metal-insulator-metal contacts: Time-dependent configuration interaction singles calculations for a jellium model. *Physical Review B*, 68:425421, 2003.

[9] A. D. McLachlan. A variational solution of the time-dependent Schrödinger equation. *Molecular Physics*, 8:39–44, 1964.

[10] G. Pöplau and U. van Rienen. A self-adaptive multigrid technique for 3D space charge calculations. *IEEE Transactions on Magnetics*, 44:1242, 2008.

[11] N. Rohringer, A. Gordon, and R. Santra. Configuration-interaction-based time-dependent orbital approach for ab initio treatment of electronic dynamics in a strong optical laser field. *Physical Review A*, 74:043420, 2006.

[12] H. B. Schlegel, S. M. Smith, and X. Li. Electronic optical response of molecules in intense fields: Comparison of TD-HF, TD-CIS, and TD-CIS(D) approaches. *Journal of Chemical Physics*, 126:244110, 2007.

[13] I. A. Solov'yov, A. V. Solov'yov, and W. Greiner. Structure and properties of small sodium clusters. *Physical Review A*, 65:053203, 2002.

[14] S. Sukiasyan, C. McDonald, C. Destefani, M. Y. Ivanov, and T. Brabec. Multielectron correlation in high-harmonic generation: A 2D model analysis. *Physical Review Letters*, 102:223002, 2009.

[15] M. Uiberacker, Th. Uphues, M. Schultze, A. J. Verhoef, V. Yakovlev, M. F. Kling, J. Rauschenberger, N. M. Kabachnik, H. Schröder, M. Lezius, K. L. Kompa, H.-G. Muller, M. J. J. Vrakking, S. Hendel, U. Kleineberg, U. Heinzmann, M. Drescher, and F. Krausz. Attosecond real-time observation of electron tunnelling in atoms. *Nature*, 446:627, 2007.

[16] W. C. Topp and J. J. Hopfield. Chemically motivated pseudopotential for sodium. *Physical Review B*, 7:1295–1303, 1973.

[17] J. B. Watson, A. Sanpera, D. G. Lappas, P. L. Knight, and K. Burnett. Nonsequential double ionization of helium. *Physical Review Letters*, 78:1884–1887, 1997.

Stefan Pabst and Robin Santra

VI Time-dependent configuration interaction singles

Many-body effects are becoming more prominent and omnipresent as strong-field phenomena are studied in larger and more complex multielectron systems. Many-body effects such as multiorbital ionization or collective excitations can dramatically change the system response requiring a theoretical description that goes beyond the single-particle picture. In this chapter, we discuss a practically useful many-body method, time-dependent configuration interaction singles (TDCIS), and its specific implementation (grid representation, etc.) to describe strong-field processes. TDCIS captures a variety of many-body effects and at the same time keeps much of the single-particle picture making TDCIS a computationally very attractive approach.

1 Introduction

The absorption and emission of electromagnetic radiation have been the method of choice for over a century to access the quantum world of atoms, molecules, and solids. Since the invention of the laser in the 1960s and the possibility to produce well-defined pulses of radiation (laser pulses), the underlying structure and mechanisms in these microscopic systems can be studied in unprecedented detail. By knowing exactly the spectrum of the radiation pulse hitting the system, a direct connection between the final pulse spectrum (after passing through the system) and the internal structure of the system can be established. The kinetic energy distribution of electrons expelled from the system (the photoelectron spectrum) can be measured and contains quite complementary information on the system.

Atoms, molecules, and solids are all multielectron systems. The Coulomb interaction among the electrons causes them to become correlated, i.e., the behavior of any given electron can generally not be specified without specifying the behavior of all other electrons in the system. Even though the importance of electron correlations for physical or chemical properties has been realized in the early years of quantum mechanics, it remains a challenge to rigorously describe them. For any realistic

Stefan Pabst: ITAMP, Harvard-Smithsonian Center for Astrophysics, Cambridge, MA 02138, USA; Department of Physics, Harvard University, Cambridge MA 02138, USA; email: stefan.pabst@cfa.harvard.edu
Robin Santra: Center for Free-Electron Laser Science, DESY, Notkestraße 85, 22607 Hamburg, Germany; Department of Physics, University of Hamburg, Jungiusstraße 9, 20355 Hamburg, Germany; email: robin.santra@cfel.de

De Gruyter Graduate – Computational Strong-Field Quantum Dynamics, Volume 5, 2017, pp. 169–201.
DOI 10.1515/9783110417265-006

system, the exact N-electron wavefunction needs to be approximated to make it computationally tractable. Even for a helium atom, the simplest multielectron system with two electrons, the exact wavefunction is analytically not known.

Many theories have been developed over the last 80 years to cope with this problem. The most popular theories are mean-field theories, where the N electrons are approximated as being independent but experiencing each other via a common mean-field potential. The most common mean-field theory is Hartree–Fock (HF), where the N-electron wavefunction is approximated as a single Slater determinant. Mean-field theories, by definition, do not include electron correlations as they treat each electron as if it were an independent particle.

In many systems (e.g., open-shell atoms), already the ground state can contain a significant amount of correlation. Additionally, electronic correlations can quite dramatically change the character of excited states. One example is autoionizing states, which become unstable and decay only because of electronic correlations. Most (wavefunction-based) many-body theories [1] that include electron correlation use the HF ground state as a reference state to build up a multiconfiguration wavefunction consisting of many Slater determinants. These theories are, therefore, also called post-HF methods. Correlation effects beyond the independent particle (mean-field) picture are included through the multiconfigurational nature of the post-HF wavefunction.[1] However, quantum many-body theories have the disadvantage that they quickly become too hard to be solvable on a computer, and the amount of correlation that can be described has to be restricted. Hence, most many-body theories are used for static problems such as calculating the ground state and its properties.

A dynamical or time-dependent description is generally needed to describe nonstationary situations, particularly when the perturbation causing the dynamics of the system is so strong that perturbation theory breaks down. This is the case for strong-field processes: The large intensities of strong-field pulses generate electromagnetic forces that are comparable with the Coulomb forces within the system. Consequently, the light-matter interaction does not just perturb but significantly alters the electronic structure of the system. Perturbation theory cannot be applied as the assumption that the light-matter interaction is weak is not fulfilled anymore. In this situation, the electric field of the pulse[2] is typically so strong that an electron can be pulled out of the system, leading to strong-field or tunnel ionization.

Because of its strength, a strong-field pulse can affect more than just a single electron. Electronic correlations that may have been negligible in the initial ground state can now become quite important as the system is strongly driven by the laser pulse. The presence of a strong-field pulse can significantly alter the properties and

1 See Chapter V on multiconfigurational time-dependent HF.

2 The magnetic component can normally be ignored in the nonrelativistic regime; see Chapter III for the relativistic regime.

the appearance of many-body effects. In the field-free scenario, many-body effects are well studied, but in the strong-field regime, they are basically unexplored.

A natural way to build strong-field many-body theories is by using well-known many-body theories from the quantum chemistry community [1] and adopting them for the strong-field regime. In strong-field calculations, it is necessary to describe a wide range of continuum states, since an electron can be freed from the system and may even return to the system at a later time. Highly delocalized continuum states do not favor an orbital description based on Gaussian-type functions, which are employed in most quantum chemistry approaches. A nonuniform grid representation, which accounts for localized bound orbitals and highly delocalized excited and continuum orbitals, is better suited for strong-field problems.

In the following, we explain why TDCIS is an ideal candidate for describing many-body effects in the strong-field regime and how it can be used to identify many-body processes. In Section 2, we derive the basic equations and properties of TDCIS. The specific implementation of TDCIS for closed-shell atoms is presented in Section 3. It includes discussions about the orbital and grid representations, spin-orbit interaction, the calculation of expectation values and the ion density matrix (IDM), the HF procedure, and the use of complex absorbing potentials (CAPs). Finally, in Section 4, we show examples of how TDCIS allows one to explain multiorbital effects and collective many-body excitations in the strong-field regime. We specifically discuss in Section 4.1 the creation of spin-orbit hole motion via strong-field ionization and how to measure it with attosecond transient absorption spectroscopy (ATAS). Another example, in Section 4.2, will focus on the appearance of collective excitations in high-harmonic generation (HHG) and the dramatic consequences for the HHG spectrum.

Throughout this chapter, we use atomic units ($\hbar = |q_e| = 1/(4\pi\epsilon_0) = a_0 = 1$), which are natural units when dealing with the electronic structure of atoms, molecules, and solids.

2 Basics of TDCIS

2.1 TDCIS wavefunction

Let us start at the very beginning, i.e., with the main concept behind configuration-interaction (CI). CI is based on the idea that an N-electron state that can be easily determined is used as a reference to describe the full N-electron state Ψ. When this reference state is a good approximation of Ψ, then only very few additional terms are needed to describe the full state Ψ. The reference state in CI is normally the HF ground state,

$$|\Phi_0\rangle = \prod_i \hat{c}_i^\dagger |0\rangle , \tag{1}$$

where the index i runs over N one-particle orbitals (including electron spin), which are the orbitals occupied by the N electrons, and $|0\rangle$ is the vacuum state with no electrons present. CI is one of the post-HF methods.[3] After we have found the reference state Φ_0, we can build Ψ in terms of Φ_0:

$$\begin{aligned} |\Psi\rangle &= \Big(\alpha_0 + \sum_{n=1}^{N} \sum_{\substack{a_1<\ldots<a_n \\ i_1<\ldots<i_n}} \alpha_{i_1\ldots i_n}^{a_1\ldots a_n} \hat{c}_{a_1}^\dagger \hat{c}_{i_1} \ldots \hat{c}_{a_n}^\dagger \hat{c}_{i_n}\Big) |\Phi_0\rangle \\ &= \alpha_0 |\Phi_0\rangle + \sum_{a_1,i_1} \alpha_{i_1}^{a_1} |\Phi_{i_1}^{a_1}\rangle + \sum_{\substack{a_1<a_2 \\ i_1<i_2}} \alpha_{i_1 i_2}^{a_1 a_2} |\Phi_{i_1 i_2}^{a_1 a_2}\rangle + \ldots , \end{aligned} \tag{2}$$

where $\hat{c}_a^\dagger$ and $\hat{c}_i$ are the creation and annihilation operators of an electron in orbitals a and i, respectively. The n-particle-n-hole (np–nh) configurations, $|\Phi_{i_1\ldots i_n}^{a_1\ldots a_n}\rangle = \hat{c}_{a_1}^\dagger \hat{c}_{i_1} \ldots \hat{c}_{a_n}^\dagger \hat{c}_{i_n} |\Phi_0\rangle$, represent N-electron configurations where n electrons are moved from their initially occupied orbitals i_k (leaving holes behind) to initially unoccupied (virtual) orbitals a_k with $k = 1, \ldots, n$. Note that the indices i, j refer always to occupied orbitals, and a, b refer to virtual (unoccupied) orbitals. The indices p, q, r, s are used when referring to any (occupied or virtual) orbitals.

If all possible np–nh configurations are included, Ψ is the exact N-electron wavefunction including all electronic correlations. In this case, the method is called full CI. Including only singly excited configurations is called CI-singles (CIS). Similarly, including only doubly excited configurations is called CI-doubles, and including singly and doubly excited configurations is called CI-singles-doubles.

The size of the np-nh-configuration space scales as $(N_a N_i)^n$, where N_a and N_i are the numbers of unoccupied and occupied orbitals, respectively. Since N_a is quite large (in principle infinitely large), it is computationally not feasible to go to very high excitation classes. To calculate exact ground-state properties (e.g., expectation values of suitable observables), it is not necessary to include states that are very delocalized and extend far from the atom/molecule. Hence, N_a may be relatively small, and higher np-nh-excitations classes and, therefore, higher-order correlations can be included in the calculations. In strong-field physics, where it is important to describe a wide range of the continuum states, N_a can easily be a 5-digit number. Including doubly or triply excited configurations quickly becomes infeasible. Consequently, we limit ourselves to CIS and focus on many-body and correlation effects within the CIS configuration space. As we will see in Section 4, already with CIS, a variety of new effects can be included that are not possible to describe with a single-particle picture.

3 HF is explained in Section 3.5.

By turning the CIS coefficients α_0 and α_i^a into time-dependent coefficients, $\alpha_0(t)$ and $\alpha_i^a(t)$, we obtain the TDCIS wavefunction:

$$|\Psi(t)\rangle = \alpha_0(t)\,|\Phi_0\rangle + \sum_{a,i} \alpha_i^a(t)\,|\Phi_i^a\rangle\,. \tag{3}$$

This wavefunction ansatz still makes use of the fact that most strong-field processes in atoms are one-electron–dominated processes, but it also acknowledges (via the sum over i) that the active electron may not just come from the least weakly bound (outermost) occupied orbital as it is usually assumed in single-active electron (SAE) calculations. TDCIS also takes into account that the active electron can influence the state of the parent ion (the state of the hole) leading to correlations between the excited electron and the hole. In comparison with single-particle approaches, TDCIS does not use model potentials to mimic the interaction of the active electron with the other electrons. It describes the exact electron-electron interaction within the CIS configuration space. For these reasons, CIS is a good candidate for a many-body theory in the strong-field regime.

2.2 The *N*-body Hamiltonian

The exact nonrelativistic N-electron Hamiltonian in an external electric field reads as

$$\hat{H}(t) = \sum_{n=1}^{N} \frac{\hat{\mathbf{p}}_n^2}{2} - \sum_{n=1}^{N} \frac{Z}{|\hat{\mathbf{r}}_n|} + \frac{1}{2} \sum_{\substack{n,m=1 \\ n \neq m}}^{N} \frac{1}{|\hat{\mathbf{r}}_n - \hat{\mathbf{r}}_m|} + \sum_{n=1}^{N} \mathbf{E}(t) \cdot \hat{\mathbf{r}}_n, \tag{4}$$

with the kinetic energy $\frac{\hat{\mathbf{p}}_n^2}{2}$, the attractive nuclear potential $-\frac{Z}{|\hat{\mathbf{r}}_n|}$, the repulsive electron-electron interaction $|\hat{\mathbf{r}}_n - \hat{\mathbf{r}}_m|^{-1}$, and the light-matter interaction $\mathbf{E}(t) \cdot \hat{\mathbf{r}}_n$. The bold symbols represent vectors, and $\cdot$ is the scalar product. For TDCIS, it is convenient to partition the Hamiltonian into three parts,

$$\hat{H}(t) = \hat{H}_0 + \hat{H}_1 + \hat{H}_{\mathrm{LM}}(t) - E_{\mathrm{HF}}, \tag{5}$$

which have an intuitive physical meaning:

$\hat{H}_0$ is the Fock operator containing the kinetic term, the nuclear potential, and the effective one-particle mean-field potential $\hat{V}_{\mathrm{HF}}$.

$\hat{H}_1$ is the residual electron-electron interaction including everything that goes beyond the mean-field potential $\hat{V}_{\mathrm{HF}}$ included in $\hat{H}_0$.

$\hat{H}_{\mathrm{LM}}(t) = \sum_{n=1}^{N} \mathbf{E}(t) \cdot \hat{\mathbf{r}}_n$ is the light-matter interaction in the electric-dipole approximation using the length form.

E_{HF} $= \langle \Phi_0 | \hat{H}_0 + \hat{H}_1 | \Phi_0 \rangle$ is the HF ground-state energy. This energy shift is only done out of convenience to shift all energies such that the HF ground state Φ_0 has zero energy.

The Fock operator $\hat{H}_0$ includes all one-particle potentials and defines the one-particle orbitals φ_p, which are the eigenstates of $\hat{H}_0$ with the eigenvalues ε_p. Consequently, we can write

$$\hat{H}_0 = \sum_p \varepsilon_p \, \hat{c}_p^\dagger \hat{c}_p. \tag{6}$$

The residual Coulomb interaction $\hat{H}_1$ is the only two-body operator in (5) and captures all effects beyond an independent-particle picture. The detailed expression for $\hat{H}_1$ reads as

$$\hat{H}_1 = \hat{V}_c - \hat{V}_{\mathrm{HF}}, \tag{7a}$$

$$\hat{V}_c = \frac{1}{2} \sum_{i \neq j} \frac{1}{|\hat{\mathbf{r}}_i - \hat{\mathbf{r}}_j|} = \frac{1}{2} \sum_{pqrs} v_{pqrs} \, \hat{c}_p^\dagger \hat{c}_q^\dagger \hat{c}_s \hat{c}_r, \tag{7b}$$

$$\hat{V}_{\mathrm{HF}} = \sum_{pq} \Big(\sum_i v_{piqi} - v_{piiq} \Big) \hat{c}_p^\dagger \hat{c}_q, \tag{7c}$$

where $\hat{V}_c$ is the exact electron-electron interaction and $\hat{V}_{\mathrm{HF}}$ is the HF mean-field potential. In these equations, we changed from a general representation to an orbital representation (defined by $\hat{H}_0$). In the orbital representation, the matrix elements of the electron-electron interaction read as

$$v_{pqrs} = \iint d^3 r_1 \, d^3 r_2 \, \frac{\varphi_p^\star(\mathbf{r}_1) \varphi_r(\mathbf{r}_1) \, \varphi_q^\star(\mathbf{r}_2) \varphi_s(\mathbf{r}_2)}{|\mathbf{r}_1 - \mathbf{r}_2|} \, \delta_{\sigma_p, \sigma_r} \, \delta_{\sigma_q, \sigma_s}, \tag{8}$$

where σ_p is the spin of the electron in orbital p.

The light-matter interaction $\hat{H}_{\mathrm{LM}}(t)$ is expressed in the length form $\sum_n \mathbf{E}(t) \cdot \hat{\mathbf{r}}_n$. An equivalent and also important way to express the light-matter interaction in the electric-dipole approximation is the velocity form $\sum_n \mathbf{A}(t) \cdot \hat{\mathbf{p}}_n$, where $\mathbf{A}(t)$ is the vector potential and $\hat{\mathbf{p}}_n$ is the canonical momentum operator.[4] The spatial dependence of the electric field, $\mathbf{E}(t)$, is neglected (this is the essence of the electric-dipole approximation) because strong-field pulses normally have wavelengths around 1 μm, i.e., orders of magnitude larger than the size of atoms, which is a few Bohr radii ($a_0 = 0.529$). Spatial variations of the electric field across the atom are negligibly small. In the following, we restrict ourselves to linearly polarized pulses where we can exploit azimuthal symmetry, greatly reducing the computational effort. The laser

4 See also Section 1.5.5 in Chapter I.

polarization axis is chosen to be the z-axis, $\mathbf{E}(t) = \mathcal{E}(t)\mathbf{e}_z$. The light-matter interaction in the orbital representation reads as

$$\hat{H}_{\mathrm{LM}}(t) = \mathcal{E}(t) \sum_{pq} z_{pq}\, \hat{c}_p^\dagger \hat{c}_q, \tag{9}$$

where $z_{pq} = \int d^3r\, \varphi_p^*(r)\, z\, \varphi_q(r)\, \delta_{\sigma_p,\sigma_q}$ is the dipole transition matrix element between orbitals p and q. Note that $\hat{H}_{\mathrm{LM}}(t)$ does not change the spin of the electron.

2.3 Equations of motion

In Sections 2.1 and 2.2, we discussed the N-body wavefunction and the N-body Hamiltonian. After inserting (3) and (5) into the time-dependent Schrödinger equation (TDSE)

$$\hat{H}(t)\,|\Psi(t)\rangle = i\frac{\partial}{\partial t}\,|\Psi(t)\rangle\,, \tag{10}$$

we need to project the TDSE on each configuration in Ψ, i.e., Φ_0 and the Φ_i^a, to obtain the equation of motion for each corresponding time-dependent coefficient:

$$i\dot{\alpha}_0(t) = \mathcal{E}(t) \sum_{a,i} \langle \Phi_0 | \hat{z} | \Phi_i^a \rangle\, \alpha_i^a(t), \tag{11a}$$

$$\begin{aligned} i\dot{\alpha}_i^a(t) = {} & \left(\langle \Phi_i^a | \hat{H}_0 | \Phi_i^a \rangle - E_{\mathrm{HF}} \right) \alpha_i^a(t) \\ & + \mathcal{E}(t) \Big[\langle \Phi_i^a | \hat{z} | \Phi_0 \rangle\, \alpha_0(t) + \sum_{b,j} \langle \Phi_i^a | \hat{z} | \Phi_j^b \rangle\, \alpha_j^b(t) \Big] \\ & + \sum_{b,j} \langle \Phi_i^a | \hat{H}_1 | \Phi_j^b \rangle\, \alpha_j^b(t). \end{aligned} \tag{11b}$$

Note that the configurations Φ_0 and Φ_i^a form an orthonormal set in the N-body Hilbert space, and $\hat{H}_0$ is diagonal in this set: $\langle \Phi_0 | \hat{H}_0 | \Phi_i^a \rangle = 0$ and $\langle \Phi_i^a | \hat{H}_0 | \Phi_j^b \rangle = \left(\varepsilon_a - \varepsilon_i + \langle \Phi_0 | \hat{H}_0 | \Phi_0 \rangle \right) \delta_{a,b} \delta_{i,j}$. Furthermore, we use the fact that $\hat{H}_1$ does not lead to couplings between the ground state and the singly excited states, i.e., $\langle \Phi_0 | \hat{H}_1 | \Phi_i^a \rangle = 0$. This nontrivial statement is known as Brillouin's theorem and can be shown by using (1) and (7). Consequently, the HF ground state is an eigenstate – in fact, the ground state – within the CIS configuration space. Only when including double excitations, the ground state will differ from the HF ground state.

The light-matter interaction can couple via the terms $\langle \Phi_0 | \hat{z} | \Phi_i^a \rangle$ and $\langle \Phi_i^a | \hat{z} | \Phi_0 \rangle$ the ground state with the singly excited states. Without any external field, the atom remains in the ground state. Singly excited states Φ_i^a are coupled with each other via the residual Coulomb interaction $\hat{H}_1$ and via the light-matter interaction $\mathcal{E}(t)\hat{z}$. The light-matter transition matrix elements $\langle \Phi_i^a | \hat{z} | \Phi_j^b \rangle$ separate into two independent single-particle transition matrix elements: one between occupied orbitals, z_{ji}, and one between virtual orbitals, z_{ab}, see (17).

The residual Coulomb interaction $\hat{H}_1$ can change particle and hole states at the same time. The matrix elements where the hole state is changed, $\langle \Phi_i^a | \hat{H}_1 | \Phi_j^b \rangle$ with $i \neq j$, are known as *interchannel* coupling matrix elements. A channel is characterized by the ionic state that is left behind after photoelectron emission, which in the case of CIS is simply given by the hole index i. Note that for CIS, the ionic states $|\Phi_i\rangle = \hat{c}_i |\Phi_0\rangle$, which are $(N-1)$-electron states, do not couple with each other—$\langle \Phi_i | \hat{H}_0 + \hat{H}_1 | \Phi_j \rangle = 0$ for $i \neq j$, making them ionic eigenstates within the CIS configuration space. This is another nontrivial statement and is the essence of Koopmans' theorem. Therefore, the hole index i is at the same time the channel index.

By setting the matrix elements $\langle \Phi_i^a | \hat{H}_1 | \Phi_j^b \rangle$ with $i \neq j$ to zero, the interchannel interactions are switched off. In this case, only *intrachannel* interactions, $\langle \Phi_i^a | \hat{H}_1 | \Phi_i^b \rangle$, are considered. Switching on and off specific interaction makes it possible to systematically identify and study mechanisms that may cause a given physical phenomenon.

The intrachannel interactions have mainly the effect to correct the charge of the ionized parent ion, as seen by the excited electron. The virtual HF orbitals φ_a are scattering states for an electron added to the N-electron system in its HF ground state. To describe excited N-electron states, the virtual HF orbitals are not optimal because the excited electron would not experience the Coulomb potential of the hole that is left behind. This is corrected with intrachannel interactions. If the excited electron is far away from the parent ion, the intrachannel potential $\langle \Phi_i^a | \hat{H}_1 | \Phi_i^b \rangle$ turns for all hole states i into a simple Coulomb potential, i.e., $-\langle a | 1/r | b \rangle$, as the shape of the hole becomes more and more irrelevant.

To mimic SAE calculations with TDCIS, we can restrict ourselves to intrachannel interactions and allow the electron to originate only from one specific orbital. Nevertheless, there are differences to common SAE approaches. Most notably, the model potentials in SAE are for atoms spherically symmetric and local, i.e., of the form $V(r)$.[5] The intrachannel potential is neither. If the electron is close to the ion, it experiences the spatial shape of the hole, thus rendering the potential nonspherical (unless the hole has angular momentum $l = 0$). Because of the antisymmetrization of the full N-electron wavefunction, the intrachannel potential is also nonlocal.

2.4 Limitations

TDCIS is a multichannel theory where only one electron can be emitted. Hence, multiple ionization processes cannot be described by TDCIS. Because of the residual Coulomb interaction, the outgoing electron can alter the ionic state. The degree of

5 See also Section 2.1 in Chapter IV.

freedom of the ionic state is, however, quite limited. The only allowed ionic states are one-hole configurations $|\Phi_i\rangle = \hat{c}_i |\Phi_0\rangle$, where an electron is removed from the HF ground state. In these configurations, all other electrons are frozen. The only ionic motion allowed is the hopping of the hole between occupied orbitals. More complicated dynamics require at least one more electron to be active leading to 2p-2h configurations $|\Phi^{a_1,a_2}_{i_1,i_2}\rangle$ and beyond. The comparison with higher-order approaches like the complete active space self-consistent field (CASSCF) method has shown that polarizability is underestimated by CIS [21], indicating that the response of the remaining electrons may be important to obtain a more accurate picture of the ionic system.

3 Implementation of TDCIS

TDCIS can be applied to a wide variety of systems, ranging from atoms to molecules and to solids. Our focus is on noble gas atoms, which are the most experimentally and theoretically studied systems in the strong-field and attosecond regimes.

The TDCIS method presented here has first been presented by Rohringer, Gordon, and Santra [24] and has later been extended to include spin-orbit splitting for the occupied orbitals [21, 23] and the exact residual Coulomb interaction $\hat{H}_1$ [9].

In Section 2, the orbital index, p, refers to the spin and to the spatial degrees of freedom of the electron. This will change in the following because the spin is treated differently than the spatial components. Therefore, the spin degree of freedom σ will be stated separately, i.e.,

$$\varphi_p(\mathbf{r}, \sigma) \to \varphi_{p,\sigma}(\mathbf{r}) := \varphi_p(\mathbf{r})\chi_\sigma, \tag{12}$$

where χ_σ is the spinor corresponding to the spin σ. The total spin of the electron is always $1/2$ and need not be explicitly mentioned. From now on, when we refer to the orbital p, we mean only the spatial component $\varphi_p(\mathbf{r})$.

3.1 Symmetries and orbital representations

Closed-shell atomic systems are spherically symmetric. It is therefore advantageous to use spherical coordinates (r, θ, ϕ). Furthermore, for spherically symmetric systems, the one-particle orbitals φ_p have a well-defined angular momentum l and can be written in terms of spherical harmonics [33] $Y_{l,m}(\theta, \phi)$. The magnetic quantum number m indicates in which direction the angular momentum is pointing. Hence, the one-particle basis φ_p can be factorized into a radial and an angular part,

$$\varphi_p(\mathbf{r}) = \langle \mathbf{r} | \varphi_p \rangle = \frac{u_{n_p,l_p}(r)}{r} Y_{l_p,m_p}(\theta, \phi), \tag{13}$$

where n_p is the radial quantum number of the orbital p. Often, the principal quantum number $n^P = n + l + 1$ instead of n is used to characterize atomic orbitals. This factorization has several advantages. It separates the radial degree of freedom from the angular degrees of freedom. The angular part is not just analytically known, also all integrals and derivatives involving angles can be analytically calculated (see Section 3.2). From an implementation point of view, this is important because the dimensionality of the problem that has to be solved just dropped from 3D to coupled 1D (radial) problems—each electron with a different angular momentum (l, m) experiences a different radial potential.

All N-electron eigenstates have well-defined angular momenta as well. For noble gas atoms, the neutral ground state is a singlet state with no overall spin, $S = 0$, and no overall angular momentum, $L = M = 0$. Only by absorbing or emitting a photon, the N-electron state can change its angular momentum by ± 1. Linearly polarized pulses can only change the total angular momentum L but not the magnetic quantum number M, which is conserved throughout the light-matter interaction. This has beneficial consequences for computation as the configuration space that needs to be captured is greatly reduced, and all configurations with $M \neq 0$ can be ignored. In CIS, only one electron can be excited, and the restriction to $M = 0$ means that for each Φ_i^a configuration, the magnetic quantum number m_a of the excited particle has to be the same as for the hole, $m_i = m_a$.

The overall spin state as well as the spin of each electron cannot be changed by the light-matter interaction, as it is not spin sensitive. Since the ground state is a singlet state $S = 0$, all excited states are singlet states too. It is therefore desirable to write each CIS configuration as a singlet state [24],

$$|\Phi_i^a\rangle_S = \frac{1}{\sqrt{2}} \left(|\Phi_{i,\uparrow}^{a,\uparrow}\rangle + |\Phi_{i,\downarrow}^{a,\downarrow}\rangle \right), \tag{14}$$

where $|\Phi_{i,\sigma}^{a,\sigma}\rangle$ describes an excitation of an electron with spin σ.

A similar transformation can be made for the m state of the excited electron [20] because linearly polarized light is used, and an electron with magnetic number m_a behaves exactly the same way as an electron with $-m_a$. This reduces the configuration space by almost a factor 2 (similar to the spin consideration). The implementation can be found in [20] and looks very similar to the spin symmetrization done in (14). However, we will not perform and use this transformation in our discussion here to keep equations more compact.

3.2 Evaluating matrix elements

It is time to evaluate the matrix elements for $\hat{z}$ and for $\hat{H}_1$ in (11) explicitly. The dipole operator $\hat{z}$ is a one-particle operator, whereas $\hat{H}_1$ is a two-particle operator. A one-particle operator changes the state of only one particle in an N-particle wavefunction, whereas a two-particle operator can change two particles at the same time. The goal is to turn the general expressions in (11), e.g., $\langle\Phi_0|\hat{z}|\Phi_i^a\rangle_S$, into expressions that depend only on the one-particle orbitals φ_p. For the light-matter interaction that couples the ground state to the singly excited states, we find

$$\langle\Phi_0|\hat{z}|\Phi_i^a\rangle_S = \frac{1}{\sqrt{2}}\sum_{\sigma}\left\langle\Phi_0\middle|\hat{z}\middle|\Phi_{i,\sigma}^{a,\sigma}\right\rangle = \sqrt{2}\,z_{ia}, \tag{15a}$$

$$\left\langle\Phi_0\middle|\hat{z}\middle|\Phi_{i,\sigma}^{a,\sigma}\right\rangle = \sum_{pq,\sigma'} z_{pq}\left\langle\Phi_0\middle|\hat{c}_{p,\sigma'}^{\dagger}\hat{c}_{q,\sigma'}\hat{c}_{a,\sigma}^{\dagger}\hat{c}_{i,\sigma}\middle|\Phi_0\right\rangle = z_{ia}, \tag{15b}$$

where the expression for the dipole operator is taken from (9), and the anticommutator relations,

$$\{\hat{c}_{p,\sigma},\hat{c}_{q,\sigma'}^{\dagger}\} = \delta_{pq}\,\delta_{\sigma,\sigma'}, \tag{16a}$$

$$\{\hat{c}_{p,\sigma},\hat{c}_{q,\sigma'}\} = \{\hat{c}_{p,\sigma}^{\dagger},\hat{c}_{q,\sigma'}^{\dagger}\} = 0, \tag{16b}$$

are used in addition to $\hat{c}_{i,\sigma}^{\dagger}|\Phi_0\rangle = \hat{c}_{a,\sigma}|\Phi_0\rangle = 0$. One interesting consequence of (16a) is that $\{\hat{c}_{i,\sigma},\hat{c}_{a,\sigma'}^{\dagger}\} = \{\hat{c}_{a,\sigma},\hat{c}_{i,\sigma'}^{\dagger}\} = 0$ because the virtual orbital space is orthogonal to the occupied orbital space. Note that in (15a), a factor $\sqrt{2}$ appears because we are using CIS configurations with spin singlet character. For the dipole matrix elements between distinct singly excited states, we get

$${}_S\langle\Phi_i^a|\hat{z}|\Phi_j^b\rangle_S = \frac{1}{2}\sum_{\sigma,\sigma'}\left\langle\Phi_{i,\sigma}^{a,\sigma}\middle|\hat{z}\middle|\Phi_{j,\sigma'}^{b,\sigma'}\right\rangle = z_{ab}\,\delta_{ij} - z_{ji}\,\delta_{ab}, \tag{17a}$$

$$\begin{aligned}\left\langle\Phi_{i,\sigma}^{a,\sigma}\middle|\hat{z}\middle|\Phi_{j,\sigma'}^{b,\sigma'}\right\rangle &= \sum_{pq,\tilde{\sigma}} z_{pq}\left\langle\Phi_0\middle|\hat{c}_{i,\sigma}^{\dagger}\hat{c}_{a,\sigma}\,\hat{c}_{p,\tilde{\sigma}}^{\dagger}\hat{c}_{q,\tilde{\sigma}}\,\hat{c}_{b,\sigma'}^{\dagger}\hat{c}_{j,\sigma'}\middle|\Phi_0\right\rangle \\ &= \sum_{q}\left\langle\Phi_0\middle|z_{aq}\,\hat{c}_{i,\sigma}^{\dagger}\hat{c}_{j,\sigma'}\,\hat{c}_{q,\sigma}\hat{c}_{b,\sigma'}^{\dagger} - z_{jq}\,\hat{c}_{i,\sigma}^{\dagger}\hat{c}_{q,\sigma'}\,\hat{c}_{a,\sigma}\hat{c}_{b,\sigma'}^{\dagger}\middle|\Phi_0\right\rangle \\ &= z_{ab}\,\delta_{ij}\,\delta_{\sigma,\sigma'} - z_{ji}\,\delta_{ab}\,\delta_{\sigma,\sigma'}.\end{aligned} \tag{17b}$$

Because the dipole operator is a one-particle operator, it can only change the state of the hole or the state of the excited electron but not both at the same time. Furthermore, the dipole operator changes the angular momentum of the orbital by ±1, as we will see when evaluating z_{pq} in (19). Consequently, we find $z_{pp} = 0$ and $\left\langle\Phi_{i,\sigma}^{a,\sigma}\middle|\hat{z}\middle|\Phi_{i,\sigma'}^{a,\sigma'}\right\rangle = 0$.

In a similar way, the matrix elements of $\hat{H}_1$ can be determined by using (7), which read as

$${}_S\left\langle\Phi_i^a\middle|\hat{H}_1\middle|\Phi_j^b\right\rangle_S = 2v_{ajib} - v_{ajbi} + \left\langle\Phi_0\middle|\hat{H}_1\middle|\Phi_0\right\rangle\delta_{ab}\,\delta_{ij}. \tag{18}$$

With the orbital ansatz from (13), we can evaluate z_{pq} and v_{pqrs} and separate the radial integrals from the angular integrals. For the one-particle operator $\hat{z}$, the calculation of the matrix elements z_{pq} involves a 3D integral,

$$
\begin{aligned}
z_{pq} &= \int d^3r\, \varphi_p^\star(\mathbf{r})\, z\, \varphi_q(\mathbf{r}) \qquad (19)\\
&= \int_0^\infty dr\, u^\star_{n_p,l_p}(r)\, r\, u_{n_q,l_q}(r) \int_0^\pi d\theta \sin\theta \int_0^{2\pi} d\phi\, Y^\star_{l_p,m_p}(\theta,\phi)\cos(\theta)\, Y_{l_q,m_q}(\theta,\phi)\\
&= R^{n_p,l_p}_{n_q,l_q} \sqrt{\frac{2l_q+1}{2l_p+1}}\, C^{l_p,m_p}_{l_q,m_q;1,0}\, C^{l_p,0}_{l_q,0;1,0},
\end{aligned}
$$

where $R^{n_p,l_p}_{n_q,l_q}$ stands for the radial integral, which is made more explicit in Section 3.4 when the radial grid is introduced. The angular integral can be analytically evaluated[6] by expressing it in terms of Clebsch–Gordan coefficients, $C^{l_3,m_3}_{l_1,m_1;l_2,m_2} = \langle l_1,m_1;l_2,m_2|l_3,m_3\rangle$ [33]. We also used $z = r\cos\theta$ and $Y_{1,0}(\theta,\phi) = \sqrt{\frac{3}{4\pi}}\cos\theta$. From the property $C^{l_3,-m_3}_{l_1,-m_1;l_2,-m_2} = (-1)^{l_1+l_2-l_3} C^{l_3,m_3}_{l_1,m_1;l_2,m_2}$, we find that an electron with quantum number m behaves the same way as an electron with $-m$ because the second Clebsch–Gordan coefficient in (19), i.e., $C^{l_3,0}_{l_1,0;l_2,0}$, enforces $(-1)^{l_1+l_2-l_3} = 1$. For the dipole operator ($l_2 = 1$), this relation means the angular momentum of the orbital has to change by ± 1. Keep in mind that $|l_1 - l_3| \le l_2$ has to be fulfilled.

Before we can write down an explicit expression for v_{pqrs}, we need to perform a multipole expansion of the electron-electron interaction,

$$
\frac{1}{|\mathbf{r}_1-\mathbf{r}_2|} = \sum_L \frac{r_<^L}{r_>^{L+1}} \frac{4\pi}{2L+1} \sum_{M=-L}^{L} (-1)^M Y_{LM}(\Omega_1) Y_{L-M}(\Omega_2), \qquad (20)
$$

where $\Omega = (\theta,\phi)$ combines both angular coordinates, and $r_> = \max(r_1,r_2)$ and $r_< = \min(r_1,r_2)$. Inserting (20) and (13) in (8) yields

$$
\begin{aligned}
v_{pqrs} &= \sum_{L,M} \frac{4\pi}{2L+1}(-1)^M \iint dr_1\, dr_2\, \frac{r_<^L}{r_>^{L+1}} u^\star_{n_p,l_p}(r_1)\, u_{n_r,l_r}(r_1)\, u^\star_{n_q,l_q}(r_2)\, u_{n_s,l_s}(r_2)\\
&\quad \times \int d\Omega_1\, Y^\star_{l_p m_p}(\Omega_1) Y_{LM}(\Omega_1) Y_{l_r m_r}(\Omega_1) \int d\Omega_2\, Y^\star_{l_q m_q}(\Omega_2) Y_{L-M}(\Omega_2) Y_{l_s m_s}(\Omega_2)\\
&= \sqrt{\frac{(2l_r+1)(2l_s+1)}{(2l_p+1)(2l_q+1)}} (-1)^{m_p-m_r}\, \delta_{m_p+m_q,m_r+m_s}\\
&\quad \times \sum_L R^{[L]}_{pqrs}\, C^{l_p m_p}_{l_r m_r;LM}\, C^{l_p 0}_{l_r 0;L0}\, C^{l_q m_q}_{l_s m_s;L-M}\, C^{l_q 0}_{l_s 0;L0}, \qquad (21)
\end{aligned}
$$

6 See also Section 1.5.6 in Chapter I.

where $R^{[L]}_{pqrs}$ stands for the 2D radial integral. Electrons with large angular momenta $l_a \gg 1$ are normally located far away from the atom because of the centrifugal barrier $\frac{l_a(l_a+1)}{2r^2}$, and bound electrons are located very near the atom. By using this information, we can simplify the terms v_{ajib} and v_{ajbi} in (18) for excited electrons with $l_a \gg 1$. The exchange term v_{ajib} disappears as the overlap between electron φ_a and the hole φ_i vanishes. For the direct term v_{ajbi}, the monopole term $L = 0$ becomes dominant as it decreases most slowly with $r_>$. At the same time, the integral over the bound orbitals can be performed, which is then nothing else than the overlap between the two bound orbitals φ_i and φ_j. Since all orbitals are orthogonal to each other, we get δ_{ij}, and the Coulomb term simplifies to a long-range Coulomb potential, $v_{ajbi} \to \delta_{ij} \langle \varphi_a | 1/\hat{r} | \varphi_b \rangle$.

3.3 Spin-orbit interaction

Spin-orbit interaction is usually a small effect in the nonrelativistic regime. In the outermost p-shell of noble gas atoms, which is the shell that is predominantly affected by a strong-field pulse, spin-orbit coupling leads to energy splittings of up to ~ 1 eV. To have a more accurate description of strong-field processes, one should include spin-orbit effects, particularly since spin-orbit effects lead to multiorbital tunnel ionization in noble gas atoms, see Section 4.1. Spin-orbit interactions couple the spatial degrees of freedom to the spin degree of freedom, which means a simple product between the two degrees of freedom as in (13) is not sufficient anymore. In the nonrelativistic limit, the spin-orbit interaction reads as [5]

$$\hat{H}_{\text{so}} = \frac{\alpha^2}{2} \frac{1}{r} \frac{dV(r)}{dr} \sum_n \hat{\mathbf{l}}_n \cdot \hat{\mathbf{s}}_n, \tag{22}$$

where $\hat{\mathbf{l}}_n$ and $\hat{\mathbf{s}}_n$ are the orbital angular momentum operator and the spin operator of the nth electron. The potential $V(r)$ denotes the mean-field potential plus the nuclear Coulomb potential. For hydrogen, $V(r)$ includes only the nuclear Coulomb potential and one finds $\hat{H}_{\text{so}} \propto r^{-3}$. Spin-orbit interaction does not change the magnitude of the orbital angular momentum l and of the spin $s_i = \frac{1}{2}$ of the electron. Only the spin and orbital momentum projections σ and m_l get mixed with each other. Note that (22) is an approximative extension (from a one-electron atom) to a many-electron atom neglecting explicit two-body spin-orbit terms.

For many-electron systems, the mean-field potential (including intrachannel corrections) scales as $-Z/r$ for small radii and as $-1/r$ for large radii. In both cases, the radial dependence of the spin-orbit interaction goes as $r^{-1}\,\partial_r V(r) \propto r^{-3}$. Applying this result to Rydberg and continuum states, which are quite delocalized and on average far away from the nucleus, we find that the strength of the spin-orbit interaction is strongly reduced in comparison to occupied orbitals and, therefore, can be neglected.

As a result, the virtual orbitals φ_a can be directly taken from the nonrelativistic HF calculations.

After rewriting the spin-orbit operator, $2\hat{\mathbf{l}} \cdot \hat{\mathbf{s}} = \hat{\mathbf{j}}^2 - \hat{\mathbf{l}}^2 - \hat{\mathbf{s}}^2$, in terms of the total angular momentum $\hat{\mathbf{j}}$, it becomes clear that the new one-particle orbitals are eigenstates of the operators $\hat{\mathbf{j}}^2, \hat{\mathbf{l}}^2, \hat{\mathbf{s}}^2$, and $\hat{j}_z$, which constitute the LS-coupled basis. The new spin-orbit–coupled occupied orbitals φ_i^{SO}, expressed in terms of the uncoupled orbital basis, read [18, 23]

$$|\varphi_i^{\text{SO}}\rangle := |\varphi_{n_i,l_i,j_i,m_i^J}\rangle = \begin{pmatrix} C^{j_i,m_i^J}_{l_i,m_i^J-\frac{1}{2};s_i,+\frac{1}{2}} & |\varphi_{n_i,l_i,m_i^J-\frac{1}{2}}\rangle \\ C^{j_i,m_i^J}_{l_i,m_i^J+\frac{1}{2};s_i,-\frac{1}{2}} & |\varphi_{n_i,l_i,m_i^J+\frac{1}{2}}\rangle, \end{pmatrix} \tag{23}$$

where the upper entry is the spin-up component, and the lower entry is the spin-down component. To arrive at this results, we made a few assumptions. We ignored spin-orbit interactions in the Hamiltonian when performing the HF procedure to find the uncoupled orbitals φ_i, which are degenerate in σ_i and m_i^L. Only after we have determined φ_i, we apply the spin-orbit interaction using degenerate-state perturbation theory to capture the spin-orbit coupling. As a consequence, we ignore the influence on the radial part of the orbital wavefunction, which generally depends also on j_i on top of n_i and l_i. As long as the spin-orbit effect is small, the radial components for the two possible j_i momenta are similar.

Finally, we only need to find the orbital energies of the new spin-orbit–coupled orbitals. Instead of determining them with perturbation theory, it is convenient to set the orbital energies to the experimental values. This has the advantage that we have the exact ionization potential, which is important for accurate tunnel ionization dynamics.

The spin symmetry that has been used for deriving the spin-singlet CIS configurations (14) cannot be used anymore, since the spin is coupled to the orbital angular momentum. But this does not mean that there is no symmetry left. The symmetry exists now between the m^J and $-m^J$ orbitals and reads as

$$|\Phi_i^a\rangle_{g/u} = \frac{1}{\sqrt{2}}\left(|\Phi_{+i}^{+a}\rangle \pm (-1)^{l_i+s_i-j_i} |\Phi_{-i}^{-a}\rangle \right), \tag{24}$$

where $\pm i$ refers to the orbital with $m_i^J \gtrless 0$, and $\pm a$ refers to the corresponding virtual orbital fulfilling $m_a^L = m_i^J - \sigma_a$. For linearly polarized light, only the *gerade* (g) CIS configuration gets populated. Depending on whether the total angular momentum, j_i, is larger or smaller than l_i, a sign jump occurs between the two equivalent configurations $|\Phi_{+i}^{+a}\rangle$ and $|\Phi_{-i}^{-a}\rangle$.

3.4 Grid representation

The specific representation of the radial degree of freedom has not been chosen so far. In the strong-field regime, the electron can travel far away from the atom and may even return. At the same time, we have very localized bound electrons. A nonuniform radial grid is, therefore, desirable with many grid points around the nucleus and fewer grid points far away from it[7]. By introducing the mapping,

$$x \mapsto r(x) = \frac{R_{\max}\zeta}{2}\frac{1+x}{1-x+\zeta}, \qquad \text{with } \zeta \neq 0, \tag{25}$$

we map the original grid, $r \in [0, R_{\max}]$, onto $x \in [-1, 1]$. We can now choose a uniform grid in x and obtain a nonuniform grid in r. The mapping parameter ζ controls the grid-point density near the origin as compared to further outside.

The question which basis or representation we choose for the radial degree of freedom is not answered with this mapping. We only transferred the question from r to x. So let us do that now. We use a pseudospectral grid representation based on Gauss–Lobatto quadrature [3, 9]. For Gauss–Lobatto, the interval is $x \in [-1, 1]$. The grid points are given by the two end points $x_0 = -1, x_{N_g} = 1$ and the roots of the first derivative of the N_g-th–order Legendre polynomial obeying $P'_{N_g}(x_k) = 0$. The number of grid points is $N_g + 1$. Now, we approximate any function $f(x)$ on this interval in terms of Legendre polynomials $P_k(x)$,

$$f(x) \approx f_{N_g}(x) = \sum_{k=0}^{N_g} a_k P_k(x), \tag{26}$$

which may also be written in terms of cardinal functions, $g_k(x)$,

$$f_{N_g}(x) = \sum_{k=0}^{N_g} f(x_k) g_k(x), \tag{27}$$

where we require that the approximation yields the exact value of the function $f(x_k)$ at the grid points x_k. The cardinal functions have the form [31],

$$g_k(x) = -\frac{1}{N_g(N_g+1)P_{N_g}(x_k)} \frac{(1-x^2)P'_{N_g}(x)}{x - x_k}, \tag{28}$$

and satisfy the unique property $g_k(x_l) = \delta_{kl}$.

7 TDCIS implementations based on Gaussian basis sets have been used as well for strong-field processes [12, 27].

The second derivative of $g_k(x)$ at $x = x_l$, which is needed to calculate the radial kinetic energy, can be written as $g_k''(x_l) = d^{(2)}_{lk} \frac{P_{N_g}(x_l)}{P_{N_g}(x_k)}$ with

$$d^{(2)}_{ij} = \begin{cases} \dfrac{N_g(N_g+1)(N_g(N_g+1)-2)}{24} & \text{for } i = j \text{ and } i,j \in \{0, N_g\}, \\ \dfrac{N_g(N_g+1)-2}{4} & \text{for } i \neq j \text{ and } i,j \in \{0, N_g\}, \\ -\dfrac{N_g(N_g+1)}{3(1-x_i^2)} & \text{for } i = j \text{ and } i,j \notin \{0, N_g\}, \\ -\dfrac{2}{(x_i - x_j)^2} & \text{for } i \neq j \text{ and } i,j \notin \{0, N_g\}. \end{cases} \tag{29a}$$

With the spherical orbital ansatz in (13), we know that $u_{n,l}(r = 0) = 0$ for all n and l. That means we can ignore the grid point $x_0 = -1$. Also the last point, $x_{N_g} = +1$, is ignored because we enforce that the wavefunction has to vanish at this point, $u_{n,l}(R_{\max}) = 0$. In other words, the electron is bound between $r = 0$ and $r = R_{\max}$. When the electron reaches $r = R_{\max}$, it will be reflected. This reflection is of course artificial, and in Section 3.6, we describe how we treat this problem.

Now, we can give an explicit expression for the radial integrals in Section 3.2, which turns into a sum over the grid points x_k:

$$\int_0^{R_{\max}} dr\, f(r) = \int_{-1}^{1} dx\, r'(x) f[r(x)] \approx \sum_{k=0}^{N_g} \frac{2\, r'(x_k)}{N_g(N_g+1) P^2_{N_g}(x_k)} f[r(x_k)]. \tag{30}$$

It is convenient to absorb the factor $\frac{r'(x_k)}{P^2_{N_g}(x_k)}$ in the orbitals by introducing

$$A^k_{n,l} = \frac{\sqrt{r'(x_k)}}{P_{N_g}(x_k)} u_{n,l}(x_k), \tag{31}$$

where the derivative $r'(x)$ stays always positive so that the square root is always well defined. Thus, the radial integral involving $u_{n,l}(r)$ becomes

$$\int_0^{R_{\max}} dr\, u^*_{n_p,l_p}(r) f(r)\, u_{n_q,l_q}(r) \approx \frac{2}{N_g(N_g+1)} \sum_k [A^k_{n_p,l_p}]^* f[r(x_k)]\, A^k_{n_q,l_q}. \tag{32}$$

3.5 Hartree–Fock

HF is the very first step in most many-body theories [1, 30]. HF defines not just the reference state for CIS but also the (uncoupled) orbitals φ_p in which each electron is represented.

HF is based on finding an N-electron wavefunction Φ_0, which can be written in terms of a single Slater determinant and has an energy closest to the exact ground-state energy E. To find this HF ground-state Φ_0, we use the Ritz variational principle,

$$\delta \langle \Phi_0 | \hat{H} - \lambda \mathbb{1} | \Phi_0 \rangle = 0, \tag{33}$$

where $\hat{H}$ is the exact field-free many-body Hamiltonian. We impose that the norm of Φ_0 is constant by using a Lagrange multiplier λ. The variational principle leads to N independent equations

$$\begin{aligned} \varepsilon_i \varphi_{i,\sigma_i}(\mathbf{r}) = & \left(\frac{\hat{\mathbf{p}}^2}{2} - \frac{Z}{r} + \int d^3r' \sum_{j,\sigma_j} \frac{|\varphi_{j,\sigma_j}(\mathbf{r}')|^2}{|\mathbf{r}-\mathbf{r}'|} \right) \varphi_{i,\sigma_i}(\mathbf{r}) \\ & - \int d^3r' \sum_j \frac{\varphi^*_{j,\sigma_i}(\mathbf{r}')\varphi_{j,\sigma_i}(\mathbf{r})}{|\mathbf{r}-\mathbf{r}'|} \varphi_{i,\sigma_i}(\mathbf{r}'). \end{aligned} \tag{34}$$

These equations have to be solved self-consistently because the last two terms, which involve electron-electron interactions, depend on the orbital solutions themselves. The exchange term, i.e., the last term in (34), acts as a nonlocal potential because the influence of the exchange potential on $\varphi_{i,\sigma_i}(\mathbf{r})$ depends on $\varphi_{i,\sigma_i}(\mathbf{r}')$. The exchange term, however, is important because it ensures that the electron does not see its own Coulomb potential because the Coulomb term (third term on the right-hand side of (34)) sums over all N electrons.[8]

For closed-shell HF, (34) reduces to a set of equations involving only the spatial orbitals φ_i, as we can exploit that each spatial orbital is doubly occupied. Furthermore, we use the spherical orbital representation (13) to reduce (34) to an equation depending only on the radial components $u_{n,l}(r)$. After expressing $u_{n,l}(r)$ in the pseudospectral-grid representation (31), we obtain the final HF questions as we would solve them on a computer [9],

$$\begin{aligned} \varepsilon_i A^k_{n_i,l_i} = & -\frac{1}{2} \sum_{k'} \frac{1}{r'(x_k)} d^{(2)}_{k,k'} \frac{1}{r'(x_{k'})} A^{k'}_{n_i,l_i} + \left(\frac{l_i(l_i+1)}{2r^2(x_k)} - \frac{Z}{r(x_k)} \right) A^k_{n_i,l_i} \\ & + \frac{2}{N_g(N_g+1)} \sum_j (4l_j+2) \sum_{k'} \frac{|A^{k'}_{n_j,l_j}|^2}{r_>(k,k')} A^k_{n_i,l_i} \\ & - \frac{2}{N_g(N_g+1)} \sum_{j,L} \left[C^{l_j0}_{l_i0,L0} \right]^2 \sum_{k'} \frac{r^L_<(k,k')}{r^{L+1}_>(k,k')} [A^{k'}_{n_j,l_j}]^* A^k_{n_j,l_j} A^{k'}_{n_i,l_i}, \end{aligned} \tag{35}$$

8 Compare to Section 2.2.2 in Chapter IV.

where $r^L_<(k,k') = \min[r(x_k), r(x_{k'})]$ and $r^L_>(k,k') = \max[r(x_k), r(x_{k'})]$. The HF equation (35) can be written as a matrix eigenvalue problem, $\varepsilon_i A_{n_i,l_i} = H_0 A_{n_i,l_i}$, where $A_{n_i,l_i} = (A^0_{n_i,l_i}, \cdots, A^{N_g}_{n_i,l_i})^T$ is the eigenvector and H_0 is the Fock operator expressed in the grid representation as a matrix.

Once all occupied orbitals φ_i are determined, $\hat{H}_0$ can be fully diagonalized. The $N/2$ orbitals with the lowest orbital energies are the occupied orbitals, φ_i. The rest are the virtual orbitals. Note that (35) does not depend on m. As a result, all $2l_i + 1$ orbitals with quantum number n_i and l_i are energetically degenerate. After fully diagonalizing $\hat{H}_0$, also the virtual orbitals φ_a are found, and $\hat{H}_0$ can be rewritten in terms of the eigenstates, which reads, using creation and annihilation operators, $\hat{H}_0 = \sum_{p,\sigma_p} \varepsilon_p\, \hat{c}^\dagger_{p,\sigma_p} \hat{c}_{p,\sigma_p}$.

3.6 Complex absorbing potential

For ionization scenarios, the emitted electron separates from the ion often with great speed. Before the pulse is over, the electron may have traveled several hundreds or thousands of Bohr radii. Such large grids are computationally not feasible, and smaller grids are chosen. However, in order to avoid artificial reflections from the grid boundary, a CAP is introduced [22], which absorbs the outgoing electron just before the grid ends.[9] By putting the absorbing potential at the end of the grid, the absorbed electron is far away from the atom such that bound and low excited states are not influenced by the CAP. Hence, the absorbing potential does not influence the physics near the atom/ion.

Unfortunately, introducing a CAP results in a non-Hermitian Hamiltonian, $\hat{H}_0 \to \hat{H}_0 - i\eta\hat{W}$, where η is the CAP strength and $\hat{W}$ is the CAP operator. One example of a CAP realization is [9]

$$W(r) = (r - r_c)^2\, \Theta(r - r_c), \tag{36}$$

where $\Theta(x)$ is the Heaviside function. It absorbs the electron only after it has reached r_c. Having a non-Hermitian $\hat{H}_0$ also means that the orbital energies ε_p become complex. For a non-Hermitian Hamiltonian, the norm of the wavefunction is not conserved. The norm loss has also a physical meaning. The electrons that are absorbed would be normally outside of the box. The loss in norm is, therefore, a direct measure of the amount of ionization.

Another, more technical consequence of the non-hermiticity of $\hat{H}_0$ is that the left eigenstates in $(\varphi_p|\hat{H}_0 = (\varphi_p|\varepsilon_p$ are not anymore the Hermitian conjugate of the right eigenstates in $\hat{H}_0|\varphi_p) = \varepsilon_p|\varphi_p)$. Since $\langle\varphi_p|$ refers to the Hermitian conjugate of

9 Such potentials were also used in Chapters II and V.

$|\varphi_p\rangle$, the left eigenstates of $\hat{H}_0$ are written as $(\varphi_p|$. For consistency, we write the right eigenstates $|\varphi_p) = |\varphi_p\rangle$. In fact, $\hat{H}_0$ is complex symmetric and the left eigenstate $(\varphi_p|$ in position space reads as

$$(\varphi_p|\mathbf{r}) = \frac{u_{n_p,l_p}(r)}{r} Y^{\star}_{l_p,m_p}(\theta,\phi), \tag{37}$$

where the radial part is the same as for the right eigenstate and only the angular part is complex conjugated. The reason for this mix is that the CAP only influences the radial coordinate but not the angular ones. The inner (or scalar) product for the orbitals and the CIS configurations read now

$$(\varphi_p|\varphi_q) = \delta_{pq}, \tag{38a}$$

$$_S\left(\Phi_i^a|\Phi_j^b\right)_S = \delta_{ab}\,\delta_{ij}, \tag{38b}$$

which we refer to as the complex inner product opposed to the Hermitian inner product. Since the CAP does not affect the occupied orbitals φ_i, the orthogonality relation between different φ_i survives in the Hermitian inner product, which for these orbitals coincides with the complex symmetric inner product, because the radial part of φ_i is a purely real function. The HF procedure for the occupied orbitals according (35) is not affected by the CAP, because the CAP is placed far away from the nucleus. Only when the virtual orbitals are calculated, the CAP has to be included in the Fock operator, $\hat{H}_0 \to \hat{H}_0 - i\eta\hat{W}$.

To arrive at the equations of motion (11), the TDSE has to be projected on the left eigenstates using the complex symmetric inner product. Therefore, the matrix elements in (11) have to be slightly modified. For example, the new dipole matrix elements are

$$z_{(pq)} = (\varphi_p|\hat{z}|\varphi_q) = \int dr\, u_{n_p,l_p}(r)\, r\, u_{n_q,l_q}(r) \int d\Omega\, Y^{\star}_{l_p,m_p}(\Omega)\cos\theta\, Y_{l_q,m_q}(\Omega). \tag{39}$$

Note that we have used parentheses in the subscript to indicate the complex symmetric inner product. The new Coulomb matrix elements $v_{(pqrs)}$ can be expressed in a similar fashion.

3.7 Expectation values

All physical and measurable quantities can be expressed through Hermitian operators $\hat{A}$. Hermitian operators have real eigenvalues, and, as a result, all expectation values $\left\langle \hat{A} \right\rangle$ are real as well. The calculation of an expectation value for an N-body wavefunction is a bit more involved than for a one-electron wavefunction. By expressing the wavefunction with creation and annihilation operators as in (3) and using the anticommutator relations in (16), we get the expression

$$\begin{aligned}\left\langle \hat{A} \right\rangle(t) &= \langle \Psi(t)|\hat{A}|\Psi(t)\rangle = |\alpha_0(t)|^2 \langle \Phi_0|\hat{A}|\Phi_0\rangle \\ &+ \sum_{ai} \left[\alpha_0^*(t)\alpha_i^a(t)\langle \Phi_0|\hat{A}|\Phi_i^a\rangle_S + c.c.\right] + \sum_{ai,bj} [\alpha_i^a(t)]^* \alpha_j^b(t)\,{}_S\langle \Phi_i^a|\hat{A}|\Phi_j^b\rangle_S \\ &= 2|\alpha_0(t)|^2 \sum_i A_{ii} + 2\sqrt{2}\sum_{ai} \mathrm{Re}\left[\alpha_0^*(t)\alpha_i^a(t)A_{ia}\right] \\ &+ \sum_{ai,bj} [\alpha_i^a(t)]^* \alpha_j^b(t)(A_{ab}\delta_{ij} - A_{ji}\delta_{ab} + 2\delta_{ab}\delta_{ij}\sum_{i'} A_{i'i'})\end{aligned} \tag{40}$$

for the expectation value of a one-particle operator ($c.c.$ stands for the complex conjugate). In the case of no CAP, the norm of $\Psi(t)$ is conserved, i.e., $|\alpha_0(t)|^2 + \sum_{ai} |\alpha_i^a(t)|^2 = 1$, and (40) can be further simplified.

One very important expectation value in strong-field physics is the dipole moment, $\hat{A} = \hat{\mathbf{d}} = q_e \hat{\mathbf{r}}$, where $q_e = -1$ is the charge of the electron. From electrodynamics [10], we know that the dipole acceleration $\frac{d^2}{dt^2}\langle \mathbf{d}\rangle(t)$ is directly related to the radiation spectrum [20]

$$S(\omega) = \frac{1}{20}\frac{1}{3\pi c^3}\left|\int_{-\infty}^{\infty} dt \left[\frac{d^2}{dt^2}\langle z\rangle(t)\right] e^{-i\omega t}\right|^2 , \tag{41}$$

which the atoms emits when driven by a strong-field pulse polarized along the z-axis. As we will see in Section 4.2, the radiation spectrum can contain frequencies that are hundred times higher than the driving frequency of the strong-field pulse. This process is known as HHG and addressed in several chapters of this book.

3.8 Ion density matrix

In strong-field and attosecond physics, the ionization dynamics and particularly its subcycle dynamics are of high interest, as it contains information on how electrons move within a laser pulse on femtosecond or even attosecond time scales. Particularly interesting is the question how the ion or the hole is formed during ionization.

To gain access to this quantity, it is desirable to partition the full N-electron system into two subsystems: the excited electron and the parent ion containing the remaining $N - 1$ electrons. To calculate the reduced density matrix of a subsystem, the trace of the full N-electron density matrix $\hat{\rho}(t) = |\Psi(t)\rangle\langle\Psi(t)|$ has to be performed over all unobserved degrees of freedom [2]. In our case, the unobserved degrees of freedom are those of the excited electron. The reduced density matrix of the ionic subsystem reads as

$$\hat{\rho}^{\mathrm{IDM}}(t) = \mathrm{Tr}_a\left[\hat{\rho}(t)\right], \tag{42a}$$

$$\rho^{\mathrm{IDM}}_{i,j}(t) = \sum_a \langle \Phi_i^a | \Psi(t)\rangle \langle \Psi(t) | \Phi_j^a \rangle, \tag{42b}$$

which is also called the IDM. The IDM uniquely characterizes the state of the ion. Since more than one occupied orbital (channel) can contribute to ionization, it is possible to create a superposition of ionic eigenstates. This results in nonzero off-diagonal elements in the IDM. Therefore, the IDM is an ideal quantity to study coherences in the ionic subsystem. After the atom has been ionized, the parent ion is normally not in a fully coherent state [19]. As we will show in Section 4.1, transient absorption spectroscopy can probe the IDM including its off-diagonal elements, making them experimentally accessible.

The introduction of a CAP as explained in Section 3.6 results in a loss of norm for the N-electron wavefunction and also for the IDM. The probability of emitting an electron from orbital i, which is equivalent to the hole population $\rho^{\mathrm{IDM}}_{ii}(t)$, should not be affected by the absorption of the photoelectron by the CAP. Therefore, we need to introduce an IDM correction [9],

$$\rho^{\mathrm{IDM}}_{i,j}(t) \to \rho^{\mathrm{IDM}}_{i,j}(t) + 2\eta\, e^{i(\varepsilon_i-\varepsilon_j)t} \int_{-\infty}^{t} dt' \sum_a \langle \Phi_i^a | \hat{\rho}(t') \hat{W} | \Phi_j^a \rangle\, e^{-i(\varepsilon_i-\varepsilon_j)t'}, \tag{43}$$

which ensures that all IDM entries—also the off-diagonal ones—are not affected by the CAP.

4 Strong-field applications of TDCIS

After discussing the basics of TDCIS and its specific implementation for closed-shell atoms, it is time to discuss applications of TDCIS in the strong-field regime. The subcycle dynamics of the multiorbital ionization in krypton and its characterization via transient absorption spectroscopy is presented in Section 4.1. In Section 4.2, we investigate the influence of multiorbital and collective-excitation effects in xenon as it is driven by a strong-field pulse to generate high-harmonics with photon energies up to 170 eV.

4.1 Subcycle ionization dynamics and coherent hole motion

Pulses in the nonrelativistic strong-field regime have intensities in the range 10^{13}–10^{15} W/cm^2, wavelengths in the near-infrared (NIR) ($\lambda \simeq 1\,\mu$m, i.e., $\omega \simeq 1$ eV), and pulse durations of a few to tens of femtosecond [4]. The main mechanism through which strong-field pulses ionize the system is tunnel ionization (also known as strong-field ionization). The light-matter interaction is pictured, using the dipole

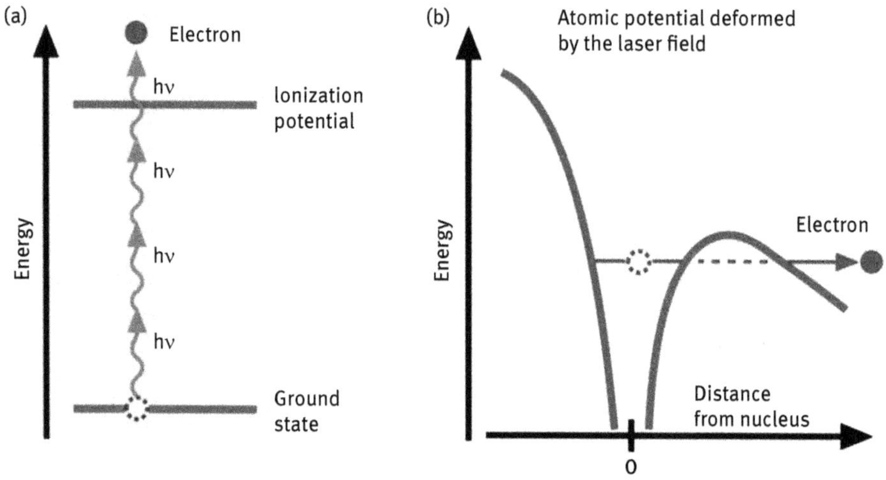

Fig. 1. The perturbative (a) and the nonperturbative (b) multiphoton regimes are illustrated. This figure is taken from http://www.desy.de. Copyright © 2007 DESY.

length form, as a local potential that strongly distorts the Coulomb potential, and already after a few Bohr radii, the field-induced potential starts to dominate the motion of the electron. This distortion creates a potential barrier, which can be overcome by the electron by tunneling through this barrier and, consequently, out of the system, see Figure 1(b). Once the electron has tunneled to the outer side of the barrier, the Coulomb potential becomes negligible and the dynamics are governed by the laser pulse. This picture is, normally, applicable when the light-matter interaction creates a potential that is comparable with the atomic Coulomb potential, and the ionization potential, I_p, of the system is much larger than the laser frequency ω.

If only one or a few photons are needed to ionize the system, the ionization can be described with perturbation theory. Based on the kinetic energy of the photoelectron, one can determine how many photons have been absorbed by the photoelectron, see Figure 1(a). This is the perturbative multiphoton regime. Tunnel ionization, on the other hand, is a nonperturbative multiphoton process where the number of photons absorbed is ill-defined.

Tunnel ionization is one of the most fundamental processes in strong-field physics. Even though it has been intensively studied in the last 50 years [7], the dynamics and especially the subcycle behavior of ionization have only recently been studied with the emergence of attosecond physics [13]. Before attosecond physics, only the final state of the system after the pulse could be studied. Attosecond pulses and the controlled delay of two pulses relative to each other, with an accuracy of a few attoseconds, have made it possible to resolve the electronic dynamics during a strong-field pulse. Using a second pulse to probe the dynamics triggered by a first pulse is known as *pump-probe* spectroscopy and is the method of choice in attosecond physics to probe dynamics. There are many variants of pump-probe studies [13, 16] but the basic idea is always the same.

By applying a strong-field pump pulse to tunnel ionized atomic krypton, it has been theoretically predicted [23] and experimentally shown [8] that within the $4p$ shell, a (partially) coherent hole wave packet can be launched, which is driven by spin-orbit interaction. For xenon, however, theory predicted that the coherence within the ion is very low even though the hole populations are quite similar to krypton. This shows that the degree of coherence cannot be determined from the hole populations and must be calculated or measured separately. Theoretically, the full IDM, including the off-diagonal elements, is needed to exactly know the ionic state after ionization.

With an attosecond probe pulse in the extreme-ultraviolet (XUV) regime, the $4p_j$ hole in krypton can be probed by promoting an electron from the $3d_j$ shells into the $4p_j$ shells. Promoting an electron from $3d$ into $4p$ requires the absorption of one photon. Measuring the spectrum of the transmitted probe pulse and comparing it to its original spectrum tells us how probable it was to excite a $3d_j$ electron into the $4p_j$ shell. For a one-photon dipole transition, there are three possibilities how a $3d_j$ electron can be excited into the $4p$ shell: $3d_{5/2} \to 4p_{3/2}$, $3d_{3/2} \to 4p_{3/2}$, and $3d_{3/2} \to 4p_{1/2}$. Each of them requires a slightly different energy, which makes them distinguishable in the absorption spectrum, see Figure 2.

The transition probability of each transition is directly proportional to the hole population in the $4p_j$ orbitals because, due to Pauli blocking, a transition can only occur when an electron has been removed from $4p_j$. The strength of the absorption signal is, therefore, a direct measure of the hole population. The coherence between $[4p_{3/2}^{1/2}]^{-1}$ and $[4p_{1/2}^{1/2}]^{-1}$ can be measured because both ionic states can be excited by

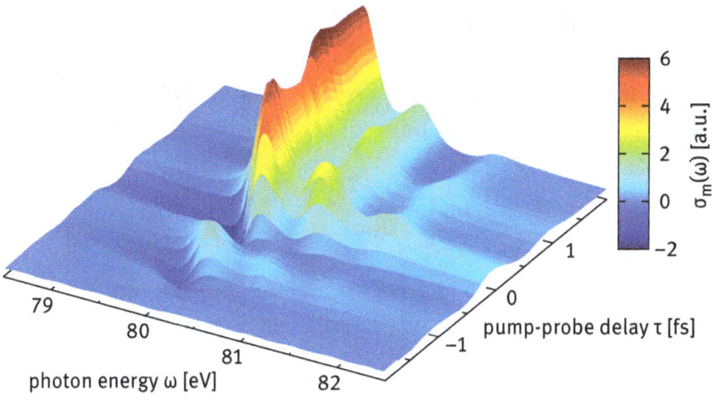

Fig. 2. An attosecond transient absorption spectrum of krypton during tunnel ionization, calculated using TDCIS, is shown as a function of the probe photon energy ω and the pump-probe delay τ. The figure is taken from [21]. Copyright © 2012 American Physical Society (APS).

the XUV probe pulse to $[3d_{3/2}^{1/2}]^{-1}$. This creates interferences that lead to oscillations in the transient absorption spectrum as a function of the pump-probe delay [25]. The term $[4p_j^m]^{-1}$ refers to the ionic state that has a hole (with respect to the neutral ground state) in the $4p_j$ orbital with $m^J = m$.

This type of probing the system by measuring the transient absorption signal is known as *ATAS*. The attosecond character comes in because the probe pulse is an attosecond pulse providing a broad bandwidth. The delay between the pump and probe pulses can be controlled within a few attoseconds allowing one to resolve subcycle dynamics. Note that an optical ($\lambda \simeq 1\ \mu$m) cycle has a period of around 3 fs.

Using this probing scheme during the strong-field ionization, we can obtain information about the subcycle ionization dynamics [32]. The described method, where a direct connection between the transition strength and the population is established, is only valid as long as pump and probe pulses do not temporally overlap. For overlapping pulses, as it is the case when we want to probe strong-field ionization dynamics taking place during the pump pulse, this direct connection is not true anymore because the emitted electron and the strong-field pulse influence the ion state during the probe step. Therefore, it is not clear to which extent the instantaneous IDM can be probed during the ionization process [14].

With TDCIS, we can study the subcycle dynamics of these multiorbital ionization processes including the probe step, which involves the deeply bound $3d$ shell. Before we start discussing the transient absorption spectrum and what information it contains, we need to understand the basic mechanism behind transient absorption spectroscopy and why only for nonoverlapping pulses, the exact IDM can be probed.

The modification of the number of transmitted photons can be described in a semiclassical picture, where the electric field is not quantized but treated as a classical field. In classical electrodynamics, the Larmor formula describes the generation of radiation due to the acceleration of charged particles [10]. Quantum mechanically, the electron motion is captured in the dipole moment $\langle \hat{d} \rangle(t)$ with $\hat{d} = -\hat{z}$, see Section 3.7. The generated and the absorbed radiation of an ion is, therefore, determined by the dynamics of the ionic dipole moment $\langle \hat{d} \rangle_{\text{ion}}(t)$. The probe field-independent photoabsorption cross section $\sigma(\omega)$ reads in terms of the ionic dipole moment

$$\sigma(\omega) = 4\pi\alpha\,\omega\,\text{Im}\left[\frac{\langle \hat{d} \rangle_{\text{ion}}(\omega)}{E(\omega)}\right], \tag{44}$$

where ω is the photon energy and $E(\omega)$ is the spectrum of the incident probe electric field. The ionic dipole is only sensitive to the off-diagonal elements of $\rho^{\text{IDM}}(t)$,

$$\langle \hat{d} \rangle_{\text{ion}}(t) = \text{Tr}[\hat{d}\hat{\rho}^{\text{IDM}}(t)] = \sum_{ij} z_{ij}\rho_{ij}^{\text{IDM}}(t). \tag{45}$$

By focusing on the three $3d^{-1} \to 4p^{-1}$ transitions, only the matrix elements $\rho^{\mathrm{IDM}}_{4p^m_j, 3d^m_{j'}}(t)$ with $m = m^J = 1/2$ are probed[10]. Knowing that only the XUV probe pulse $\mathcal{E}(t) = \mathcal{E}_0\delta(t-\tau)$ at a pump-probe delay τ is able to access the $3d$ shell via resonant photoexcitation, which can be treated perturbatively, the off-diagonal elements $\rho^{\mathrm{IDM}}_{4p^m_j, 3d^m_{j'}}(t)$ can be expressed in terms of the ionic state at time τ, $\hat{\rho}^{\mathrm{IDM}}(t > \tau) \propto -i\mathcal{E}_0 e^{-i\hat{H}(t-\tau)}[\hat{z}, \hat{\rho}^{\mathrm{IDM}}(\tau)]e^{i\hat{H}(t-\tau)}$ where $\rho^{\mathrm{IDM}}_{4p^m_j, 3d^m_{j'}}(\tau) = \rho^{\mathrm{IDM}}_{3d^m_j, 3d^m_{j'}}(\tau) = 0$. Using this information, ignoring the counter-rotating terms, and assuming the ion evolves field-free after the probe step, the ionic dipole moment reads as [25]

$$\begin{aligned}\langle\hat{d}\rangle_{\mathrm{ion}}(t) = &-2\mathcal{E}_0 \sum_{j,j',m} z^2_{3d^m_j, 4p^m_{j'}}\, \rho^{\mathrm{IDM}}_{4p^m_{j'}, 4p^m_{j'}}(\tau) \sin[(\varepsilon_{4p_{j'}} - \varepsilon_{3d_j})(t-\tau)] \\ &-2\mathcal{E}_0 \sum_{j,j',m} \sum_{j'' \neq j'} z_{3d^m_j, 4p^m_{j'}} z_{3d^m_j, 4p^m_{j''}}\, \mathrm{Im}\left[\rho^{\mathrm{IDM}}_{4p^m_{j'}, 4p^m_{j''}}(\tau)\, e^{i(\varepsilon_{4p_{j'}} - \varepsilon_{3d_j})(t-\tau)}\right],\end{aligned} \tag{46}$$

where j, j', j'' stand for the total angular momenta. The dipole transitions introduce the restrictions $|j - j'|, |j'' - j| \leq 1$. We see now that each transition $[3d^m_j]^{-1} \to [4p^m_{j'}]^{-1}$ probes different entries of $\hat{\rho}^{\mathrm{IDM}}(\tau)$. Particularly interesting are the two possible transitions to $[3d^{1/2}_{3/2}]^{-1}$, which are sensitive to the off-diagonal element $\rho^{\mathrm{IDM}}_{4p^m_{1/2}, 4p^m_{3/2}}(\tau)$ with $m = 1/2$. This is exactly what we need to probe the hole coherence within the $4p$ shell. Equation (46) assumes field-free propagation of the ion after the probe pulse. This makes the equation only exact for nonoverlapping pump-probe pulses.

In the case of absorption, the ionic dipole $\langle\hat{d}\rangle_{\mathrm{ion}}(t) \propto \cos(\omega_0 t + \pi/2) = -\sin(\omega_0 t)$ oscillates $\pi/2$ out of phase with respect to $E(t) \propto \cos(\omega_0 t)$, and the cross section is positive. If the dipole oscillates with a phase shift of $-\pi/2$, the probe electric field gets enhanced, leading to an emitting behavior, and the cross section becomes negative. The energetically lowest absorption line $[4p^m_{3/2}]^{-1} \to [3d^m_{5/2}]^{-1}$ in Figure 2 shows a purely absorbing (Lorentzian) behavior when pump and probe pulses do not overlap[11]. The widths of the transition lines are determined by the lifetime of the $3d^{-1}$ ionic states and by the energy resolution of the XUV detector.

When the phase shift ϕ in the oscillating ionic dipole $[\langle\hat{d}\rangle(t) \propto \cos(\omega_0 t + \phi)]$ is not $\pm\pi/2$, the transition line shapes are not Lorentzian anymore, see Figure 3(a) for $\tau = 0$. Figure 3(a) shows cuts of the transient absorption spectrum of Figure 2 for the pump-probe delays $\tau = 0, 2.4$ fs. At $\tau = 0$ fs, the magnitude of the NIR field has a maximum, and all three transition lines are strongly deformed.

From the calculated ATAS spectrum in Figure 2, the dipole strength and the dipole phase for each pump-probe delay can be extracted. The strength of the

10 The $m^J = -1/2$ component behaves identically as the $m^J = 1/2$ component, see Section 3.3.

11 The other two absorption lines do not show purely absorbing behavior due to coherences between the ionic states $[4p^{\pm 1/2}_{1/2}]^{-1}$ and $[4p^{\pm 1/2}_{3/2}]^{-1}$.

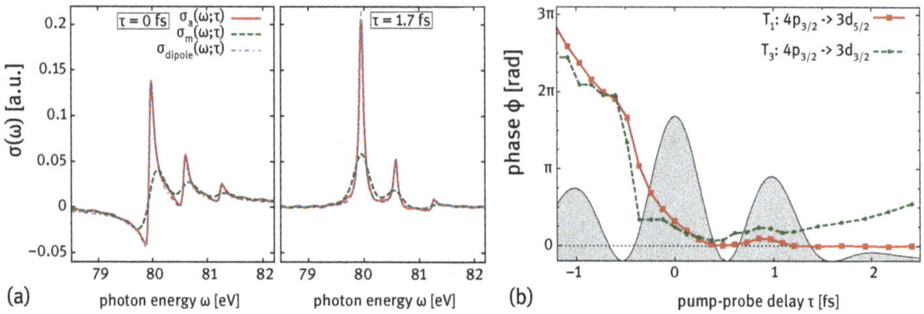

Fig. 3. (a) The calculated atomic cross sections (red solid line), the measured cross sections (green dashed line), and the fit obtained from a simplified oscillating dipole model (yellow dotted line) are shown for the pump-probe delays $\tau = 0$ fs and $\tau = 2.4$ fs, respectively. The measured cross section is taken from [32]. (b) Ionic dipole phase $\phi(\tau)$ obtained from the fits for the transition lines $4p^{-1}_{3/2} \to 3d^{-1}_{5/2}$ (red solid line) and $4p^{-1}_{3/2} \to 3d^{-1}_{3/2}$ (green dashed line) is shown. The figures are taken from [21]. Copyright © 2012 American Physical Society (APS).

dipole oscillations is a direct measure of the ionic hole populations in the case of nonoverlapping pulses, see (46). The comparison between the hole populations extracted from the calculated transient absorption spectrum and the corresponding instantaneous hole populations revealed that in the case of overlapping pulses, the obtained hole populations match quite well the instantaneous ones but are not exactly the same. A delay in the extracted hole motion of up to 200 as was found [21].

Besides the question of population dynamics, a new phenomenon has been identified during the ionization process, which we want to focus on here. The transition lines do not just rise in strength as the hole populations do. They also show strong deformations in their shape, see Figure 2, signaling a rapid change in the ionic dipole phase $\phi(\tau)$.

The ionic phase shifts are shown in Figure 3(b) for the energetically lowest and highest transition lines. They are obtained from the TDCIS results by fitting (44) with $\langle \hat{d} \rangle_{\text{ion}}(t) = z_0 \sin(\omega t + \phi')$ for each pump-probe delay.[12] The increasing phase shift in the $4p^{-1}_{3/2} \to 3d^{-1}_{3/2}$ transition for large τ is due to the coherent superposition of the ionic states $4p^{-1}_{3/2}$ and $4p^{-1}_{1/2}$. When pump and probe pulses overlap, the phase changes for all transitions quite dramatically. There are three mechanisms that could explain this rapid change in ϕ:

- the quadratic Stark shift of the ionic energy levels due to the high electric field strength of the ionizing pulse,
- the residual Coulomb interaction between the ion and the electron,

12 Please note that the ionic dipole phase ϕ' is here redefined such that an absorbing behavior corresponds to $\phi' = \phi - \pi/2 = 0$.

– field-driven coupling between the freed electron and the parent ion via the neutral ground state Φ_0.

The polarizability for all ionic states with a $3d$ or $4p$ hole has values around 10 atomic units.[13] The peak field intensity in these studies was $4.8 \cdot 10^{14}$ W/cm^2, corresponding to an electric field of 0.08 atomic units. Even though the resulting ionic Stark shifts can be as large as 1 eV, the relative energy differences do not change by more than 100 meV. The maximum phase shift that can be expected is $\simeq \frac{\pi}{10}$. The phase shift shown in Figure 3(b) is, however, much larger.

Since TDCIS allows us to turn on and off specific effects, we can test to which extent the residual Coulomb interaction is responsible for the phase shift. Ignoring residual Coulomb interaction leads to basically unchanged line deformations, see Figure 9 in [21], indicating that electron-electron interactions are unimportant for this effect even though the emitted electron is still quite close to the ion during the probe step.

The large phase shifts disappear when the field-driven coupling to the neutral ground state is immediately switched off after the ion has been probed, see Figure 10 in [21]. Interestingly, the electric field cannot directly couple the ion with the emitted electron. Only via the neutral ground state (i.e., $\langle \Phi_0 | \hat{z} | \Phi_i^a \rangle$), it is possible to create a field-driven interaction between the ion and the photoelectron.

4.2 Multiorbital and collective excitations in HHG

Multiorbital and interchannel-coupling effects are well known in one-photon photoionization [29]. They lead to strong modifications in the partial cross sections of valence and inner-shell orbitals. In the strong-field regime and specifically for tunnel ionization, the electron is normally ejected from the outermost orbital because the tunneling rate decreases exponentially with the ionization potential. This is especially true for atoms where the next occupied orbitals are more strongly bound by tens of electronvolts. One would, therefore, expect that multiorbital effects do not exist in the strong-field regime, except in the case of spin-orbit effects, see Section 4.1. Many strong-field theories make use of this fact [7]. With TDCIS, we can test this approximation as we can freeze and unfreeze occupied orbitals.

Another fundamental process in strong-field physics is HHG [11]. HHG is used to generate isolated attosecond pulses (or trains of attosecond pulses) with photon energies in the XUV range. The mechanism behind HHG is well explained by a semi-classical model called the three-step model [6, 26]. It factorizes the HHG mechanism into three separate steps, see Figure 4. In the first step, the outermost electron tunnels

13 These values are based on CASSCF calculations.

out. In step two, the electron moves in the presence of the electric field and, within an optical cycle, the electric field changes sign and drives the electron back toward the ion. In the third step, the electron can recombine with the ion, thereby emitting a high-energy photon. The photon energy is given by the ionization potential I_p plus the amount of energy that the electron gained in the NIR field. The maximum emitted photon energy is $E_{\text{cutoff}} = I_p + 3.17\ U_p$, where $U_p = \frac{\mathcal{E}^2}{4\omega^2}$ is the ponderomotive potential, i.e., the cycle-averaged quiver energy of a free electron in an electric field with amplitude $\mathcal{E}$ and frequency ω. Characteristic for HHG is the plateau region, where the harmonics extend up to the cut-off energy without decreasing in strength.

The first step in HHG is tunnel ionization, so the usual picture in HHG for noble gas atoms is that only the electron in the p_z orbital is active and all other electrons are unaffected. In argon, we tested this statement for a multielectron system [20]. By activating all $3p$ orbitals or only the $3p_z$ orbital within TDCIS, we found that indeed only the $3p_z$ orbital gets effectively ionized, but including the other orbitals leads to small changes because the N-electron state is allowed to move in a larger configuration space. This affects the short-range part of the potential the emitted electron experiences.

Tunnel ionization is, however, only the first step of HHG. Multiorbital and many-body effects can also occur when the electron is driven back to the atom. A very prominent example is the giant dipole resonance in xenon. The giant dipole resonance originates from the $4d$ shell, which is bound by 68 eV. In no way would one intuitively expect that such a deeply bound orbital affects a strong-field-driven process. The electron that is driven back to the atom can have hundreds of electronvolt of energy and, hence, can trigger complex many-body excitations.[14] This is exactly the case in xenon, where the electron triggers collective excitations that are mainly located in the $4d$ shell but through interchannel interaction influence also the $5s$ and $5p$ shells. As a result, the signatures of the giant dipole resonance appear in the recombination step and, therefore, also in the HHG spectrum. A schematic of the underlying processes is shown in Figure 4.

Within the TDCIS theory, interchannel interactions, which are responsible for the collective excitation effect, see step 3.2 in Figure 4, can be turned off or on. In this way, the electron is allowed to recombine either only via step 3.1 or via both steps 3.1 and 3.2, respectively. Besides the physical mechanisms, the active orbitals participating in the HHG process can also be controlled. Let us consider in particular three different scenarios:

- all orbitals in the $4d, 5s$, and $5p$ shells are active, and all interchannel and intrachannel interactions (included in TDCIS) are allowed,
- only the orbitals aligned with the laser polarization ($4d_0, 5s, 5p_0$) are active, and all interchannel and intrachannel interactions are allowed,

14 Compare to the case of HHG in C_{60} in Section 3.1.2 of Chapter IV.

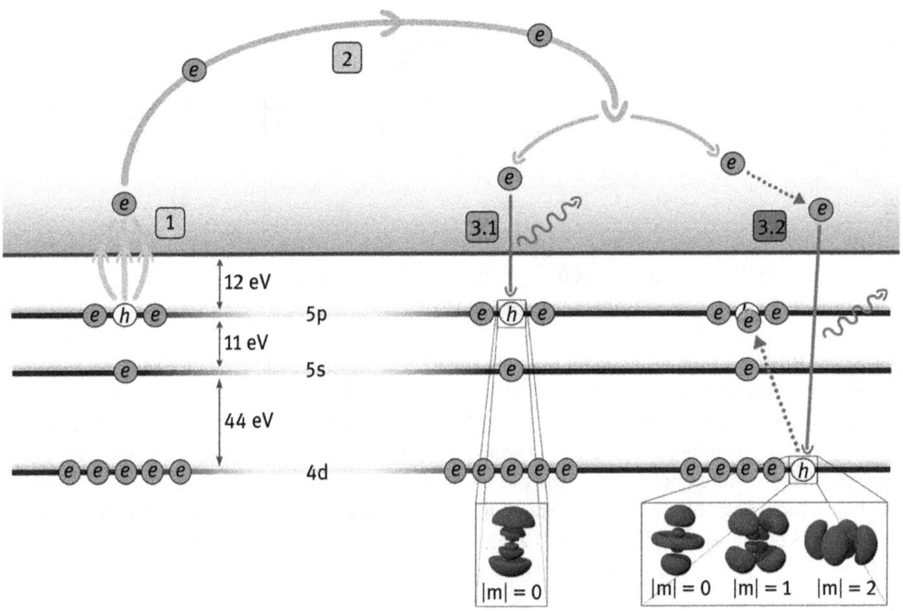

Fig. 4. Schematic illustration of the HHG process in xenon (spin states are excluded). (1) The electron tunnels mainly from the $5p_0$ orbital because of the strong-field driving pulse. (2) The electron is driven back to the ion by the oscillating electric field. In the third step, the electron recombines with the ion in two different ways: (3.1) the electron recombines with the very same hole that was generated in step 1, or (3.2) the electron exchanges energy with the ion by promoting an inner-shell electron from the $4d$ shell into the $5p_0$ hole via Coulomb interaction before recombining in a more tightly bound $4d$ orbital. All five $4d_m$ orbitals ($m = -2, \dots, 2$) contribute to the Coulomb interaction. The figure is taken from [17]. Copyright © 2013 American Physical Society (APS).

- all orbitals in the $4d, 5s$, and $5p$ shells are active, and only intrachannel interactions are allowed.

Even though the last two scenarios are artificially simplified by ignoring certain interactions or orbitals, they are of high educational utility since the direct comparison enables us to identify the underlying mechanisms and orbitals involved.

Figure 5 shows the HHG spectra for these three cases. When all orbitals are active and only intrachannel interactions are included (dotted red line), the HHG yield has the typical form of an atomic HHG spectrum with a flat plateau region up to the cut-off energy—no enhancement can be seen. This result does not change when only the $5p$ shell is considered (not shown) indicating that the direct contributions from $4d$ and $5s$ are negligible.

By including the many-body interchannel interactions in the calculations, a strong enhancement of up to one order of magnitude in the HHG yield appears. Depending on which orbitals are active, the width and center of the enhancement

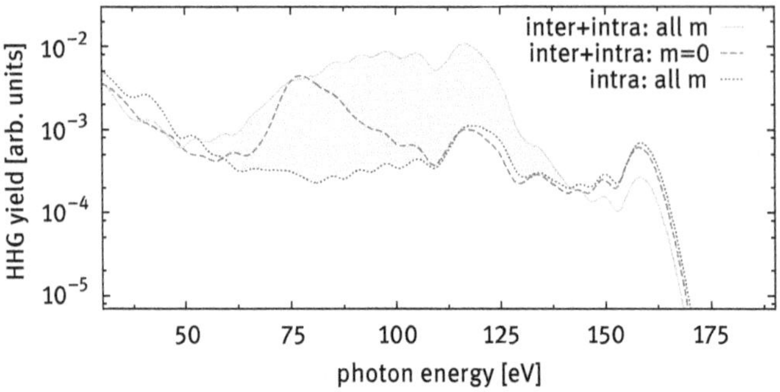

Fig. 5. The HHG spectrum of xenon for different theoretical models: (solid) the full TDCIS model (intrachannel + interchannel interactions) with all orbitals in the $4d$, $5s$, and $5p$ shells active, (dashed) the full TDCIS model with only $m = 0$ electrons in the $4d$, $5s$, and $5p$ shells active, and (dotted) the TDCIS model, excluding interchannel coupling, with all orbitals in the $4d$, $5s$, and $5p$ shells active. The difference between the two models including interchannel-coupling effects and the model including only intrachannel coupling is highlighted by the corresponding shaded areas. The figure is taken from [17]. Copyright © 2013 American Physical Society (APS).

region are quite different. When only laser-aligned $4d_0$, $5s$, and $5p_0$ orbitals are active, the enhancement is located in the energy region of 60–90 eV compared with an enhancement ranging from 60 to 125 eV when all electrons in the $4d$, $5s$, and $5p$ shells are active. This has been confirmed experimentally where a strong enhancement in the HHG yield was also seen around 100 eV [28].

The giant dipole resonance in the $4d$ shell couples via interchannel coupling to other shells and also to the $5p_0$ orbital, which got ionized in step 1. Unlike the interaction with the light field, where the magnetic quantum number m of each electron cannot change, the electron-electron interaction can change the m state. Only the overall quantum number, $M = m_a - m_i$, is conserved. We see this consequence already in the HHG spectrum because when only the $4d_0$ orbital is considered, the HHG enhancement is too narrow and too low in energy, see Figure 5.

Using TDCIS, we can also test the quantitative rescattering theory (QRS) [15] in the presence of this collective many-body effect. QRS assumes that the spectrum of the returning electron is universal and not system dependent, and all the system-dependence is included in the recombination cross section. This sounds reasonable because the electron, when moving in the field, is far away from the atom and "feels" only the charge of the ion. However, recombination happens very close to the atom, where many-body effects like interchannel interactions may be important. This short-range potential is highly system and state dependent and may affect the spectrum of the returning electron.

With TDCIS, we can calculate within the same theory the HHG spectrum and the recombination cross section, and we can test to which extent the returning photoelectron spectrum is really system independent. To this end, we can compare

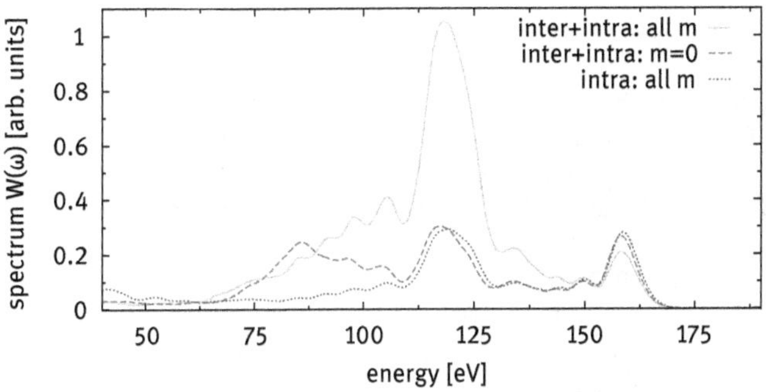

Fig. 6. The returning-electron spectrum $W(\omega)$, calculated by using the factorization $W(\omega) = S(\omega)/\sigma(\omega)$, where $S(\omega)$ is the HHG spectrum and $\sigma(\omega)$ is the photoionization cross section. The electron spectra shown for the three different theoretical models, see Figure 5, illustrate the system dependence of the returning electron wave packet, especially when interchannel interactions are not negligible. The figure is taken from [17]. Copyright © 2013 American Physical Society (APS).

the spectrum of the returning electron for the three different CIS models. Each CIS model describes different electron-electron interactions and, consequently, different short-range potentials. For large distances, all three models turn into a long-range Coulomb potential $-1/r$.

The three returning-electron spectra $W(\omega) = S(\omega)/\sigma(\omega)$ shown in Figure 6 agree well with each other above and below the giant resonance. In the region of the giant enhancement (60–140 eV), the spectra start to differ from each other and reach a maximum around 120 eV. The enhancement seen in the HHG spectra and in the cross sections persists also in the returning-photoelectron spectrum indicating that the enhancement in the HHG yield cannot be solely explained by the modified cross sections. It demonstrates that the modified electron-electron interaction influences not just the recombination matrix elements but also the electron spectrum.

Bibliography

[1] R. J. Bartlett and J. F. Stanton. Applications of post-Hartree–Fock methods: A tutorial. In *Reviews in Computational Chemistry*, vol. 5 (pp. 65–169). John Wiley & Sons, Inc., Hoboken, NJ, USA, 2007.

[2] K. Blum. *Density Matrix Theory and Applications*, 3rd edn. Springer, Berlin, 2012.

[3] J. P. Boyd. *Chebyshev and Fourier Spectral Methods*. Dover, New York, 2001.

[4] T. Brabec and F. Krausz. Intense few-cycle laser fields: Frontiers of nonlinear optics. *Reviews of Modern Physics*, 72(2):545, 2000.

[5] C. Cohen-Tannoudji, B. Diu, and F. Laloe. *Quantum Mechanics*, vol. 2, 1st edn. Wiley, New York, 1991.

[6] P. B. Corkum. Plasma perspective on strong field multiphoton ionization. *Physical Review Letters*, 71(13):1994–1997, 1993.

[7] F. H. M. Faisal. *Theory of Multiphoton Processes*. Springer Science+ Business Media, New York, 1987.

[8] E. Goulielmakis, Z.-H. Loh, A. Wirth, R. Santra, N. Rohringer, V. S. Yakovlev, S. Zherebtsov, T. Pfeifer, A. M. Azzeer, M. F. Kling, S. R. Leone, and F. Krausz. Real-time observation of valence electron motion. *Nature*, 466(7307):739–743, 2010.

[9] L. Greenman, P. J. Ho, S. Pabst, E. Kamarchik, D. A. Mazziotti, and R. Santra. Implementation of the time-dependent configuration-interaction singles method for atomic strong-field processes. *Physical Review A*, 82(2):023406, 2010.

[10] J. D. Jackson. *Classical Electrodynamics*, 3rd edn. Wiley, New York, 1998.

[11] M. C. Kohler, T. Pfeifer, K. Z. Hatsagortsyan, and C. H. Keitel. Frontiers of atomic high-harmonic generation. *Advances in Atomic, Molecular, and Optical Physics*, 61:159–208, 2012.

[12] P. Krause, T. Klamroth, and P. Saalfrank. Time-dependent configuration-interaction calculations of laser-pulse-driven many-electron dynamics: Controlled dipole switching in lithium cyanide. *Journal of Chemical Physics*, 123(7):074105, 2005.

[13] F. Krausz and M. Ivanov. Attosecond physics. *Reviews of Modern Physics*, 81:163–234, 2009.

[14] S. R. Leone, C. W. McCurdy, J. Burgdorfer, L. S. Cederbaum, Z. Chang, N. Dudovich, J. Feist, C. H. Greene, M. Ivanov, R. Kienberger, U. Keller, M. F. Kling, Z.-H. Loh, T. Pfeifer, A. N. Pfeiffer, R. Santra, K. Schafer, A. Stolow, U. Thumm, and M. J. J. Vrakking. What will it take to observe processes in 'real time'? *Nature Photonics*, 8(3):162–166, 2014.

[15] T. Morishita, A.-T. Le, Z. Chen, and C. D. Lin. Accurate retrieval of structural information from laser-induced photoelectron and high-order harmonic spectra by few-cycle laser pulses. *Physical Review Letters*, 100:013903, 2008.

[16] S. Pabst. New theoretical approaches to atomic and molecular dynamics triggered by ultrashort light pulses on the atto- to picosecond time scale. *European Physical Journal: Special Topics*, 221:1–72, 2013.

[17] S. Pabst and R. Santra. Strong-field many-body physics and the giant enhancement in the high-harmonic spectrum of xenon. *Physical Review Letters*, 111:233005, 2013.

[18] S. Pabst and R. Santra. Spin–orbit effects in atomic high-harmonic generation. *Journal of Physics B: Atomic Molecular and Optical Physics*, 47(12):124026, 2014. 10.1088/0953-4075/47/12/124026.

[19] S. Pabst, L. Greenman, P. J. Ho, D. A. Mazziotti, and R. Santra. Decoherence in attosecond photoionization. *Physical Review Letters*, 106(5):053003, 2011.

[20] S. Pabst, L. Greenman, D. A. Mazziotti, and R. Santra. Impact of multichannel and multipole effects on the cooper minimum in the high-order-harmonic spectrum of argon. *Physical Review A*, 85:023411, 2012.

[21] S. Pabst, A. Sytcheva, A. Moulet, A. Wirth, E. Goulielmakis, and R. Santra. Theory of attosecond transient-absorption spectroscopy of krypton for overlapping pump and probe pulses. *Physical Review A*, 86:063411, 2012.

[22] U. V. Riss and H.-D. Meyer. Calculation of resonance energies and widths using the complex absorbing potential method. *Journal of Physics B: Atomic Molecular and Optical Physics*, 26 (23):4503, 1993. 10.1088/0953-4075/26/23/021.

[23] N. Rohringer and R. Santra. Multichannel coherence in strong-field ionization. *Physical Review A*, 79(5):053402, 2009.

[24] N. Rohringer, A. Gordon, and R. Santra. Configuration-interaction-based time-dependent orbital approach for ab initio treatment of electronic dynamics in a strong optical laser field. *Physical Review A*, 74(4):043420, 2006.

[25] R. Santra, V. S. Yakovlev, T. Pfeifer, and Z.-H. Loh. Theory of attosecond transient absorption spectroscopy of strong-field-generated ions. *Physical Review A*, 83:033405, 2011.

[26] K. J. Schafer, B. Yang, L. F. DiMauro, and K. C. Kulander. Above threshold ionization beyond the high harmonic cutoff. *Physical Review Letters*, 70(11):1599–1602, 1993.
[27] H. B. Schlegel, S. M. Smith, and X. Li. Electronic optical response of molecules in intense fields: Comparison of TD-HF, TD-CIS, and TD-CIS(D) approaches. *Journal of Chemical Physics*, 126(24):244110, 2007.
[28] A. D. Shiner, B. E. Schmidt, C. Trallero-Herrero, H. J. Wörner, S. Patchkovskii, P. B. Corkum, J-C. Kieffer, F. Legare, and D. M. Villeneuve. Probing collective multi-electron dynamics in xenon with high-harmonic spectroscopy. *Nature Physics*, 7(6):464–467, 2011.
[29] A. F. Starace. Theory of atomic photoionization. In *Encyclopedia of Physics, vol. 31: Corpuscles and Radiation in Matter I* (pp. 1–121). Springer, Berlin, 1982.
[30] A. Szabo and N. S. Ostlund. *Modern Quantum Chemistry*. Dover Publication Inc., Mineola, NY, 1996.
[31] D. A. Telnov and S.-I Chu. Multiphoton detachment of H^- near the one-photon threshold: Exterior complex-scaling–generalized pseudospectral method for complex quasienergy resonances. *Physical Review A*, 59:2864–2874, 1999.
[32] A. Wirth, M. Th. Hassan, I. Grguras, J. Gagnon, A. Moulet, T. T. Luu, S. Pabst, R. Santra, Z. A. Alahmed, A. M. Azzeer, V. S. Yakovlev, V. Pervak, F. Krausz, and E. Goulielmakis. Synthesized light transients. *Science*, 334(6053):195–200, 2011.
[33] R. N. Zare. *Angular Momentum*. Wiley, New York, 1988.

Dejan B. Milošević

VII Strong-field approximation and quantum orbits

Strong-field physics is the area of science that explores the interaction of intense electromagnetic fields with matter. The advancements of ultrafast laser science and strong-field physics have led to the possibility of tracking electronic and structural dynamics on the subfemtosecond time scale, which has enabled the development of a new area of research: attoscience (see, for example, the review articles [1, 2, 4, 5, 19, 22, 24, 36, 39, 41, 44, 49, 50, 52, 54] and references therein). In this context, particularly important are two laser-induced processes: high-order harmonic generation (HHG) and above-threshold ionization (ATI). In the ATI process, the atom or molecule is ionized by a strong laser field. During the ATI, more photons are absorbed from the laser field than are necessary for ionization, and the emitted electron spectrum consists of peaks separated by the photon energy $\hbar\omega$. If the oscillating laser field is linearly polarized, then the emitted electron can be driven back by the laser field to the parent ion. If this electron recombines with the ion, one high-energy photon can be emitted in the HHG process. The returned electron can also elastically rescatter off the parent ion, move away from it, and reach the detector with a higher energy, which is the high-order ATI (HATI) process. The so-called three-step model (ionization, propagation, and rescattering or recombination with high-harmonic emission) [8] was crucial for the development of strong-field physics and attoscience.

Atomic and molecular processes in a strong field are quantum mechanical in nature. For a complete description of these processes, one has to solve the time-dependent Schrödinger equation (TDSE), which is a difficult task to which most chapters of this book are devoted. Fortunately, the fact that the laser field is strong may be used to simplify the theory. As early as in 1964, Keldysh [16] has introduced an approximation in which he neglected the influence of the atomic ion on the emitted electron and has supposed that, since the field is strong, the electron moves only in this field on its way to the detector. In this approximation, it was crucial that the solution of the TDSE for the electron under the influence of the laser field only was known in analytical form as the so-called Gordon–Volkov solution [12, 53]. This theory was further developed in [9, 45, 47] and now is known as direct strong-field approximation (SFA) or Keldysh–Faisal–Reiss theory. The SFA is applied also to other strong-field processes, and it usually consists in neglecting the influence of the atomic

Dejan B. Milošević: Faculty of Science, University of Sarajevo, Zmaja od Bosne 35, 71000 Sarajevo, Bosnia and Herzegovina, Max-Born-Institut, Max-Born-Strasse 2a, 12489 Berlin, Germany, and Academy of Sciences and Arts of Bosnia and Herzegovina, Bistrik 7, 71000 Sarajevo, Bosnia and Herzegovina; email: milo@bih.net.ba

De Gruyter Graduate – Computational Strong-Field Quantum Dynamics, Volume 5, 2017, pp. 203–225.
DOI 10.1515/9783110417265-007

or ionic potential on the electron in the final state or intermediate states. For example, for HHG, the SFA is introduced by the so-called Lewenstein model [25]. In this chapter, we will deal with various versions of the SFA.

Having in mind the wave-particle duality of electrons and photons, the above-described classical electron trajectories in reality should be replaced by electron wave packets. In fact, Feynman's path integral approach is more suitable for the description of HHG and HATI processes [33–35, 48]. A realization of this approach is the so-called quantum-orbit theory [4, 20, 21, 31]. In this formalism, the electron does not have to start with zero velocity and follow only one classical trajectory and return to the parent ion. In Feynman's path integral approach and quantum-orbit theory, the transition amplitude is a coherent sum of many different paths, expressed as $\sum_s A_s \exp(iS_s)$, with S_s the action along the sth path. In laser-induced processes, the ionization time and the recombination (for HHG) or rescattering (for HATI) time are complex and so are the corresponding trajectories. A visualization of such processes will be the second main theme in this chapter.

In this chapter, we first formulate the S-matrix theory in Section 1, with particular emphasis on HHG and ATI processes. In Section 2, the S-matrix elements are simplified using the SFA. For a periodic laser field, the observable quantities are expressed via the emission rate. In Section 3, we present explicit expressions for the harmonic generation rate and the ionization rate. For numerical calculations, it is necessary to define precisely the atomic ground state wave function and the rescattering potential. This problem is considered in Section 4. In addition, in this section, we explain how multielectron effects are taken into account in our approach. Numerical results for harmonic photon and electron energy spectra, obtained by numerical integration of the SFA matrix element, are presented in Section 5. The application of the saddle-point method (SPM) to strong-field processes is examined in Section 6. Explicit expressions for the emission rates are given with particular emphasis on the modified SPM and the uniform approximation. An example of the classification of the saddle-point solutions is presented in Section 7. In Section 8, we show numerical results for the HATI spectra obtained using the SPM and the uniform approximation. Section 9 is devoted to the quantum-orbit solutions and to the visualization of electron trajectories. Finally, a summary is given in Section 10. Atomic units ($\hbar = e = m = 4\pi\varepsilon_0 = 1$) are used unless indicated otherwise.

1 *S*-matrix elements

In this section, we will derive expressions for the S-matrix element of the HHG and ATI processes. Our starting point is a general form of the S-matrix (see [39])

$$S_{\mathrm{fi}} = i \lim_{t'\to\infty} \lim_{t\to-\infty} \langle \Phi_{\mathrm{out}}(t') | G_{\mathrm{tot}}^{(+)}(t', t) | \Phi_{\mathrm{in}}(t) \rangle, \tag{1}$$

where the total Green's operator $G_{\mathrm{tot}}^{(+)}$ corresponds to the Hamiltonian

$$H_{\mathrm{tot}}(t) = H(t) + \mathbf{r}\cdot\mathbf{E}_{\mathrm{har}}(t), \qquad H(t) = H_0 + V_{\mathrm{le}}(t) + V(\mathbf{r}). \tag{2}$$

Here, $\mathbf{r}\cdot\mathbf{E}_{\mathrm{har}}(t)$ is the interaction of the atom with the quantized high-harmonic field

$$\mathbf{E}_{\mathrm{har}}(t) = \sum_{\mathbf{K}} c_{\mathbf{K}} \left(a_{\mathbf{K}}^{\dagger} e^{i\omega_{\mathbf{K}}t} \hat{\mathbf{e}}_{\mathbf{K}}^{*} - a_{\mathbf{K}} e^{-i\omega_{\mathbf{K}}t} \hat{\mathbf{e}}_{\mathbf{K}} \right) \tag{3}$$

(in length gauge and dipole approximation), where $a_{\mathbf{K}}$ and $a_{\mathbf{K}}^{\dagger}$ are the annihilation and creation operators of the high-harmonic field photons corresponding to the wavevectors $\mathbf{K}$, frequencies $\omega_{\mathbf{K}}$, and complex unit polarization vectors $\hat{\mathbf{e}}_{\mathbf{K}}$. The interaction $\mathbf{r}\cdot\mathbf{E}_{\mathrm{har}}(t)$ is absent for the ATI process. The parts of the Hamiltonian $H(t)$ in (2) are $H_0 = -\boldsymbol{\nabla}^2/2$, with $\boldsymbol{\nabla} \equiv \partial/\partial\mathbf{r}$, $V_{\mathrm{le}}(t) = \mathbf{r}\cdot\mathbf{E}(t)$ the laser-field electron interaction, with $\mathbf{E}(t) = -d\mathbf{A}(t)/dt$ the electric field vector, and $V(\mathbf{r}) = V_C(r) + V_{\mathrm{sh}}(r)$, with $V_C(r) = -Z/r$ the Coulomb interaction ($Z = 1$ for atoms and $Z = 0$ for negative ions) and $V_{\mathrm{sh}}(r)$ a short-range interaction. The interaction with the laser field is off for the *in* and *out* states. The number of harmonic photons is zero for the *in* state, so that $|\Phi_{\mathrm{in}}(t)\rangle = |\psi_i(t)\rangle|0_{\mathbf{K}}\rangle$. For the *out state,* we have one high-harmonic photon for HHG so that $|\Phi_{\mathrm{out}}(t)\rangle = |\psi_f(t)\rangle|1_{\mathbf{K}}\rangle$, while for the ATI process, we have one electron with the asymptotic momentum $\mathbf{p}$ in the final state and zero harmonic photons, i.e., $|\Phi_{\mathrm{out}}(t)\rangle = |\psi_{\mathbf{p}}(t)\rangle|0_{\mathbf{K}}\rangle$. Here, $|\psi_i(t)\rangle = |\psi_i\rangle e^{-iE_it}$ and $|\psi_f(t)\rangle = |\psi_f\rangle e^{-iE_ft}$ (for HHG) or $|\psi_f(t)\rangle = |\psi_{\mathbf{p}}\rangle e^{-iE_{\mathbf{p}}t}$ (for ATI) are the solutions of the laser-free TDSE with the Hamiltonian $H_V = H_0 + V(\mathbf{r})$. Atomic or negative ion binding energy is E_i, while the final electron kinetic energy is $E_{\mathbf{p}} = \mathbf{p}^2/2$. The states $|\psi_i\rangle$ and $|\psi_{\mathbf{p}}\rangle$ are mutually orthogonal eigenstates of the Hamiltonian H_V. Using the Lippmann–Schwinger equation

$$G_{\mathrm{tot}}^{(+)}(t,t') = G^{(+)}(t,t') + \int dt''\, G_{\mathrm{tot}}^{(+)}(t,t'')\mathbf{r}\cdot\mathbf{E}_{\mathrm{har}}(t'')G^{(+)}(t'',t'), \tag{4}$$

as well as the relation $\langle 1_{\mathbf{K}}|G_{\mathrm{tot}}^{(+)}|1_{\mathbf{K}}\rangle = G^{(+)}$, the S-matrix for the HHG process can be further simplified [39]. The field-free boundary conditions have the form

$$i \lim_{t'\to\infty} \langle\psi_f(t')|G^{(+)}(t',t'') = \langle\Phi_f^{(-)}(t'')|, \tag{5}$$

$$i \lim_{t\to-\infty} G^{(+)}(t'',t)|\psi_i(t)\rangle = |\Phi_i^{(+)}(t'')\rangle, \tag{6}$$

where the states $|\Phi_j^{(\pm)}(t)\rangle$, $j = i,f$, satisfy the TDSE with the Hamiltonian $H(t)$. In this case, the S-matrix element for the emission of one high-harmonic photon, having the wavevector $\mathbf{K}$, frequency $\omega_{\mathbf{K}}$, and complex unit polarization vector $\hat{\mathbf{e}}_{\mathbf{K}}$, can be written as

$$S_{fi}^{\mathrm{HHG}} = -ic_{\mathbf{K}} \int_{-\infty}^{\infty} dt\, e^{i\omega_{\mathbf{K}}t} \hat{\mathbf{e}}_{\mathbf{K}}^{*}\cdot\mathbf{d}_{fi}(t), \tag{7}$$

where $\mathbf{d}_{fi}(t)$ is the time-dependent dipole matrix element between the initial and final laser-dressed states,

$$\mathbf{d}_{fi}(t) = \langle \Phi_f^{(-)}(t)|\mathbf{r}|\Phi_i^{(+)}(t)\rangle. \tag{8}$$

Taking into account the Lippmann–Schwinger equation

$$G^{(+)}(t,t') = G_V^{(+)}(t,t') + \int dt'' \, G^{(+)}(t,t'')V_{\mathrm{le}}(t'')G_V^{(+)}(t'',t') \tag{9}$$

and the orthogonality of the *in* and *out* states, we obtain from (1) the S-matrix element for the ATI process

$$S_{fi}^{\mathrm{ATI}} = -i \int_{-\infty}^{\infty} dt \, \langle \Phi_f^{(-)}(t)| V_{\mathrm{le}}(t)|\psi_i(t)\rangle. \tag{10}$$

2 Strong-field approximation

The states $|\Phi_j^{(\pm)}(t)\rangle$, $j = i, f$, satisfy the Lippmann–Schwinger equation

$$|\Phi_j^{(\pm)}(t)\rangle = |\psi_j(t)\rangle + \int dt' \, G^{(\pm)}(t,t')V_{\mathrm{le}}(t')|\psi_j(t')\rangle. \tag{11}$$

Inserting this into (8), we obtain four terms. Each of these terms has a clear physical meaning. The dominant one is the term for which, after the laser field-electron interaction at time t' when photons are absorbed from the laser field, the system propagates until time t, when a harmonic photon is emitted during the transition to the final state $|\psi_f(t)\rangle$. This term is consistent with the three-step model in which the photons are first absorbed from the laser field. Denoting this term with the superscript "a," we have

$$\mathbf{d}_{fi}(t) \simeq \mathbf{d}_{fi}^{\mathrm{a}}(t) = \langle \psi_f(t)| \int dt' \, G^{(+)}(t,t')V_{\mathrm{le}}(t')|\psi_i(t')\rangle. \tag{12}$$

Furthermore, it can be shown [39] that in the Fourier transform (7), instead of $\mathbf{d}_{fi}^{\mathrm{a}}(t)$, one can take $2\mathrm{Re}\,\mathbf{d}_{fi}^{\mathrm{a}}(t)$, which simplifies the calculations.

We will now apply the SFA and suppose that, in the intermediate states, the atomic interaction $V(\mathbf{r})$ can be neglected in comparison with the laser field, so that the Green's operator $G^{(+)}$ can be approximated by the Volkov–Green's operator (in length gauge)

$$G_{\mathrm{le}}(t,t') = -i \int d\mathbf{q} \, |\mathbf{q} + \mathbf{A}(t)\rangle\langle \mathbf{q} + \mathbf{A}(t')| \exp[-iS_{\mathbf{q}}(t) + iS_{\mathbf{q}}(t')], \tag{13}$$

where $dS_{\mathbf{q}}(t)/dt = \left[\mathbf{q}+\mathbf{A}(t)\right]^2/2$, and $|\mathbf{q}\rangle$ is a plane-wave ket vector such that $\langle\mathbf{r}|\mathbf{q}\rangle = (2\pi)^{-3/2}\exp(i\mathbf{q}\cdot\mathbf{r})$. Substituting (13) into (12) and solving the three-dimensional integral over the intermediate electron momenta using the SPM, we obtain [39]

$$\mathbf{d}_{fi}^{\mathrm{a}}(t) \simeq -i\int_0^\infty d\tau \left(\frac{2\pi}{i\tau}\right)^{3/2} \langle\psi_f(t)|\mathbf{r}|\mathbf{q}_{\mathrm{st}}+\mathbf{A}(t)\rangle\langle\mathbf{q}_{\mathrm{st}}+\mathbf{A}(t_0)|V_{\mathrm{le}}(t_0)|\psi_i\rangle e^{iS_{\mathbf{q}_{\mathrm{st}}i}(t,t_0)}, \quad (14)$$

where $t_0 = t-\tau$ and

$$\mathbf{q}_{\mathrm{st}} \equiv -\frac{1}{\tau}\int_{t_0}^{t} dt'\,\mathbf{A}(t'),\quad S_{\mathbf{q}i}(t,t_0) \equiv -S_{\mathbf{q}}(t)+S_{\mathbf{q}}(t_0)-E_i t_0. \quad (15)$$

The stationary momentum $\mathbf{q}_{\mathrm{st}}$ is obtained as the solution of the saddle-point equation $\nabla_{\mathbf{q}}S_{\mathbf{q}i}(t,t_0) = \mathbf{0}$, where $S_{\mathbf{q}i}(t,t_0)$ is the relevant part of the action. According to the three-step model, the physical meaning of (14) is the following: the electron appears in the continuum at the ionization time t_0 and propagates in the field, during the travel time τ, up to the time t, when a high-harmonic photon is emitted in the transition to the final state $|\psi_f(t)\rangle$ at time t.

The Green's operator $G^{(+)}$ satisfies also the Lippmann–Schwinger equation

$$G^{(+)}(t,t') = G_{\mathrm{le}}(t,t') + \int dt''\,G^{(+)}(t,t'')V(\mathbf{r})G_{\mathrm{le}}(t'',t'). \quad (16)$$

Inserting (5) and (16) into (10), we obtain

$$S_{fi}^{\mathrm{ATI}} = -i\lim_{t'\to\infty}\lim_{t\to-\infty}\langle\psi_{\mathbf{p}}(t')|\int_{-\infty}^{\infty} dt\Big[G_{\mathrm{le}}(t',t) + \int dt''\,G^{(+)}(t',t'')V(\mathbf{r})G_{\mathrm{le}}(t'',t)\Big]V_{\mathrm{le}}(t)|\psi_i(t)\rangle. \quad (17)$$

The usual procedure in the SFA for ATI is to approximate $G^{(+)}$ with G_{le} and to replace $\langle\psi_{\mathbf{p}}(t')|G_{\mathrm{le}}(t',t)$ with the Volkov state $\langle\chi_{\mathbf{p}}(t)| = \langle\mathbf{p}+\mathbf{A}(t)|\exp[iS_{\mathbf{p}}(t)]$. In this way, we obtain

$$S_{fi}^{\mathrm{ATI}} \simeq S_{fi}^{\mathrm{dir}} + S_{fi}^{\mathrm{res}}, \quad (18)$$

where the so-called direct and rescattering S-matrix elements are

$$S_{fi}^{\mathrm{dir}} = -i\int_{-\infty}^{\infty} dt_0\,\langle\chi_{\mathbf{p}}(t_0)|V_{\mathrm{le}}(t_0)|\psi_i(t_0)\rangle, \quad (19)$$

$$S_{fi}^{\mathrm{res}} = -i\int_{-\infty}^{\infty} dt\int_{t}^{\infty} dt''\,\langle\chi_{\mathbf{p}}(t'')|V(\mathbf{r})G_{\mathrm{le}}(t'',t)V_{\mathrm{le}}(t)|\psi_i(t)\rangle. \quad (20)$$

For the rescattering matrix element, we obtain, using (13) and solving the integral over $d\mathbf{q}$ with the SPM, similarly as for (14),

$$S_{fi}^{\mathrm{res}} = (-i)^2 \int_{-\infty}^{\infty} dt \int_0^{\infty} d\tau \left(\frac{2\pi}{i\tau}\right)^{3/2} e^{iS_{\mathbf{p}}(t)} \langle \mathbf{p}|V(\mathbf{r})|\mathbf{q}_{\mathrm{st}}\rangle$$

$$\times \langle \mathbf{q}_{\mathrm{st}} + \mathbf{A}(t_0)|V_{\mathrm{le}}(t_0)|\psi_i\rangle e^{iS_{\mathbf{q}_{\mathrm{st}}i}(t,t_0)}. \tag{21}$$

The physical meaning of the direct S-matrix element is that, after the ionization at the time t_0 via the interaction $V_{\mathrm{le}}(t_0)$, the electron directly goes to the continuum and reaches the detector with the asymptotic momentum $\mathbf{p}$. Instead, for the rescattering S-matrix element, the electron, after the ionization at the time t_0, propagates in the laser field up to the time t, when it returns to the nucleus and elastically rescatters off the potential $V(\mathbf{r})$ and then goes to the detector. This is in accordance with the three-step model in which the third step (recombination with the harmonic emission) is replaced by the rescattering event.

3 Harmonic generation rate and ionization rate

For a periodic laser field with the period $T = 2\pi/\omega$ and the fundamental frequency ω, it is possible to introduce the T-matrix determined by

$$S_{fi}^{\mathrm{HHG}} = -2\pi i \sum_n \delta(\omega_{\mathbf{K}} + E_f - E_i - n\omega) T_{fi}^{\mathrm{HHG}}(n), \tag{22}$$

$$S_{fi}^{\mathrm{ATI}} = -2\pi i \sum_n \delta(E_{\mathbf{p}} - E_i + U_p - n\omega) T_{fi}^{\mathrm{ATI}}(n). \tag{23}$$

Here, n is the number of photons absorbed from the laser field, the δ function expresses the energy conservation, and $U_p = \int_0^T \mathbf{A}^2(t)\,dt/(2T)$ is the ponderomotive energy. The T-matrix element for the HHG process is

$$T_{fi}^{\mathrm{HHG}}(n) = \int_0^T \frac{dt}{T} e^{i\omega_{\mathbf{K}} t} \hat{\mathbf{e}}_{\mathbf{K}}^{*} \cdot \mathbf{d}_{fi}(t). \tag{24}$$

Knowing the T-matrix element, we define the rate (probability per unit time) of emission of a harmonic photon (having the frequency $\omega_{\mathbf{K}}$ and polarization $\hat{\mathbf{e}}_{\mathbf{K}}$) into a solid angle $d\Omega_{\hat{\mathbf{K}}}$ by

$$w_{fi}(\omega_{\mathbf{K}}, \hat{\mathbf{e}}_{\mathbf{K}}) = \frac{\omega_{\mathbf{K}}^3}{2\pi c^3} \left| T_{fi}^{\mathrm{HHG}}(n) \right|^2, \quad \omega_{\mathbf{K}} + E_f = n\omega + E_i. \tag{25}$$

The corresponding harmonic intensity (power) is $I_n = \omega_{\mathbf{K}} w_{fi}(\omega_{\mathbf{K}}, \hat{\mathbf{e}}_{\mathbf{K}})$.

The T-matrix element for the direct ATI is given by

$$T_{fi}^{\mathrm{dir}}(n) = \int_0^T \frac{dt_0}{T} \langle \mathbf{p} + \mathbf{A}(t_0) | V_{\mathrm{le}}(t_0) | \psi_i \rangle e^{i[S_{\mathbf{p}}(t_0) - E_i t_0]}. \tag{26}$$

Using the energy-conserving condition $n\omega = E_{\mathbf{p}} - E_i + U_p$, we can rewrite the exponent in (26) as $S_{\mathbf{p}}(t_0) - E_i t_0 = \mathbf{p} \cdot \boldsymbol{\alpha}(t_0) + \mathcal{U}_1(t_0) + n\omega t_0$, where $\mathbf{A}(t) = d\boldsymbol{\alpha}(t)/dt$ and $\mathcal{U}_1(t) = \int^t dt' \, \mathbf{A}(t')^2/2 - U_p t$. For the rescattering T-matrix element, we obtain

$$T_{\mathbf{p}i}^{\mathrm{res}}(n) = -i \int_0^T \frac{dt}{T} \int_0^\infty d\tau \left(\frac{2\pi}{i\tau} \right)^{3/2} e^{iS_{\mathbf{p}}(t)} \langle \mathbf{p} | V(\mathbf{r}) | \mathbf{q}_{\mathrm{st}} \rangle$$
$$\times \langle \mathbf{q}_{\mathrm{st}} + \mathbf{A}(t_0) | V_{\mathrm{le}}(t_0) | \psi_i \rangle e^{iS_{\mathbf{q}_{\mathrm{st}}i}(t,t_0)}. \tag{27}$$

The term $S_{\mathbf{p}}(t) - E_i t_0$ can be replaced by $\mathbf{p} \cdot \boldsymbol{\alpha}(t) + \mathcal{U}_1(t) + n\omega t + E_i \tau$. The differential ionization rate for ATI with absorption of n photons from the laser field is

$$w_{fi}(n) = 2\pi p \left| T_{fi}^{\mathrm{dir}}(n) + T_{fi}^{\mathrm{res}}(n) \right|^2. \tag{28}$$

4 Ground-state wavefunctions, rescattering potential, and multielectron effects

Our ground-state wavefunctions ψ_i are given in the form of an ith atomic orbital, obtained by solving the Roothaan–Hartree–Fock equations [6, 7, 27]. In spherical coordinates, with the z-axis as axis of quantization, the orbitals are given by

$$\psi_{ilm}(\mathbf{r}) = \langle r\theta\phi | \psi_{ilm} \rangle = R_{il}(r) Y_{lm}(\theta, \phi), \tag{29}$$

where $Y_{lm}(\theta, \phi) = \langle \theta\phi | lm \rangle$ are normalized spherical harmonics in complex form, while the radial wavefunctions $R_{il}(r)$ are expanded in terms of basis functions (for example, Slater-type orbitals [6, 7, 27, 46]) or represented by asymptotic wavefunctions [11, 13, 46]. In the case of the expansion in Slater-type orbitals, the radial wavefunction is

$$R_{il}(r) = \sum_a C_a \frac{(2\zeta_a)^{n_a+1/2}}{\sqrt{(2n_a)!}} r^{n_a-1} e^{-\zeta_a r}, \tag{30}$$

where n_a and l are the quantum numbers of the electron and the parameters C_a and ζ_a characterize the radial distribution of the electron density and can be found tabulated in [6, 7, 27, 46]. For the inert gases considered here, we will use the orbitals given in Table 1. Since we are using length gauge, the interaction $\mathbf{r} \cdot \mathbf{E}(t)$ emphasizes large distances where the bound-state wavefunctions have well-defined asymptotic

Tab. 1. Ground-state configurations of valence electrons (the number of Slater-type orbitals used is given in parenthesis), ionization potentials, and the asymptotic expansion coefficients A for inert atomic gases [17, 46, 51].

Atom	Configuration	I_p (eV)	A (a.u.)
He	1s(5)	24.59	2.87
Ne	2p(4)	21.56	2.1
Ar	2p(2) + 3p(2)	15.76	2.51
Kr	3p(2) + 4p(2)	14.00	2.59
Xe	3p(1) + 4p(2) + 5p(2)	12.13	2.72

behavior. Hence, the approximation

$$R_{il}(r) \simeq A r^{\nu-1} \exp(-\kappa r), \quad r \gg 1, \tag{31}$$

is well justified [13]. Here, the constant A is tabulated in [17, 46, 51], $\nu = Z/\kappa$, and $\kappa = \sqrt{2I_p}$, with I_p the ionization potential. For inert gases, the corresponding values of I_p and A are given in Table 1.

For the rescattering potential, we use two approaches. The first is based on the independent-particle model potential represented by the short-range double Yukawa potential, see (21) and Table 1 in [14]: $V(r) = -Ze^{-r/D}[1 + (H - 1)e^{-Hr/D}]/(Hr)$, with $H = DZ^{0.4}$. The second enables calculation of the electron-ion rescattering, see the Appendix B in [15] and (21) and Table 2 in [37]: $V(r) = -\left(1 + a_1 e^{-a_2 r} + a_3 r e^{-a_4 r} + a_5 e^{-a_6 r}\right)/r$ (for the Kr atom there was a misprint: parameters a_3 and a_4 should be interchanged with the parameters a_5 and a_6).

We are using the single-active-electron approximation. However, different electrons from the ground-state configuration of an atom (or negative ion) may play the role of this active electron. Denoting by N_e the number of equivalent electrons in the ionizing shell of the target and averaging over the possible values of m, we obtain that (28) should be replaced by

$$\bar{w}_{\mathbf{p}i}(n) = \frac{N_e}{2l+1} \sum_{m=-l}^{l} w_{\mathbf{p}ilm}(n). \tag{32}$$

We usually consider the case of closed subshells specified by the orbital quantum number l. This subshell has $2l+1$ orbitals specified by m. Each orbital can be occupied by two electrons having different values of the spin projection. Therefore, in this case, we have $N_e = 2(2l+1)$ in (32). For the He atom, we have $l = m = 0$, while for other inert gases, we have $l = 1$, so that three terms ($m = 0, \pm 1$) have to be taken into account in (32).

Let us now consider multielectron effects for HHG. For the p ground state, the magnetic quantum number is $m = 0, \pm 1$. The initial (final) state is characterized only by the quantum number m_i (m_f). For atoms with closed electron shells (Ne, Ar, Kr, Xe), the m-changing transitions are forbidden by the Pauli exclusion principle. Namely, for an outer electron configuration np^6, the electron emitted initially from an m_i state cannot end up in an $m_f \neq m_i$ state since these states are occupied by other np-electrons. If we neglect the influence of the spin (i.e., we consider that the pairs of electrons with the spin $m_s = \pm 1/2$ from the np^6 configuration interact with the laser field in the same way), only three electrons are active and we have

$$\mathbf{d}_{fi}^{\mathrm{a}}(t) = \sum_{m=0,\pm 1} \mathbf{d}_m(t). \tag{33}$$

According to (29), we have $\psi_j(\mathbf{r}) \propto Y_{lm_j}(\hat{\mathbf{r}})$, $j = i, f$, and $E_f = E_i = -I_p$. The matrix elements in (14) can be calculated analytically, taking into account that $\langle \mathbf{q}|\mathbf{r}|\psi_{n_a lm}\rangle = i\partial\tilde{\psi}_{n_a lm}(\mathbf{q})/\partial\mathbf{q}$, where $\tilde{\psi}_{n_a lm}(\mathbf{q})$ are the Slater-type orbitals in momentum space. It is important that in the multielectron theory, we have a coherent sum of the time-dependent dipoles $\mathbf{d}_m(t)$. However, in the one-electron theory, the nth harmonic power, summed over all final states and averaged over all initial states, is given by

$$I_n^{(1e)} = \frac{(n\omega)^4}{2\pi c^3} \frac{1}{2l+1} \sum_{m_i, m_f} \left| \int_0^T \frac{dt}{T} \mathbf{d}_{fi}^{(1e)}(t) e^{in\omega t} \right|^2, \tag{34}$$

where $\mathbf{d}_{fi}^{(1e)}(t)$ is the (one-term) time-dependent dipole matrix element for the transition from the initial state m_i to the final state m_f. In the case of the s ground state, we have $m_i = m_f = 0$, and the multi- and one-electron theories give the same result.

5 Numerical examples for harmonic and electron spectra

In this section, we present two examples of photon and electron spectra obtained using the SFA. The corresponding T-matrix elements are calculated numerically. The integral over the travel time τ, (14) and (27), goes from zero to infinity. However, it is enough to take the upper limit to be five optical cycles ($5T$), since the probability of the process is low for long travel times. An exception is the case of resonant-like enhancements in HHG and HATI (see [31, 32]) where the long travel times are important and one has to take a larger upper limit, say $10T$. A standard Gauss–Legendre quadrature can be used for the calculation of this integral. Depending on the laser parameters, a few hundreds points can be enough. Various numerical tricks can be used. For example, one can divide the integration interval into many small subintervals and use Gauss–Legendre quadrature with a smaller number of points (say six) on each of these subintervals. Also, one can calculate the integral up to $5T$ and up to $6T$ and use one half of the sum of these integrals as the final result. The lower limit of the integral over τ can be problematic in some cases where the subintegral matrix element remains finite for $\tau \to 0$. In this case, one can set the lower limit to be, say, $0.05T$. This procedure does not affect the plateau of the spectra. For the low-energy electrons, the contribution of the direct SFA is dominant. Exceptions are the so-called low-energy structures [28]. The integral over the time t_0 in (26) and the integral over the time t in (24) and (27) can also be solved using appropriate Gauss–Legendre quadrature with a few hundreds points. In the case of HHG, it is convenient to use the FFT to calculate the integral over t. Also, Filon's method for highly oscillatory integrals can be used.

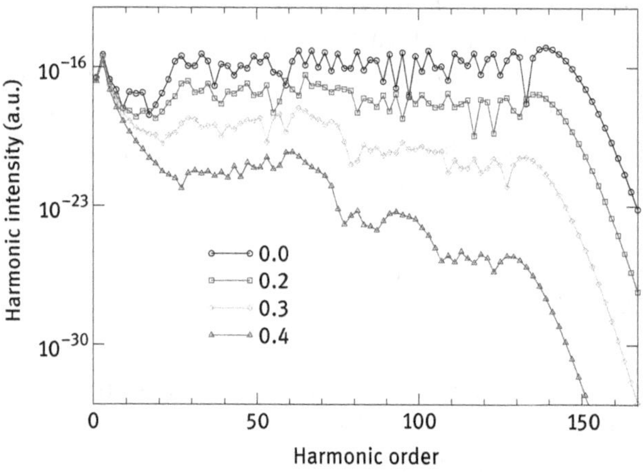

Fig. 1. Harmonic intensity as a function of the harmonic order for HHG by He atoms exposed to an elliptically polarized laser field of intensity 10^{15} W/cm^2, wavelength 800 nm, and ellipticity $\varepsilon = 0.0$, 0.2, 0.3, or 0.4, as denoted in the legend.

In Figure 1, we show the harmonic intensity as a function of the harmonic order for HHG by He atoms and an elliptically polarized laser field given by $\mathbf{E}(t) = E_0(\hat{\mathbf{e}}_x \sin\omega t - \varepsilon\hat{\mathbf{e}}_y \cos\omega t)/\sqrt{1+\varepsilon^2}$, where ε is the ellipticity. The laser intensity and the wavelength are 10^{15} W/cm^2 and 800 nm, respectively. We see that for the linearly polarized field case ($\varepsilon = 0$), the plateau extends up to $3U_p$ (a more precise semiclassical cutoff law is $n_{\max}\omega = 3.173U_p + 1.325I_p$ [25]). With the increase of the ellipticity, the plateau height decreases and a multiplateau structure starts to appear. In fact, the plateau becomes a staircase having three steps, and each of these steps has its own cutoff, which can be determined semiclassically using the quantum-orbit theory [38]. This behavior of the HHG process for elliptical polarization is in accordance with the three-step model: the probability of the electron return decreases with the increase of ε, and for circular polarization, the atomic HHG is completely suppressed. In general, the field of the emitted high harmonics is elliptically polarized, and furthermore, the corresponding polarization ellipse is rotated vs the laser-field polarization ellipse by an offset angle. A method of calculating the harmonic ellipticity and the offset angle can be found in [43] and references therein.

For the ATI process, besides the emitted electron energy, we have the electron emission angle θ with respect to the laser polarization axis as an additional parameter. In this case, it is convenient to present the electron spectra in the momentum plane as a false-color plot in which the color bar denotes the differential ionization rate on a logarithmic scale. Such an example is shown in Figure 2 where the spectrum for Xe atoms ionized by a linearly polarized laser field $\mathbf{E}(t) = E_0\hat{\mathbf{e}}_x \sin \omega t$ of intensity 4.5×10^{13} W/cm^2 and wavelength 1800 nm is presented. The results are obtained using (26) and (27) with a modification of the rescattering matrix element, which is called the low-frequency approximation in [15]. Namely, in (27), the rescattering amplitude $\langle \mathbf{p}|V|\mathbf{q}_{\mathrm{st}}\rangle$ is taken into account in first Born approximation, which cannot properly describe the exact rescattering amplitude for all angles. This is clearly visible in Figure 2, where characteristic minima for particular values of the angle θ appear. These minima are atom specific, and similar results have already been presented for short-wavelength lasers in [15] (see also [30] for a more rigorous treatment of the low-frequency approximation; the method of calculation of the electron-atom (ion) scattering amplitude can be found in [3]). The spectrum obtained using (26) and (27) without the mentioned modification is very similar to that shown in Figure 2 (except the discussed minima). The direct amplitude (26) dominates the low-energy part of the spectrum and is visible in the form of the most intense ellipsoid-shaped central region, which is elongated in the p_x direction. The high-energy part of the spectrum

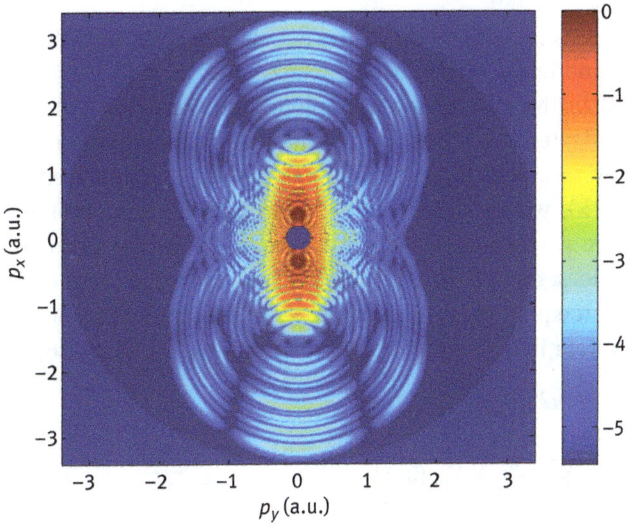

Fig. 2. The logarithm of the differential ionization rate of Xe atoms, presented in false colors in the electron momentum plane, for ionization by a linearly polarized laser field of intensity 4.5×10^{13} W/cm^2 and wavelength 1800 nm. The results are obtained using the SFA for the direct process and the low-frequency approximation for rescattering.

is determined by the rescattering amplitude. The spectrum obeys the symmetry $\theta \leftrightarrow 180^\circ - \theta$. The characteristic cutoff at $E_{\mathbf{p}} \simeq 10U_p$ for $\theta = 0^\circ$ [4], as well as the fork-like structures perpendicular to the polarization axis (the so-called off-axis low-energy structures, [40]), is noticeable.

6 Saddle-point method

We first consider the direct part of the ATI T-matrix element (26). The integral over the direct ionization time t_0 can be approximately solved using the SPM. The application of the SPM is justified by the high laser intensity for which the phase of the integrand is large. This is also in accordance with the SFA, which assumes high intensities. The saddle-point equation $d[S_{\mathbf{p}}(t_0) - E_i t_0]/dt_0 = 0$ is equivalent to the energy-conserving condition at the ionization time t_0

$$\frac{1}{2}\left[\mathbf{p} + \mathbf{A}(t_0)\right]^2 = E_i. \tag{35}$$

Since $E_i < 0$, the solutions t_{0s} of (35), which we distinguish by the index s, are complex. In saddle-point approximation, the T-matrix element for direct ATI assumes the form

$$T_{fi}^{\mathrm{dir,SP}}(n) = \frac{1}{T}\sum_s \left|\frac{2\pi}{S_{\mathbf{p}}''(t_{0s})}\right|^{1/2} \langle \mathbf{p} + \mathbf{A}(t_{0s})|V_{\mathrm{le}}(t_{0s})|\psi_i\rangle e^{i(S_s + \delta\pi/4)}, \tag{36}$$

where $S_{\mathbf{p}}''(t) = -\mathbf{E}(t)\cdot[\mathbf{p} + \mathbf{A}(t)]$, $S_s \equiv S_{\mathbf{p}}(t_{0s}) - E_i t_{0s}$, and $\delta \equiv \operatorname{sgn}\left[\operatorname{Im} S_{\mathbf{p}}''(t_{0s})\right]$. Only the solutions with $0 \le \operatorname{Re} t_{0s} \le T$ and $\operatorname{Im} t_{0s} > 0$ are taken into account.

Application of the SPM in the case of the asymptotic functions (31) with $\nu = Z/\kappa = 1/\kappa \neq 0$ is more complicated [13, 18]. In short, after a partial integration over the time t_0, the matrix element $\langle \mathbf{k}|V_{\mathrm{le}}(t_0)|\psi_i\rangle$, with $\mathbf{k} = \mathbf{p} + \mathbf{A}(t_0)$, is replaced by the function $-(I_p + \mathbf{k}^2/2)\tilde{\psi}_i(\mathbf{k})$, where $\tilde{\psi}_i(\mathbf{k}) = (2\pi)^{-3/2}\int d\mathbf{r}\,\psi_i(\mathbf{r})\exp(-i\mathbf{k}\cdot\mathbf{r})$ is the momentum-space asymptotic wavefunction, which can be expressed as a product of the Gauss hypergeometric series ${}_2F_1(a, b; c; -\mathbf{k}^2/\kappa^2)$ and the solid harmonics $Y_{lm}(\hat{\mathbf{k}})(k/\kappa)^l$. The saddle-point equation implies that $\mathbf{k}^2 = -\kappa^2$, so that, with a proper choice of the integration contour in the complex plain, one obtains [13, 18, 26, 36]

$$T_{fi}^{\mathrm{dir,MSP}}(n) = i2^{-3/2}T^{-1}A\kappa^{\nu}\nu\Gamma(\nu/2) \times \sum_s \left(\frac{k_s}{i\kappa}\right)^l Y_{lm}(\hat{\mathbf{k}}_s)\left[\frac{2i}{S_{\mathbf{p}}''(t_{0s})}\right]^{(\nu+1)/2} e^{iS_s}, \tag{37}$$

where $\mathbf{k}_s \equiv \mathbf{p} + \mathbf{A}(t_{0s})$, and the superscript "MSP" stands for the modified SPM. The summation in (37) is over the complex saddle-point solutions of (35) for the ionization time t_0, which are in the upper half of the complex t_0 plane.

The HHG S-matrix element (7), with (12) and (13), as well as the ATI rescattering matrix element (20), is expressed as five-dimensional integrals. The integration is over the three-dimensional intermediate electron momentum $\mathbf{q}$ and over the ionization and rescattering times t_0 and t. These integrals can be solved by expanding the phase about the stationary points. The stationary points are determined with respect to the variables $\mathbf{q}$, t_0, and t. The first stationarity condition $\nabla_{\mathbf{q}} S_{\mathbf{q}i}(t, t_0) = \mathbf{0}$ leads to $\mathbf{q}_{\text{st}}$ given by (15). Physically, this condition ensures that the electron returns to its parent ion. The second condition $\partial S_{\mathbf{q}i}(t, t_0)/\partial t_0 = 0$ leads to

$$\frac{1}{2}\left[\mathbf{q}_{\text{st}} + \mathbf{A}(t_0)\right]^2 = E_i. \tag{38}$$

This condition represents the energy conservation at the electron tunneling time t_0. The third condition is $\partial[S_{\mathbf{q}i}(t, t_0) + E_f t]/\partial t_t = 0$ for HHG and $\partial[S_{\mathbf{q}i}(t, t_0) + S_{\mathbf{p}}(t)]/\partial t = 0$ for HATI, which gives

$$\frac{1}{2}\left[\mathbf{q}_{\text{st}} + \mathbf{A}(t)\right]^2 = \begin{cases} E_f & \text{for HHG} \\ \frac{1}{2}\left[\mathbf{p} + \mathbf{A}(t)\right]^2 & \text{for HATI} \end{cases}. \tag{39}$$

Since the third step of the three-step model is different for HHG and HATI, it is logical that condition (39), which expresses energy conservation for this third step, is also different.

Mathematically, the SPM for the five-dimensional integral leads to a five-dimensional determinant [4]. However, it is easier first to solve the three-dimensional integral over $d\mathbf{q}$ using the SPM, as we have done in (14) and (21). Then, it remains to solve the two-dimensional integral over the times t_0 and t. In this case, the following determinant appears:

$$\Delta_{\mathbf{q}fi} = \left(\frac{\partial^2 S_{\mathbf{q}fi}}{\partial t_0 \partial t}\right)^2 - \frac{\partial^2 S_{\mathbf{q}fi}}{\partial t_0^2}\frac{\partial^2 S_{\mathbf{q}fi}}{\partial t^2}, \quad S_{\mathbf{q}fi} \equiv S_{\mathbf{q}i} + \begin{cases} E_f t & \text{for HHG} \\ S_{\mathbf{p}}(t) & \text{for HATI} \end{cases}. \tag{40}$$

The final result for HHG has the form

$$\begin{aligned} T_{fi}^{\text{HHG,SP}}(n) = \omega \sum_{\{t_{0s}, t_s\}} \left(\frac{2\pi}{i\tau_s}\right)^{3/2} &\langle \psi_f | \hat{\mathbf{e}}_{\mathbf{K}}^* \cdot \mathbf{r} | \mathbf{q}_{\text{st}} + \mathbf{A}(t_s)\rangle \Delta_{\mathbf{q}fi}^{-1/2}(t_s, t_{0s}) \\ &\times \langle \mathbf{q}_{\text{st}} + \mathbf{A}(t_{0s}) | V_{\text{le}}(t_{0s}) | \psi_i \rangle e^{i[\omega_{\mathbf{K}} t_s + S_{\mathbf{q}_{\text{st}}fi}(t_s, t_{0s})]}, \end{aligned} \tag{41}$$

where $\tau_s \equiv t_s - t_{0s}$, while for HATI, we obtain

$$\begin{aligned} T_{fi}^{\text{res,SP}}(n) = \omega \sum_{\{t_{0s}, t_s\}} \left(\frac{2\pi}{i\tau_s}\right)^{3/2} &\langle \mathbf{p} | V(\mathbf{r}) | \mathbf{q}_{\text{st}}\rangle \Delta_{\mathbf{q}fi}^{-1/2}(t_s, t_{0s}) \\ &\times \langle \mathbf{q}_{\text{st}} + \mathbf{A}(t_{0s}) | V_{\text{le}}(t_{0s}) | \psi_i \rangle e^{i S_{\mathbf{q}_{\text{st}}fi}(t_s, t_{0s})}. \end{aligned} \tag{42}$$

Instead of the above-described two-dimensional SPM, sometimes a MSP is used. This method was introduced in [42] (for HHG) and in [36] (for HATI by a few-cycle laser pulse).

Finally, it should be mentioned that the SPM fails for the electron energies near and beyond the cutoff. In this case, it is better to use a uniform approximation for the case of coalescing saddle points [10, 31]. The HATI T-matrix element (42) in the uniform approximation takes the form

$$\begin{aligned} T_{fi}^{\mathrm{res,UA}}(n) &= \sum_{\alpha\beta m} A_{\alpha\beta m} \exp(iS_{\alpha\beta m}) \\ &= \sum_{\beta m} (6\pi S_-)^{1/2} e^{iS_+ + i\pi/4} \left[\frac{A_-}{\sqrt{z}} \mathrm{Ai}(-z) + \frac{iA_+}{z} \mathrm{Ai}'(-z) \right], \end{aligned} \tag{43}$$

where Ai and Ai′ are the Airy function and its first derivative, respectively (for complex arguments, they can be calculated using the subroutine ZAIRY from the Netlib library). The quantities $A_\pm$ and $S_\pm$ are related to the weights and the actions of the saddle points in (41): $A_\pm = (A_{1\beta m} \pm iA_{-1\beta m})/2$, $S_\pm = (S_{1\beta m} \pm S_{-1\beta m})/2$. In (43), beyond the cutoff, the argument $z = (3S_-/2)^{2/3}$ must be replaced by $z\exp(i2\beta\pi/3)$, in order to select the proper branch of the Airy functions, and $A_{\alpha\beta m}$ should change its sign. The cutoff is determined by the critical value $n = n_c$ for which $\mathrm{Im}\, S_{+1\beta m} = \mathrm{Im}\, S_{-1\beta m}$ (the condition for the so-called anti-Stokes transition [31, 42]). The classification of the saddle-point solutions by the multi-index $s \equiv \{\alpha, \beta, m\}$ will be explained in Section 7.

7 Classification of the saddle-point solutions

We are looking for the solutions of the system of two saddle-point equations (38) and (39) for the complex times t_0 and t. This is a system of four real equations over four real variables $\mathrm{Re}\, t_0$, $\mathrm{Im}\, t_0$, $\mathrm{Re}\, t$, and $\mathrm{Im}\, t$. It can be solved using the subroutine ZSPOW from the International Mathematics and Statistics Library (IMSL). In principle, one can scan the whole complex planes t_0 and t looking for these solutions. The problem can be simplified selecting appropriate physical solutions. In searching for the solutions $\{t_0, t\}$, we fix the real part of the rescattering time t within one cycle of the field so that $0 \le \mathrm{Re}\, t < T$ [see (24) and (27)] and look for the solutions $\{t_0, t\}$ such that $\mathrm{Re}\, t_0 < \mathrm{Re}\, t$ (ionization happens before the rescattering or recombination). We start with a low value of the harmonic photon (for HHG) or electron energy (for HATI) energy. Each solution has its own cutoff energy and we "catch" more solutions choosing a low initial energy. For this energy, we look for such solutions for which $|\mathrm{Im}\, t|$ is not too large (for example, we neglect all solutions for which $\mathrm{Im}\, t > 0.1T$). The reason is that large $|\mathrm{Im}\, t|$ is related to low probability of the process. Then, we sort all solutions according to the values of the travel time $\mathrm{Re}\,(t - t_0)$, starting from the shortest one. The solutions having large $\mathrm{Re}(t-t_0)$ can be neglected, say $\mathrm{Re}(t-t_0) > 5T$; exceptions are resonant-like enhancements of HHG and HATI, which are related to the constructive interference of contributions of many solutions including those having long travel time, see [31, 32]. Having found a set of solutions for a fixed low energy, we can continuously increase the energy using the previously found solutions as the initial condition for the subroutine ZSPOW. Then, we can present the energy as a function of $\mathrm{Re}\,t_0/T$ and $\mathrm{Re}\,t/T$ and classify solutions as it is done in Figure 3. For the linearly polarized field case, the problem can be simplified starting from the approximate analytical solutions. This procedure is described in detail in [31] for HHG and in [32] for HATI. Having found solutions for a certain parameter, we can use them as initial condition to find solutions for a slightly different value of this parameter and then iteratively find solutions for the chosen interval of interest. The parameter can be the laser field ellipticity, frequency or intensity, or the electron emission angle θ, etc.

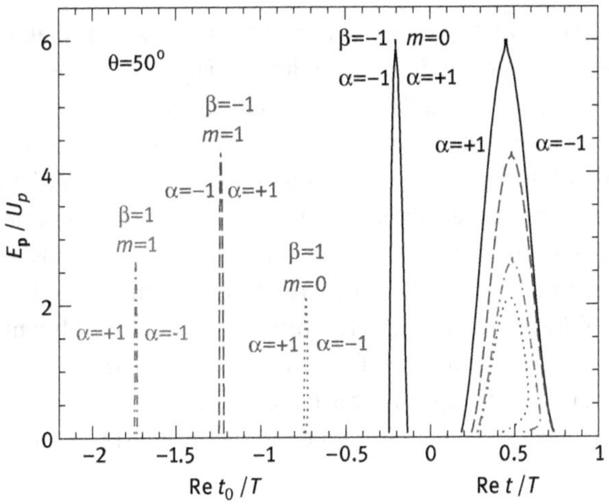

Fig. 3. The notation (α, β, m) is used to label the solutions of the system of saddle-point equations (38) and (39) for a linearly polarized laser field. The solid, dotted, long-dashed, and dotted-dashed curves in the right-hand part ($0 \leq \mathrm{Re}\,t \leq T$) specify the real part of the rescattering times for the four pairs of orbits with the shortest travel times. In the left-hand part of the figure, the counterpart of each curve identifies the corresponding real part of the ionization times t_0. The emitted electron energy in multiples of U_p is plotted on the ordinate, and horizontal lines (at constant energy) relate the real parts of the ionization and rescattering times for the respective orbits. There are infinitely many further solutions that have the real part of the ionization time beyond the left-hand margin of the figure. The curves have been calculated for Ne, for emission in the direction $\theta = 50°$, and for a linearly polarized laser field of intensity 2×10^{14} W/cm^2 and wavelength 800 nm.

The most detailed classification of the saddle-point solutions for HATI by a linearly polarized field is given in [29]. The backward-scattering solutions were classified by the multi-index (α, β, m), while for the forward-scattering solutions, the double index (ν, μ) was used. The name backward (forward) is related to the angle $\theta_s = 180°$ ($0°$), by which the recolliding electron scatters with respect to its incoming direction. We will not consider the solutions (ν, μ) since these forward-scattering solutions are related to low energies, which are not of our interest here. In order to simplify the presentation, we also do not consider the solutions $(\alpha, \beta, m) = (\pm 1, 1, 0)$, which were introduced in [29] and which correspond to short travel time and low energies.

In Figure 3, we present the saddle-point solutions $\{t_0, t\}$, obtained solving the system of (38) and (39), classified in accordance with the notation (α, β, m), $\alpha = \pm 1$, $\beta = \pm 1$, $m = 0, 1, 2, \ldots$. The physical meaning of the index m is that it gives the approximate length of the travel time in multiples of the laser period. The index β denotes the solutions within one optical cycle characterized by the index m. For each $m = 0, 1, \ldots$, there are two pairs of solutions having the real part of the ionization time t_0 between $-(m+1)T$ and mT. The travel time is longer for the pair of solutions characterized by the index $\beta = 1$.

Let us explain why we have two solutions $\beta = \pm 1$. The ionization is more probable when the absolute value of the field is close to the maximum. For a field linearly polarized along the x-axis, $E_x(t_0) = E_0 \sin \omega t_0$, if $\beta = 1$ the emitted electron is driven by the field, which decreases from a close-to-maximum positive value ($\omega \mathrm{Re}\, t_0 = -3\pi/2$, i.e., $\mathrm{Re}\, t_0/T = -3/4$ in Figure 3), while for $\beta = -1$, the situation is the opposite, i.e., the field increases from a minimum negative value, such that $\mathrm{Re}\, t_0/T = -1/4$. In one optical cycle T, the field changes along two segments of the x-axis in the opposite directions ($0°$ and $180°$). In general, we have $\beta = \mathrm{sgn}(E_x(\mathrm{Re}\, t_0))$. In addition to this, each pair of solutions having fixed β and m consists of two orbits with slightly different travel times, and we discriminate the longer ($\alpha = -1$) from the shorter orbit ($\alpha = +1$) by the index α. The terminology "short" and "long" orbits is usually used in the literature in connection with the Lewenstein model [25] of HHG where only the shortest pair of solutions is considered.

8 Numerical results for HATI spectra obtained using the SPM and uniform approximation

Let us first present numerical results for the (H)ATI differential ionization rate for Ne atoms ionized by a linearly polarized laser field of intensity 8×10^{14} W/cm^2 and wavelength 800 nm. Saddle-point solutions for this case are obtained and classified using the method described in Section 7. We see that the rates obtained using particular solutions are presented by smooth curves, which finish by abrupt cutoffs for particular values of the energy $[(E_{\mathbf{p}}/U_p)_c = \{5.6057, 2.2096, 4.1576, 2.6930, 3.8344\}$ for $\beta m = \{-10, 10, -11, 11, -12\}]$. The solutions having $\alpha = -\beta$ are unphysical (divergent) after the corresponding cutoff and should be neglected. Near the cutoff, the SPM fails so that

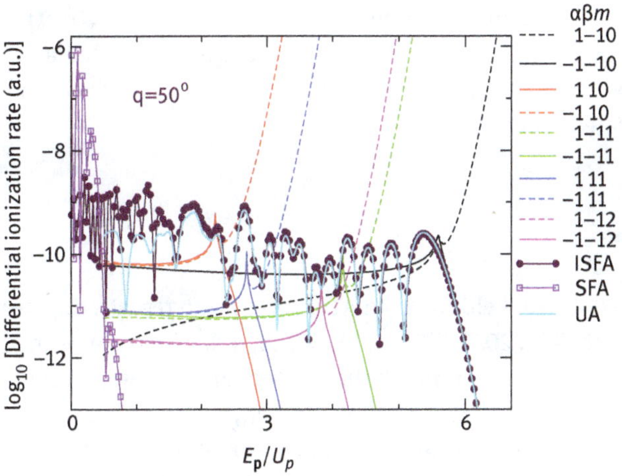

Fig. 4. The logarithm of the differential ionization rate of Ne as a function of the electron energy $E_{\mathbf{p}}$ in units of the ponderomotive energy U_p for ionization by a linearly polarized laser field of intensity 8×10^{14} W/cm^2 and wavelength 800 nm. The results are obtained using the uniform approximation with ten saddle-point solutions βm (UA cyan bold solid curve), the SFA method with one-dimensional integral over the ionization time (SFA violet line with squares), and the ISFA method with the SPM for the integral over $d\mathbf{k}$ and numerical integration over the ionization and travel times (ISFA maroon line with filled circles). The contributions of particular quantum orbits (α, β, m) are also presented, as denoted in the legend. The contributions that should be neglected after the cutoff are denoted by the dashed lines.

the procedure in which the rate is calculated as $\sum_{\alpha\beta m} A_{\alpha\beta m} \exp(iS_{\alpha\beta m})$ in which the contributions of the solutions having $\alpha = -\beta$ are neglected after $E_{\mathbf{p}_c}$ gives the spectrum with characteristic spikes at the energies $E_{\mathbf{p}_c}$. As it is explained in Section 6, in this case, it is more appropriate to use the uniform approximation (UA), which produces a continuous (and oscillatory) spectrum. An example is shown in Figure 4: the bold cyan solid line represents the spectrum obtained using ten solutions βm ($\beta = \pm 1$, $m = 0, 1, 2, 3, 4$) and the uniform approximation. In Figure 4, the spectrum obtained using the improved-SFA (ISFA) method (i.e., the SPM for the integral over $d\mathbf{k}$ and numerical integration over the ionization and travel times, as it is described in Section 2) is shown by maroon line with filled circles. The agreement with the uniform approximation is excellent for energies above $1.5U_p$ (for lower energies, the contribution of the low-energy forward-scattering solutions (ν, μ), not considered here, should be taken into account). The result obtained using the direct SFA method with one-dimensional integral over the ionization time is also presented in Figure 4 (SFA violet line with squares). This direct SFA result is dominant for energies lower than $0.6U_p$. The partial ionization rates for ten particular values of $\alpha\beta m$ are also shown in Figure 4. The partial rates, which are divergent after the cutoff, are presented by dashed lines. It is clearly visible that the high-energy part of the spectrum is determined by the coherent sum of only two contributions of the solutions $(\alpha,\beta,m) = (1,-1,0)$ and $(-1,-1,0)$. This implies that for the energies larger than $4.2U_p$, the spectrum has a simple oscillatory structure. For lower energies, more and more solutions contribute to the spectrum, and we have a more complicated oscillatory structure.

9 Quantum orbits

In order to better understand the laser field-induced ATI process with rescattering, we use the concept of quantum orbits [4, 20, 21, 29, 31, 35, 36, 38, 48], which are defined as solutions of the classical Newton's equation for the electron in the presence of the laser field, $\ddot{\mathbf{r}}(t) = -\mathbf{E}(t)$, but for complex time. For the direct SFA, we have the complex ionization time t_0, and the electron trajectories are defined as real part of $\mathbf{r}(t)$ for $t > \mathrm{Re}\, t_0$: $\mathbf{r}(t) = (t-t_0)\mathbf{p} + \boldsymbol{\alpha}(t) - \boldsymbol{\alpha}(t_0)$. For the rescattering ISFA, we have complex ionization time t_0 and rescattering time t_r for which $\mathbf{r}(t_0) = \mathbf{r}(t_r) = \mathbf{0}$. We will present the electron trajectories defined as real part of $\mathbf{r}(t)$ for t real, with

$$\mathbf{r}(t) = \begin{cases} (t-t_0)\mathbf{q}_{\mathrm{st}} + \boldsymbol{\alpha}(t) - \boldsymbol{\alpha}(t_0), & \text{if } \mathrm{Re}\, t_0 \le t \le \mathrm{Re}\, t_r, \\ (t-t_r)\mathbf{p} + \boldsymbol{\alpha}(t) - \boldsymbol{\alpha}(t_r), & \text{if } t > \mathrm{Re}\, t_r. \end{cases} \tag{44}$$

Since $\mathrm{Re}\,\mathbf{r}(\mathrm{Re}\, t_0) \neq \mathbf{0}$, the emitted electron appears in the continuum at the "exit of the tunnel," few atomic units away from the origin. After that, for the direct SFA, the electron is driven by the laser field to the detector, while for the rescattering ISFA, the electron behaves in accordance with the three-step model as described before. For HHG, the electron recombines at time t_r so that the electron trajectory is determined by (44) for $\mathrm{Re}\, t_0 \le t \le \mathrm{Re}\, t_r$.

We illustrate the quantum orbits and the corresponding real electron trajectories using the example of HATI of Ne atoms by linearly polarized laser field for the same laser parameters as in Figure 4. In the left panels of Figure 5, we present the saddle-point solutions for the ionization time t_0 (maroon curves) and the rescattering time t_r (red curves) in the complex time plane for four solutions (α,β,m) having $m = 0$, which are characterized by the short travel times. The solutions having $\beta = -1$ $(\beta = 1)$ are presented in the upper (lower) panels. In each panel, the two solutions having $\alpha = \pm 1$ are presented. The solutions for which $\alpha \neq \beta$ should be neglected after the cutoff are presented by the dotted-dashed lines. The imaginary part of the recombination time $\mathrm{Im}\, t_r$ is small for energies lower than the cutoff value for all solutions. For fixed β and m, two solutions $\alpha = \pm 1$ approach each other with the increase of the electron energy, having an avoided crossing for the energy near the cutoff value. The imaginary part of the solutions t_r for $\alpha \neq \beta$ takes large negative values with the increase of the energy beyond the cutoff, which is the cause of the divergence of the corresponding rates (see the dashed lines in Figure 4).

In the right-hand panel of Figure 5, we have presented the electron trajectories (44) for fixed electron energies, which are slightly lower than the corresponding cutoff

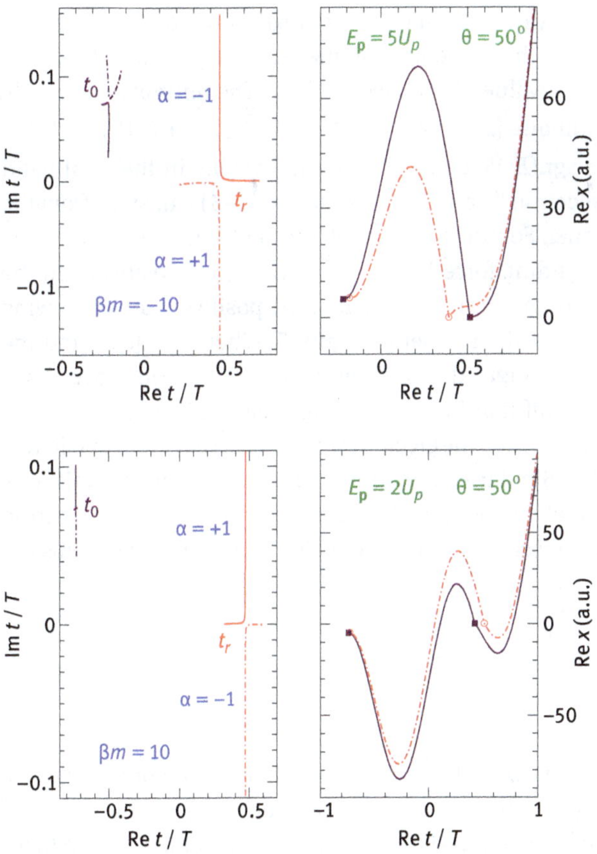

Fig. 5. The shortest rescattered quantum-orbit solutions and trajectories characterized by the multi-index (α, β, m) with $m = 0$ and $\beta = -1$ ($\beta = 1$) for upper (lower) panels. The results are for HATI of Ne atoms by a linearly polarized laser field of intensity 8×10^{14} W/cm^2 and wavelength 800 nm and for the electron emission angle $\theta = 50°$. Left-hand panels: Solutions for the complex ionization time t_0 (maroon lines) and rescattering time t_r (red lines) presented in the complex (Re t, Im t) plane, expressed in units of the optical period T. The electron energy changes as a continuous parameter along each curve from a minimum to a maximum value. The solutions $\alpha = -\beta$ whose contribution should be neglected after the cutoff are presented by dotted-dashed lines. Right-hand panels: Electron trajectories [real part of the quantum orbits $x(\mathrm{Re}\, t)$] obtained using the saddle-point solutions of the corresponding left panel, for the energies slightly below the corresponding cutoff [$E_{\mathbf{p}} = 5U_p$ ($2U_p$) for the upper (lower) panel]. The positions where the electron exits from the tunnel and where it rescatters off the core are denoted by the corresponding symbols (open symbol and dotted-dashed lines for the $\alpha = -\beta$ solutions).

values ($E_{\mathbf{p}} = 5U_p$ for $\beta = -1$ and $E_{\mathbf{p}} = 2U_p$ for $\beta = 1$). For complex time, we have $x(t_0) = 0$. However, from the left panels, we see that $\mathrm{Im}\, t_0$ is large so that $\mathrm{Re}\, x(\mathrm{Re}\, t_0) \neq 0$, i.e., the electron appears at the "tunnel exit." The position of the "tunnel exit" can be determined solving approximately the stationary Schrödinger equation for the energy $-I_p$ and for an electron bound by the Coulomb potential and the potential $E_s x$, with $E_s = E_x(\mathrm{Re}\, t_0)$ the instantaneous value of the electric field. The approximate result, obtained using parabolic coordinates [23], is $x_e = -I_p(1 + \sqrt{1 - 2E_s/I_p^2})/(2E_s) \simeq -I_p/E_s$. In our case, we obtain $x_e \simeq -\mathrm{sgn}(E_s)5$ a.u.. This is clearly visible in the right-hand panels of Figure 5 where the electron "starts" approximately 5 (−5) a.u. away from the origin for the upper (lower) panel. For the upper panel, the field at the ionization time is negative so that the corresponding force $F_x(\mathrm{Re}\, t_0) = -E_x(\mathrm{Re}\, t_0)$ is positive, and the electron moves away from the origin in the direction of the positive x-axis following the trajectory shown in the upper-right panel of Figure 5. When the field (and the corresponding force) changes the sign, the electron turns around and moves back to the parent ion and rescatters off it at the rescattering time t_r. Since $\mathrm{Im}\, t_r$ is small (negligible), we have $\mathrm{Re}\, x(\mathrm{Re}\, t_r) \simeq 0$, i.e., the electron rescatters at the origin and leaves in the direction of the detector. Similar explanations can be given for the trajectory presented in the lower-right panel. In this case, the trajectory is longer and the electron revisits the origin once before it rescatters off the parent ion when it returns to this ion for the second time.

10 Summary

We considered the SFA, an approximate quantum-mechanical theory of atomic processes in a strong laser field, which constitutes the backbone of the theory of intense laser-atom interaction. The emphasis was on the so-called ISFA, which takes into account corrections to the standard SFA, namely, an additional interaction of the emitted electron with the parent ion. In comparison with the standard SFA for which the transition matrix element can be calculated as a one-dimensional integral, in the ISFA, a five-dimensional integral appears. In accordance with the three-step model, in addition to the integral over the ionization time (first step), we have a three-dimensional integration over the intermediate electron momenta (second step) and over the electron rescattering (or recombination) time (third step). For practical purposes, it is too demanding to calculate numerically such five-dimensional integrals, and one usually solves the three-dimensional integral over the electron momenta using the SPM. The SPM can also be applied to the remaining time integrals. This approach is not only important for practical calculations. Its strength lies in the related formalism, which is called quantum-orbit theory and closely connected to Feynman's path integral approach. The quantum-orbit formalism is particularly useful for the fast calculation and the intuitive interpretation of high-order atomic processes in strong laser fields.

Bibliography

[1] P. Agostini and L. F. DiMauro. The physics of attosecond light pulses. *Reports on Progress in Physics*, 67(6):813, 2004.

[2] P. Agostini and L. F. DiMauro. Atomic and molecular ionization dynamics in strong laser fields: From optical to x-rays. In E. Arimondo, P. Berman, and C. Lin, eds, *Advances in Atomic, Molecular, and Optical Physics*, vol. 61 (pp. 117–158). Academic Press, Amsterdam, 2012.

[3] K. Bartschat, ed. *Computational Atomic Physics: Electron and Positron Collisions with Atoms and Ions*. Springer Berlin Heidelberg, Berlin, Heidelberg, 1996.

[4] W. Becker, F. Grasbon, R. Kopold, D. B. Milošević, G. G. Paulus, and H. Walther. Above-threshold ionization: From classical features to quantum effects. *Advances in Atomic, Molecular, and Optical Physics*, vol. 48 (pp. 35–98). Academic Press, Amsterdam, 2002.

[5] W. Becker, X. Liu, P. J. Ho, and J. H. Eberly. Theories of photoelectron correlation in laser-driven multiple atomic ionization. *Reviews of Modern Physics*, 84:1011–1043, 2012.

[6] C. F. Bunge, J. A. Barrientos, and A. V. Bunge. Roothaan–Hartree–Fock ground-state atomic wave functions: Slater-type orbital expansions and expectation values for Z = 2–54. *Atomic Data and Nuclear Data Tables*, 53(1):113–162, 1993.

[7] E. Clementi and C. Roetti. Roothaan–Hartree–Fock atomic wavefunctions. *Atomic Data and Nuclear Data Tables*, 14(3):177–478, 1974.

[8] P. B. Corkum. Plasma perspective on strong field multiphoton ionization. *Physical Review Letters*, 71:1994–1997, 1993.

[9] F. H. M. Faisal. Multiple absorption of laser photons by atoms. *Journal of Physics B: Atomic and Molecular Physics*, 6(4):L89, 1973.

[10] C. Figueira de Morisson Faria, H. Schomerus, and W. Becker. High-order above-threshold ionization: The uniform approximation and the effect of the binding potential. *Physical Review A*, 66:043413, 2002.

[11] A. Gazibegović-Busuladžić, D. B. Milošević, and W. Becker. High-energy above-threshold detachment from negative ions. *Physical Review A*, 70:053403, 2004.

[12] W. Gordon. Der Comptoneffekt nach der Schrödingerschen Theorie. *Zeitschrift für Physik*, 40(1):117–133, 1926.

[13] G. F. Gribakin and M. Y. Kuchiev. Multiphoton detachment of electrons from negative ions. *Physical Review A*, 55:3760–3771, 1997.

[14] E. Hasović, M. Busuladžić, A. Gazibegović-Busuladžić, D. B. Milošević, and W. Becker. Simulation of above-threshold ionization experiments using the strong-field approximation. *Laser Physics*, 17(4):376–389, 2007.

[15] A. Čerkić, E. Hasović, D. B. Milošević, and W. Becker. High-order above-threshold ionization beyond the first-order born approximation. *Physical Review A*, 79:033413, 2009.

[16] L. V. Keldysh. Ionization in the field of a strong electromagnetic wave. *Soviet Physics JETP*, 20: 1307–1314, 1965 (Russian original - ZhETF, Vol. 47, No. 5, p. 1945, 1964).

[17] T. K. Kjeldsen and L. B. Madsen. Strong-field ionization of diatomic molecules and companion atoms: Strong-field approximation and tunneling theory including nuclear motion. *Physical Review A*, 71:023411, 2005.

[18] T. K. Kjeldsen and L. B. Madsen. Strong-field ionization of atoms and molecules: The two-term saddle-point method. *Physical Review A*, 74:023407, 2006.

[19] M. C. Kohler, T. Pfeifer, K. Z. Hatsagortsyan, and C. H. Keitel. Frontiers of atomic high-harmonic generation. In E. Arimondo, P. Berman, and C. Lin, eds, *Advances in Atomic, Molecular, and Optical Physics, Advances In Atomic, Molecular, and Optical Physics*, vol. 61 (pp. 159–208). Academic Press, Amsterdam, 2012.

[20] R. Kopold, W. Becker, and M. Kleber. Quantum path analysis of high-order above-threshold ionization. *Optics Communications*, 179(1–6):39–50, 2000.
[21] R. Kopold, D. B. Milošević, and W. Becker. Rescattering processes for elliptical polarization: A quantum trajectory analysis. *Physical Review Letters*, 84:3831–3834, 2000.
[22] F. Krausz and M. Ivanov. Attosecond physics. *Reviews of Modern Physics*, 81:163–234, 2009.
[23] L. D. Landau and E. M. Lifshits. *Quantum Mechanics: Non-relativistic Theory*. Butterworth-Heinemann, Oxford, 1977.
[24] A. S. Landsman and U. Keller. Attosecond science and the tunnelling time problem. *Physics Reports*, 547:1–24, 2015.
[25] M. Lewenstein, Ph. Balcou, M. Y. Ivanov, A. L'Huillier, and P. B. Corkum. Theory of high-harmonic generation by low-frequency laser fields. *Physical Review A*, 49:2117–2132, 1994.
[26] C. P. J. Martiny. *Strong-field ionization of atoms and molecules by short femtosecond laser pulses*. PhD thesis, Department of Physics and Astronomy, Aarhus University, Aarhus, 2010.
[27] A. D. McLean and R. S. McLean. Roothaan–Hartree–Fock atomic wave functions Slater basis-set expansions for $Z = 55–92$. *Atomic Data and Nuclear Data Tables*, 26(3–4):197–381, 1981.
[28] D. B. Milošević. Reexamination of the improved strong-field approximation: Low-energy structures in the above-threshold-ionization spectra for short-range potentials. *Physical Review A*, 88:023417, 2013.
[29] D. B. Milošević. Forward- and backward-scattering quantum orbits in above-threshold ionization. *Physical Review A*, 90:063414, 2014.
[30] D. B. Milošević. Low-frequency approximation for above-threshold ionization by a laser pulse: Low-energy forward rescattering. *Physical Review A*, 90:063423, 2014.
[31] D. B. Milošević and W. Becker. Role of long quantum orbits in high-order harmonic generation. *Physical Review A*, 66:063417, 2002.
[32] D. B. Milošević, E. Hasović, M. Busuladžić, A. Gazibegović-Busuladžić, and W. Becker. Intensity-dependent enhancements in high-order above-threshold ionization. *Physical Review A*, 76:053410, 2007.
[33] D. B. Milošević. *Few-Cycle-Laser-Pulse Induced and Assisted Processes in Atoms, Molecules, and Nanostructures* (pp. 27–47). Springer International Publishing, Cham, 2016.
[34] D. B. Milošević. Phase space path-integral formulation of the above-threshold ionization. *Journal of Mathematical Physics*, 54(4):042101, 2013.
[35] D. B. Milošević, D. Bauer, and W. Becker. Quantum-orbit theory of high-order atomic processes in intense laser fields. *Journal of Modern Optics*, 53(1–2):125–134, 2006.
[36] D. B. Milošević, G. G. Paulus, D. Bauer, and W. Becker. Above-threshold ionization by few-cycle pulses. *Journal of Physics B: Atomic, Molecular and Optical Physics*, 39(14):R203, 2006.
[37] D. B. Milošević, W. Becker, M. Okunishi, G. Prümper, K. Shimada, and K. Ueda. Strong-field electron spectra of rare-gas atoms in the rescattering regime: Enhanced spectral regions and a simulation of the experiment. *Journal of Physics B: Atomic, Molecular and Optical Physics*, 43 (1):015401, 2010.
[38] D. B Milošević. Cut-off law for high-harmonic generation by an elliptically polarized laser field. *Journal of Physics B: Atomic, Molecular and Optical Physics*, 33(13):2479, 2000.
[39] D. B. Milošević and F. Ehlotzky. Scattering and reaction processes in powerful laser fields. *Advances in Atomic, Molecular, and Optical Physics*, vol. 49 (pp. 373–532). Academic Press, Amsterdam, 2003.
[40] M. Möller, F. Meyer, A. M. Sayler, G. G. Paulus, M. F. Kling, B. E. Schmidt, W. Becker, and D. B. Milošević. Off-axis low-energy structures in above-threshold ionization. *Physical Review A*, 90:023412, 2014.

[41] M. Nisoli and G. Sansone. New frontiers in attosecond science. *Progress in Quantum Electronics*, 33(1):17–59, 2009.

[42] S. Odžak and D. B. Milošević. High-order harmonic generation in the presence of a static electric field. *Physical Review A*, 72:033407, 2005.

[43] S. Odžak and D. B. Milošević. Ellipticity and the offset angle of high harmonics generated by homonuclear diatomic molecules. *Journal of Physics B: Atomic, Molecular and Optical Physics*, 44(12):125602, 2011.

[44] R. Pazourek, S. Nagele, and J. Burgdörfer. Attosecond chronoscopy of photoemission. *Reviews of Modern Physics*, 87:765–802, 2015.

[45] A. M. Perelomov, V. S. Popov, and M. V. Terent'ev. Ionization of Atoms in an alternating electric field. *Soviet Physics JETP*, 23:924, 1966. Russian original - ZhETF, Vol. 50, No. 5, p. 1393, November 1966.

[46] A. A. Radzig and B. M. Smirnov. *Reference Data on Atoms, Molecules, and Ions*. Springer Berlin Heidelberg, Berlin, Heidelberg, 1985.

[47] H. R. Reiss. Effect of an intense electromagnetic field on a weakly bound system. *Physical Review A*, 22:1786–1813, 1980.

[48] P. Salières, B. Carré, L. Le Déroff, F. Grasbon, G. G. Paulus, H. Walther, R. Kopold, W. Becker, D. B. Milošević, A. Sanpera, and M. Lewenstein. Feynman's path-integral approach for intense-laser-atom interactions. *Science*, 292(5518):902–905, 2001.

[49] P. Salières, A. Maquet, S. Haessler, J. Caillat, and R. Taïeb. Imaging orbitals with attosecond and ångström resolutions: Toward attochemistry? *Reports on Progress in Physics*, 75(6): 062401, 2012.

[50] A. Scrinzi, M. Y. Ivanov, R. Kienberger, and D. M. Villeneuve. Attosecond physics. *Journal of Physics B: Atomic, Molecular and Optical Physics*, 39(1):R1, 2006.

[51] X. M. Tong, Z. X. Zhao, and C. D. Lin. Theory of molecular tunneling ionization. *Physical Review A*, 66:033402, 2002.

[52] K. Ueda and K. L. Ishikawa. Attosecond science: Attoclocks play devil's advocate. *Nature Physics*, 7(5):371–372, 2011.

[53] D. M. Wolkow. Über eine Klasse von Lösungen der Diracschen Gleichung. *Zeitschrift für Physik*, 94(3):250–260, 1935.

[54] B. Wolter, M. G. Pullen, M. Baudisch, M. Sclafani, M. Hemmer, A. Senftleben, C. D. Schröter, J. Ullrich, R. Moshammer, and J. Biegert. Strong-field physics with mid-ir fields. *Physical Review X*, 5:021034, 2015.

Christian Peltz, Charles Varin, Thomas Brabec, and Thomas Fennel

VIII Microscopic particle-in-cell approach

A broad spectrum of scenarios resulting from the exposure of dense material to strong laser fields takes place in the realm of intense but nonrelativistic light-matter interactions. The quest for a microscopic understanding of the underlying processes is driven by both fundamental interest and various striking applications, ranging from industry-driven technologies like laser micromachining [13] and laser modification of metals and dielectric materials [48, 52] over the development of devices based on ultrafast strong-field nanoplasmonics [25, 35, 44, 54] to attosecond dynamics in solids [30, 49]. For intensities close to the ionization threshold where material is transformed from a solid into a plasma, the dynamics is particularly complicated as it is dominated by transient effects, proceeds far from equilibrium, and is strongly coupled [33, 42]. The latter aspect makes a physical understanding challenging, as the description of strong coupling is intimately connected to a correlated description of the physical many-body processes.

Modeling the interaction of laser light with strongly coupled plasmas is a challenging task even in the nondegenerate regime,[1] as the classical trajectories of all electrons and ions have to be propagated explicitly, and microscopic processes such as collisions have to be fully resolved. For small systems, where the dipole approximation is justified and field propagation effects can be neglected, this can be done efficiently with electrostatic molecular dynamics (MD) calculations [16, 43]. However, to describe macroscopic plasma volumes, an electromagnetic treatment is required that fully accounts for field propagation effects like field attenuation.

A widely used numerical method to study the interaction of light with macroscopic plasma volumes including field propagation effects is the electromagnetic particle-in-cell (PIC) approach [7, 11, 51]. In PIC, Maxwell's equations are solved on a grid along with the relativistic equations of motion for all PIC particles. Typically, these PIC particles represent an average over many physical particles and are sampled

1 For thermal energies, $k_B T$, smaller or equal to the Fermi energy, E_F, quantum effects (e.g., Pauli-blocking) become important, i.e., the plasma is degenerate, and a classical description is in general not longer justified. In terms of the degeneracy parameter, $\Theta = k_{\mathrm{B}} T / E_F \lesssim 1$, the classical approximation is valid for $\Theta > 1$.

Christian Peltz: Institute of Physics, University of Rostock, 18051 Rostock, Germany
Charles Varin, Thomas Brabec: Department of Physics, University of Ottawa, Ottawa, ON K1N 6N5, Canada
Thomas Fennel: Institute of Physics, University of Rostock, 18051 Rostock, Germany; email: thomas.fennel@uni-rostock.de

De Gruyter Graduate – Computational Strong-Field Quantum Dynamics, Volume 5, 2017, pp. 227–270.
DOI 10.1515/9783110417265-008

on a coarse grid. As a result of this averaging, all microscopic processes are lost and only the collective response is explicitly described, which precludes a meaningful description of strongly coupled plasmas.

In this chapter, we present one possible route to overcome the above limitations. The so-called microscopic particle-in-cell (MicPIC) method connects MD and PIC in a two-level approach. In MicPIC, long-range electromagnetic interactions are treated on a coarse numerical grid (PIC level). To also fully resolve microscopic processes, the short-range interactions missing in the PIC part are reintroduced via local electrostatic MD.

In the following, we begin with the description of the basic concept of the MicPIC method from a physical point of view before we examine the most important details of its numerical implementation. Subsequently, we discuss two application scenarios. First, we present a relatively simple calculation, where a solid-density foil is excited with intense laser light under normal incidence. These results might serve as a reference for your own implementation of MicPIC and can be performed easily on a desktop computer. The second scenario discusses a more sophisticated application that became numerically accessible only with the advent of MicPIC, i.e., the complete simulation of a time-resolved coherent diffractive imaging experiment.

It should be noted that the following discussion has the form of a tutorial and is in substantial parts based on previously published work. Major parts have been adapted from the PhD thesis of Christian Peltz [36].

1 Basic concept

We start with a discussion of the key idea of MicPIC from a physical point of view.

1.1 Physical problem

In a classical picture, the exact nonrelativistic dynamics of a plasma particle in the presence of electromagnetic fields is determined by the Lorentz force

$$\mathbf{f}_i = \int \rho_i(\mathbf{r}) (\mathbf{E} + \dot{\mathbf{r}}_i \times \mathbf{B}) \, d^3r, \tag{1}$$

where $\rho_i(\mathbf{r})$ and $\dot{\mathbf{r}}_i$ are the charge density distribution and velocity of the ith particle, and $\mathbf{E}$ and $\mathbf{B}$ are the microscopic electric and magnetic fields. Once these fields are known, the force on each particle can be calculated, and the dynamics of the system follows from the self-consistent integration of the classical equations of motion. The self-consistent evolution of the corresponding classical electromagnetic fields is determined by Maxwell's curl equations according to

$$\nabla \times \mathbf{E} = -\dot{\mathbf{B}}, \tag{2}$$

$$\nabla \times \mathbf{B} = \mu_0 \mathbf{j} + \mu_0 \varepsilon_0 \dot{\mathbf{E}}. \tag{3}$$

Here, the charged particles couple to the electromagnetic fields via the current density $\mathbf{j} = \sum_i \dot{\mathbf{r}}_i \rho_i$. From a physical point of view, the classical plasma dynamics is completely specified by (1)–(3). However, the full numerical solution of this set of equations becomes very demanding for many-particle systems. The reason is that the numerical effort for solving the field equations on a grid scales with $1/\Delta x^4$ (as will be shown in Section 2.6), where Δx is the mesh size (grid resolution). In order to resolve microscopic processes properly, the mesh has to be of the order of one atomic unit or even less, where the direct numerical solution becomes prohibitive in practice. The MicPIC approach offers a route to overcome this problem and is developed step by step below.

1.2 Particle representation

In MicPIC, each plasma particle (electron or ion) is represented by a charge density distribution

$$\rho_i(\mathbf{r}) = q_i g(|\mathbf{r} - \mathbf{r}_i|, w_0), \tag{4}$$

where q_i and $\mathbf{r}_i$ are the charge and position of the ith particle, and the shape function $g(r, w) = \exp(-r^2/w^2)/\pi^{3/2} w^3$ describes a normalized Gaussian distribution of width w.

The long-range interaction between two of such particles is essentially determined by the particle charges and effectively independent of the particle shape. However, their short-range interaction is strongly dependent on their exact shape and size. In general, larger particle widths yield smoother fields in their vicinity and therefore result in a softened short-range interaction. These short-range interactions primarily determine the collision dynamics of plasma particles. Therefore, the width w_0 is a key parameter that should, within technical and physical limits (see Section 2.6), be chosen as small as possible.[2] On the other hand, the particle width also determines the spatial resolution that is necessary to resolve the particle shape on a grid. In turn, a large particle width is key to an efficient solution of Maxwell's equations, as it allows the use of a coarse numerical grid and large time steps.

So far, there have been two major approaches to deal with these conflicting needs: (i) The regular PIC approach uses superparticles. A superparticle represents multiple physical particles, e.g., up to millions of electrons or ions, with large particle

2 The Coulomb singularity at zero particle distance can only be observed in the limit of vanishing width, corresponding to point like particles.

widths to describe the collective plasma dynamics but neglects particle collisions. The latter assumption is justified only in the relativistic regime or at low density [27, 40, 41]. And (ii) the collisional PIC approach, which uses the same superparticles but reintroduces the underestimated collisions via Monte-Carlo methods using approximate binary collision rates [32, 47]. Both approaches have their merits and have revolutionized our understanding of strong-field plasma physics. However, even the collisional PIC method is restricted to the regime of weak coupling, where microscopic fluctuations are negligible and microscopic interactions are limited to small-angle binary collisions. MicPIC is intended to describe the nonrelativistic dynamics of laser-driven clusters and bulk materials, which proceeds far from equilibrium and is strongly coupled. Therefore, the short-range interactions have to be taken into account explicitly. Within MicPIC, this is done in a consistent two-level approach that combines the electromagnetic treatment of the collective plasma dynamics on a PIC level with a local electrostatic MD to describe microscopic correlations (Mic).

1.3 PIC approximation

On the PIC level, particles are represented by a particle width larger than the actual particle width ($w_{\text{pic}} \gg w_0$) via the corresponding smoothed particle charge density

$$\rho_i^{\text{pic}}(\mathbf{r}) = q_i g(|\mathbf{r} - \mathbf{r}_i|, w_{\text{pic}}). \tag{5}$$

The super/subscripts "pic" indicate that the respective quantities belong to the PIC level. The corresponding PIC electric and magnetic field evolution is then given by

$$\nabla \times \mathbf{E}^{\text{pic}} = -\dot{\mathbf{B}}^{\text{pic}} \tag{6}$$

$$\nabla \times \mathbf{B}^{\text{pic}} = \mu_0 \mathbf{j}^{\text{pic}} + \mu_0 \varepsilon_0 \dot{\mathbf{E}}^{\text{pic}} \tag{7}$$

with the smoothed current density $\mathbf{j}^{\text{pic}} = \sum_i \dot{\mathbf{r}}_i \rho_i^{\text{pic}}$. Note that in this description, radiation fields are fully accounted for if w_{pic} is smaller than all relevant scales (wavelength, skin depth, etc.). However, the microscopic nature of the particles and therefore also all resulting correlation effects are lost because of the large PIC particle size. The PIC force on the ith plasma particle is given by

$$\mathbf{f}_i^{\text{pic}} = \int \rho_i^{\text{pic}} (\mathbf{E}^{\text{pic}} + \dot{\mathbf{r}}_i \times \mathbf{B}^{\text{pic}}) \, d^3 r. \tag{8}$$

1.4 MicPIC force decomposition

To identify the missing short-range forces in the PIC approximation, the actual force (1) on plasma particle i can be formally split into a microscopic part $\mathbf{f}_i^{\text{mic}}$ and a long-range

PIC part $\mathbf{f}_i^{\text{pic}}$,

$$\mathbf{f}_i = \mathbf{f}_i^{\text{mic}} + \mathbf{f}_i^{\text{pic}}, \tag{9}$$

with

$$\mathbf{f}_i^{\text{mic}} = \int \left[\rho_i(\mathbf{E} + \dot{\mathbf{r}}_i \times \mathbf{B}) - \rho_i^{\text{pic}}(\mathbf{E}^{\text{pic}} + \dot{\mathbf{r}}_i \times \mathbf{B}^{\text{pic}})\right] d^3r, \tag{10}$$

$$\mathbf{f}_i^{\text{pic}} = \int \rho_i^{\text{pic}}(\mathbf{E}^{\text{pic}} + \dot{\mathbf{r}}_i \times \mathbf{B}^{\text{pic}})\, d^3r. \tag{11}$$

Besides rearrangement of the terms, (9) is still identical to the force in (1). To show the short-range character of the microscopic contribution $\mathbf{f}_i^{\text{mic}}$, PIC and actual electric and magnetic fields have to be decomposed into their individual particle contributions. The respective total fields are then given by the sum over the field contributions created by all particles. This decomposition is justified because of the linearity of Maxwell's equations and leads to

$$\mathbf{f}_i^{\text{mic}} = \sum_j \int [\rho_i(\mathbf{E}_j + \dot{\mathbf{r}}_i \times \mathbf{B}_j) - \rho_i^{\text{pic}}(\mathbf{E}_j^{\text{pic}} + \dot{\mathbf{r}}_i \times \mathbf{B}_j^{\text{pic}})]\, d^3r. \tag{12}$$

The above sum describes the force on the ith particle, created by the fields of all other particles j. Self-force contributions cancel out automatically. For large distances between particles j and i ($r_{ij} = |\mathbf{r}_j - \mathbf{r}_i| \gg w_{\text{pic}}$), the actual and PIC fields produced in the region $\mathbf{r} \simeq \mathbf{r}_i$ are identical.[3] Therefore, remaining contributions to $\mathbf{f}_i^{\text{mic}}$ in the far field could only stem from the different actual and PIC particle densities.

The variation of the (actual and PIC) fields over the PIC particle extent can be approximated by a linear Taylor expansion around $\mathbf{r}_i$. The corresponding expansion for the actual electric field reads as

$$\mathbf{E}_j(\mathbf{r}) = \mathbf{E}_j(\mathbf{r}_i) + (\mathbf{r} - \mathbf{r}_i)\nabla\mathbf{E}_j(\mathbf{r}_i) + \cdots. \tag{13}$$

Carrying out the corresponding integration

$$\begin{aligned}\int \rho_i(\mathbf{r})\,\mathbf{E}_j(\mathbf{r})\, d^3r &= \int \rho_i(\mathbf{r})\,\mathbf{E}_j(\mathbf{r}_i)\, d^3r + \int \rho_i(\mathbf{r})\,\nabla\mathbf{E}_j(\mathbf{r}_i)(\mathbf{r} - \mathbf{r}_i)\, d^3r + \cdots \\ &\simeq \mathbf{E}_j(\mathbf{r}_i) \int \rho_i(\mathbf{r})\, d^3r \end{aligned} \tag{14}$$

gives zero for the linear field terms due to the even symmetry of the charge density. If higher-order terms are negligible, only the constant field terms remain and can be

3 Here, it is assumed that w_{pic} is much smaller than all scales of the radiated fields.

pulled out of the integral

$$\int \rho_i(\mathbf{r}) \mathbf{E}_j(\mathbf{r})\, d^3 r = \mathbf{E}_j(\mathbf{r}_i) \int \rho_i(\mathbf{r})\, d^3 r \tag{15}$$

$$= q_i \mathbf{E}_j(\mathbf{r}_i). \tag{16}$$

The remaining integral over the particle density yields the particle charge. As the total charges of actual and PIC particles are equal, their contributions cancel each other for each index j in (12), proving the short-range nature of the microscopic correction for the electric field. Analogous steps show the short-range nature of the magnetic field term. Note that the interaction of the plasma particles is described exactly with the force from (9), independent of the width of the particles on the PIC level. The value of the PIC particle width only determines the softness of the force on the PIC level and, in turn, the radius within which the Mic forces contribute.

1.5 The MicPIC approximation

So far, the above force decomposition has only formal character as everything has been derived in full generality. However, in order to facilitate the numerical evaluation of the short-range interaction, field retardation is neglected locally, i.e., within the microscopic correction. This is the only formal approximation in MicPIC. Taking the nonrelativistic, electrostatic limit of (12) by dropping magnetic fields and expressing electric fields by the respective Coulomb interaction yields

$$\mathbf{f}_i^{\text{mic}} = -\nabla_{\mathbf{r}_i} \sum_j \int\int \left[\frac{\rho_i(\mathbf{r})\rho_j(\mathbf{r}')}{4\pi\varepsilon_0|\mathbf{r}-\mathbf{r}'|} - \frac{\rho_i^{\text{pic}}(\mathbf{r})\rho_j^{\text{pic}}(\mathbf{r}')}{4\pi\varepsilon_0|\mathbf{r}-\mathbf{r}'|} \right] d^3 r' d^3 r. \tag{17}$$

For Gaussian shape functions, the above double integral can be evaluated analytically and yields the difference of the particle interaction energies for actual and PIC particles. The interaction energy of two Gaussian particles with width parameter w is given by

$$V_{ij}(r_{ij}, w) = \frac{q_i q_j}{4\pi\varepsilon_0 r_{ij}} \operatorname{erf}\left(\frac{r_{ij}}{\sqrt{2}w} \right). \tag{18}$$

Inserting this expression into (17) yields

$$\mathbf{f}_i^{\text{mic}} = -\sum_j \nabla_{\mathbf{r}_i} V_{ij}^{\text{mic}} \tag{19}$$

with

$$V_{ij}^{\text{mic}} = V_{ij}(r_{ij}, w_0) - V_{ij}(r_{ij}, w_{\text{pic}}).$$

Combining the electrostatically approximated microscopic correction in (19) and the PIC force in (8) yields the total MicPIC force

$$\mathbf{f}_i = -\sum_j \nabla_{\mathbf{r}_i} V_{ij}^{\text{mic}} + \int \rho_i^{\text{pic}} (\mathbf{E}^{\text{pic}} + \dot{\mathbf{r}}_i \times \mathbf{B}^{\text{pic}})\, d^3 r. \tag{20}$$

The complete MicPIC dynamics is now determined by the self-consistent integration of Newton's equations of motion with the force specified in (20), together with the propagation of the electromagnetic fields according to (6) and (7).

2 Numerical aspects of MicPIC

This section discusses the most important numerical aspects of the implementation of the MicPIC concept. In the first part, Section 2.1, the numerical propagation of the electromagnetic fields via the finite-difference time-domain (FDTD) algorithm is described. Special emphasis is put on the description of the field propagation algorithm on a discretized staggered spatial grid as well as the implementation of appropriate absorbing boundary conditions to emulate an infinite simulation volume. In the second part, Section 2.2, the representation of the Gaussian particles on the numerical grid is discussed in more detail. This is of particular importance, as the particle shape and its representation on the grid determine major properties of the code, like charge conservation, force anisotropy, and numerical effort. After that, the implementation and efficient evaluation of the short-range forces are discussed in Section 2.3. Subsequently, the Boris scheme for particle propagation in the presence of electromagnetic fields is described (Section 2.4), completing the basic part of the code. Finally, the implementation of atomic ionization processes into MicPIC is described in Section 2.5.

2.1 Electromagnetic field propagation with the FDTD method

The foundation of the MicPIC approach is the numerical description of the time evolution of the electric and magnetic fields on the PIC level according to (6) and (7). In this work, the FDTD method has been used for that purpose, as it solves Maxwell's equations in the time domain and is relatively simple to implement. The following discussion of the FDTD idea and its implementation follows the book of Taflove [45], which is an excellent compendium of most relevant FDTD-related information. As a starting point for our discussion, the basic equations are repeated here for convenience,

$$\nabla \times \mathbf{E} = -\dot{\mathbf{B}}, \tag{21a}$$

$$\nabla \times \mathbf{B} = \mu_0 \left(\mathbf{j} + \varepsilon_0 \dot{\mathbf{E}} \right), \tag{21b}$$

with the electric field **E**, the magnetic field **B**, the vacuum permittivity ε_0 and permeability μ_0, as well as the electric current density **j**. As fields and currents are exclusively discussed for the PIC level in this section, the superscript "pic" has been dropped. Writing out the vector components of (21a) and (21b) yields the following system of six coupled first-order differential equations:

$$\frac{\partial E_x}{\partial t} = \frac{1}{\varepsilon_0 \mu_0} \left[\frac{\partial B_z}{\partial y} - \frac{\partial B_y}{\partial z} \right] - \frac{1}{\varepsilon_0} j_x, \tag{22a}$$

$$\frac{\partial E_y}{\partial t} = \frac{1}{\varepsilon_0 \mu_0} \left[\frac{\partial B_x}{\partial z} - \frac{\partial B_z}{\partial x} \right] - \frac{1}{\varepsilon_0} j_y, \tag{22b}$$

$$\frac{\partial E_z}{\partial t} = \frac{1}{\varepsilon_0 \mu_0} \left[\frac{\partial B_y}{\partial x} - \frac{\partial B_x}{\partial y} \right] - \frac{1}{\varepsilon_0} j_z, \tag{22c}$$

$$\frac{\partial B_x}{\partial t} = \left[\frac{\partial E_y}{\partial z} - \frac{\partial E_z}{\partial y} \right], \tag{22d}$$

$$\frac{\partial B_y}{\partial t} = \left[\frac{\partial E_z}{\partial x} - \frac{\partial E_x}{\partial z} \right], \tag{22e}$$

$$\frac{\partial B_z}{\partial t} = \left[\frac{\partial E_x}{\partial y} - \frac{\partial E_y}{\partial x} \right]. \tag{22f}$$

The basic idea for the solution of this problem in the framework of the FDTD algorithm has been proposed by Kane Yee already in 1966 [53]. It is based on a specific space and time staggering of the field components in conjunction with the centered finite difference scheme to discretize the space and time derivatives.

2.1.1 Centered finite difference

To derive a finite difference expression for the spatial derivative of a scalar function $u(x, t_n)$, it is convenient to consider its Taylor series expansion around space point x_i at a fixed time t_n. For positive displacements in space, the expansion is given by

$$u(x_i + \Delta x)\,|_{t_n} = u\,|_{x_i,t_n} + \Delta x \frac{\partial u}{\partial x}\,|_{x_i,t_n} + \frac{(\Delta x)^2}{2} \frac{\partial^2 u}{\partial x^2}\,|_{x_i,t_n} + \frac{(\Delta x)^3}{6} \frac{\partial^3 u}{\partial x^3}\,|_{x_i,t_n} + \cdots$$

and for negative displacements by

$$u(x_i - \Delta x)\,|_{t_n} = u\,|_{x_i,t_n} - \Delta x \frac{\partial u}{\partial x}\,|_{x_i,t_n} + \frac{(\Delta x)^2}{2} \frac{\partial^2 u}{\partial x^2}\,|_{x_i,t_n} - \frac{(\Delta x)^3}{6} \frac{\partial^3 u}{\partial x^3}\,|_{x_i,t_n} + \cdots.$$

Subtracting the second from the first equation leads to

$$u(x_i + \Delta x)\,|_{t_n} - u(x_i - \Delta x)\,|_{t_n} = 2\Delta x \frac{\partial u}{\partial x}\,|_{x_i,t_n} + \frac{(\Delta x)^3}{3} \frac{\partial^3 u}{\partial x^3}\,|_{x_i,t_n} + \cdots. \tag{23}$$

Rearranging the terms results in the well-known centered finite difference expression for the derivative

$$\left.\frac{\partial u}{\partial x}\right|_{x_i,t_n} = \frac{u(x_i+\Delta x)\,|_{t_n} - u(x_i-\Delta x)\,|_{t_n}}{2\Delta x} + \mathcal{O}[(\Delta x)^2] \tag{24}$$

with second-order accuracy. The actual field and current components are functions of three space and one time coordinate. For convenience, the shorter notation

$$u(i\Delta x, j\Delta y, k\Delta z, n\Delta t) = u^n_{i,j,k} \tag{25}$$

is used from here on. Evaluating the Taylor expansions with half displacements $x_i \pm \frac{\Delta x}{2}$ leads to the final expressions for the space and time derivatives

$$\frac{\partial u}{\partial x}(i\Delta x, j\Delta y, k\Delta z, n\Delta t) = \frac{u^n_{i+\frac{1}{2},j,k} - u^n_{i-\frac{1}{2},j,k}}{\Delta x} + \mathcal{O}[(\Delta x)^2], \tag{26a}$$

$$\frac{\partial u}{\partial y}(i\Delta x, j\Delta y, k\Delta z, n\Delta t) = \frac{u^n_{i,j+\frac{1}{2},k} - u^n_{i,j-\frac{1}{2},k}}{\Delta y} + \mathcal{O}[(\Delta y)^2], \tag{26b}$$

$$\frac{\partial u}{\partial z}(i\Delta x, j\Delta y, k\Delta z, n\Delta t) = \frac{u^n_{i,j,k+\frac{1}{2}} - u^n_{i,j,k-\frac{1}{2}}}{\Delta z} + \mathcal{O}[(\Delta z)^2], \tag{26c}$$

$$\frac{\partial u}{\partial t}(i\Delta x, j\Delta y, k\Delta z, n\Delta t) = \frac{u^{n+\frac{1}{2}}_{i,j,k} - u^{n-\frac{1}{2}}_{i,j,k}}{\Delta t} + \mathcal{O}[(\Delta t)^2], \tag{26d}$$

which will be used in the following.

2.1.2 The Yee staggering

The positioning of the electric- and magnetic-field components on the numerical grid according to Yee [53] is shown in Figure 1. The reason for this specific staggering becomes evident when the above finite difference expressions are applied to the field equations. Doing this exemplarily for (22a) yields

$$\begin{aligned}
&\frac{E_x\,|^{n+\frac{1}{2}}_{i+\frac{1}{2},j,k} - E_x\,|^{n-\frac{1}{2}}_{i+\frac{1}{2},j,k}}{\Delta t} \\
&\quad = \frac{1}{\varepsilon_0}\left[\frac{B_z\,|^n_{i+\frac{1}{2},j+\frac{1}{2},k} - B_z\,|^n_{i+\frac{1}{2},j-\frac{1}{2},k}}{\Delta y} - \frac{B_y\,|^n_{i+\frac{1}{2},j,k+\frac{1}{2}} - B_y\,|^n_{i+\frac{1}{2},j,k-\frac{1}{2}}}{\Delta z}\right] \\
&\quad - \frac{1}{\varepsilon_0} j_x\,|^n_{i+\frac{1}{2},j,k}\,.
\end{aligned} \tag{27}$$

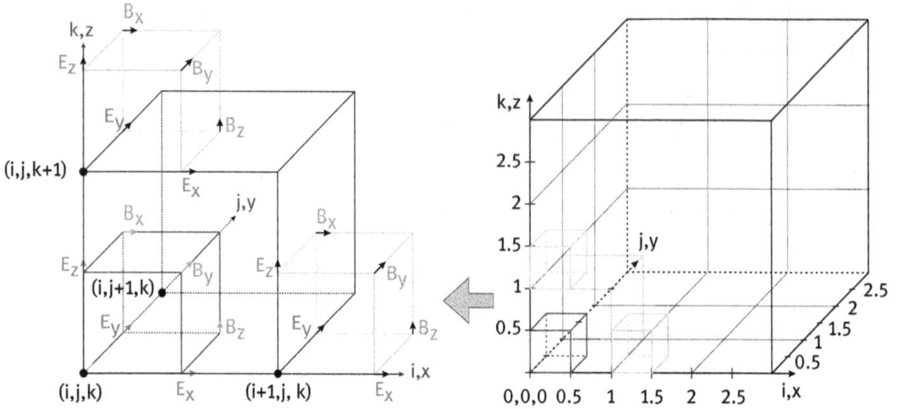

Fig. 1. Schematic illustration of the Yee field staggering [53]. Electric and magnetic field components are chosen such that the positions fit automatically to a centered finite differences solution to the curl equations.

Rearranging the terms leads to an explicit expression for the propagation of E_x

$$
\begin{aligned}
E_x\big|_{i+\frac{1}{2},j,k}^{n+\frac{1}{2}} \\
&= E_x\big|_{i+\frac{1}{2},j,k}^{n-\frac{1}{2}} \\
&\quad + \frac{\Delta t}{\varepsilon_0}\left[\frac{B_z\big|_{i+\frac{1}{2},j+\frac{1}{2},k}^{n} - B_z\big|_{i+\frac{1}{2},j-\frac{1}{2},k}^{n}}{\Delta y} - \frac{B_y\big|_{i+\frac{1}{2},j,k+\frac{1}{2}}^{n} - B_y\big|_{i+\frac{1}{2},j,k-\frac{1}{2}}^{n}}{\Delta z}\right] \\
&\quad - \frac{\Delta t}{\varepsilon_0} j_x\big|_{i+\frac{1}{2},j,k}^{n}
\end{aligned}
\tag{28}
$$

that only relies on the knowledge of all field components in the past. Comparing the positions of the required field components in (28) with the positions in Figure 1 shows that the Yee staggering is perfectly matched to this propagation scheme. Similar expressions can be derived for the other field components, completing the basic numerical recipe for the numerical solution to Maxwell's equations. Note that the electric- and magnetic-field components are also staggered in time, which results in alternating field updates for electric and magnetic fields.

2.1.3 Absorbing boundary conditions

The FDTD scheme introduced above represents a very efficient way to solve the microscopic Maxwell equations on a numerical grid. As such a grid consumes memory, it will always be limited in size so that some sort of boundary conditions have to be applied. In principle, there are three main types of boundary conditions, suitable for different physical problems. The perfect electric conductor boundary condition mimics a perfect metal surface to truncate the computational grid, reflecting all impinging electromagnetic waves. Periodic boundary conditions (PBC) emulate a periodic continuation of the computational domain. This can be very useful for the simulation of targets that are effectively homogeneous in at least one direction, e.g., bulk targets or planar surfaces. The third and for this work most important type of boundary condition is the absorbing boundary condition (ABC). It simulates the extension of the lattice to infinity and thus allows the description of open systems.

The basic idea

An obvious approach to achieve ABC's is to enclose the computational box with a highly absorbing medium layer. The main goal of such an electromagnetic absorption layer is to obtain dissipation within the layer without reflection from the interface back into the main simulation volume. From a numerical point of view, this can be achieved by using Maxwell's curl equations for a dissipative medium [24, 45]. The corresponding equations read as

$$\nabla \times \mathbf{H} = \frac{\partial \mathbf{D}}{\partial t} + \sigma \mathbf{E}, \tag{29a}$$

$$\nabla \times \mathbf{E} = -\frac{\partial \mathbf{B}}{\partial t} - \sigma^{*} \mathbf{H}, \tag{29b}$$

where **D** and **H** are the displacement field and magnetizing field, respectively. The electric conductivity σ and the equivalent magnetic loss σ^* determine the dissipative properties of the medium. For linear, isotropic, and nondispersive materials,[4] **D** and **E** as well as **B** and **H** can be related by

$$\mathbf{D} = \varepsilon_0 \varepsilon_r \mathbf{E} = \varepsilon \mathbf{E}, \tag{30a}$$

$$\mathbf{B} = \mu_0 \mu_r \mathbf{H} = \mu \mathbf{H}. \tag{30b}$$

Inserting these expressions into (29) eliminates **D** and **B** and leads to

$$\nabla \times \mathbf{H} = \varepsilon \frac{\partial \mathbf{E}}{\partial t} + \sigma \mathbf{E}, \tag{31a}$$

$$\nabla \times \mathbf{E} = -\mu \frac{\partial \mathbf{H}}{\partial t} - \sigma^{*} \mathbf{H}. \tag{31b}$$

4 The medium is assumed to have frequency independent values $\varepsilon_r(\omega) = \varepsilon_r$ and $\mu_r(\omega) = \mu_r$.

For the following derivation, it is convenient to translate these equations into the frequency domain ($\partial_t \to -i\omega$),[5] which yields a very compact notation

$$\nabla \times \breve{\mathbf{H}} = -i\omega\varepsilon\breve{\mathbf{E}} + \sigma\breve{\mathbf{E}} = -i\omega\varepsilon\left(1 - \frac{\sigma}{i\omega\varepsilon}\right)\breve{\mathbf{E}} = -i\omega\varepsilon\lambda_\varepsilon\breve{\mathbf{E}}, \tag{32a}$$

$$\nabla \times \breve{\mathbf{E}} = i\omega\mu\breve{\mathbf{H}} - \sigma^*\breve{\mathbf{H}} = i\omega\mu\left(1 - \frac{\sigma^*}{i\omega\mu}\right)\breve{\mathbf{H}} = i\omega\mu\lambda_\mu\breve{\mathbf{H}}, \tag{32b}$$

with the new material parameters λ_ε and λ_μ. Note that quantities in the frequency domain are flagged with a breve from here on.

Plane wave incident on a lossy half-space

Eventually, the main simulation volume should be completely surrounded by an absorbing medium layer. However, as a first step, the reflection properties of the interface between absorbing and nonabsorbing regions have to be evaluated. To this end, the incidence of a plane wave upon such an interface between a lossless (region 1, $x < 0$) and a lossy half-space (region 2, $x > 0$) is studied in more detail. Consider an incident wave defined by

$$\breve{\mathbf{H}}^{\text{inc}} = \mathbf{e}_z H_0\, e^{i\beta_{1_x}x + i\beta_{1_y}y}, \tag{33a}$$

$$\breve{\mathbf{E}}^{\text{inc}} = \left[\mathbf{e}_x \frac{\beta_{1y}}{\omega\varepsilon} - \mathbf{e}_y \frac{\beta_{1x}}{\omega\varepsilon}\right] H_0 e^{i\beta_{1_x}x + i\beta_{1_y}y}, \tag{33b}$$

with the wave vector

$$\beta_1 = \beta_{1_x}\mathbf{e}_x + \beta_{1_y}\mathbf{e}_y; \quad \beta_{1_x} = k_1 \cos\Theta; \quad \beta_{1_y} = k_1 \sin\Theta, \tag{34}$$

and the angle of incidence Θ. Then, according to (32), the total fields in region 1 ($\sigma = \sigma^* = 0$) are given by

$$\breve{\mathbf{H}}_1 = \mathbf{e}_z H_0 \left(1 + \Gamma e^{-2i\beta_{1_x}x}\right) e^{i\beta_{1_x}x + i\beta_{1_y}y}, \tag{35a}$$

$$\breve{\mathbf{E}}_1 = \left[\mathbf{e}_x \frac{\beta_{1y}}{\omega\varepsilon}\left(1 + \Gamma e^{-2i\beta_{1_x}x}\right) - \mathbf{e}_y \frac{\beta_{1x}}{\omega\varepsilon}\left(1 - \Gamma e^{-2i\beta_{1_x}x}\right)\right] H_0 e^{i\beta_{1_x}x + i\beta_{1_y}y}, \tag{35b}$$

where Γ is the reflection coefficient. The total fields in region 2 are given by

$$\breve{\mathbf{H}}_2 = \mathbf{e}_z H_0 \tau\, e^{i\beta_{2_x}x + i\beta_{2_y}y}, \tag{36a}$$

$$\breve{\mathbf{E}}_2 = \left[\mathbf{e}_x \frac{\beta_{2y}}{\omega\varepsilon\lambda_\varepsilon} - \mathbf{e}_y \frac{\beta_{2x}}{\omega\varepsilon\lambda_\varepsilon}\right] H_0 \tau e^{i\beta_{2_x}x + i\beta_{2_y}y}, \tag{36b}$$

5 The corresponding Fourier transforms are defined as $F(t) = \frac{1}{\sqrt{2\pi}} \int \breve{F}(\omega) e^{-i\omega t} d\omega$ and $\breve{F}(\omega) = \frac{1}{\sqrt{2\pi}} \int F(t) e^{i\omega t} dt$.

with the transmission coefficient τ. Enforcing continuity of the tangential fields across the interface at $x = 0$ yields expressions for the reflection and transmission coefficients

$$\Gamma = \frac{\frac{\beta_{1x}}{\omega\varepsilon} - \frac{\beta_{2x}}{\omega\varepsilon\lambda_\varepsilon}}{\frac{\beta_{1x}}{\omega\varepsilon} + \frac{\beta_{2x}}{\omega\varepsilon\lambda_\varepsilon}}; \quad \tau = 1 + \Gamma. \tag{37}$$

For arbitrary angles of incidence Θ, the reflection coefficient is in general nonzero. However, it can be shown that for normal incidence ($\Theta = 0$) and an appropriate choice of the parameters, i.e., $\lambda_\varepsilon = \lambda_\mu \Rightarrow \sigma^* = \sigma\mu/\varepsilon$, plane electromagnetic waves can enter the lossy half-space without reflection, irrespective of their frequency. Then, the transmitted fields given by

$$\breve{\mathbf{H}}_2 = \breve{\mathbf{H}}^{\text{inc}} e^{-\sigma\sqrt{\mu/\varepsilon}\,x}, \tag{38a}$$

$$\breve{\mathbf{E}}_2 = \breve{\mathbf{E}}^{\text{inc}} e^{-\sigma\sqrt{\mu/\varepsilon}\,x} \tag{38b}$$

resemble the incident fields but exhibit additional exponential damping in region 2. In this case, region 2 is called perfectly matched to region 1 for normal incident waves [23]. However, to be of practical use, the observed behavior is needed for arbitrary angles of incidence.

The uniaxial perfectly matched layer

A first solution to this problem has been presented by Berenger in 1994 in terms of the so-called perfectly matched layer (PML) [6]. Berenger used a split-field formulation of Maxwell's equations, which leads to a set of 12 coupled first-order differential equations. With an appropriate choice of the loss parameters, a perfectly matched interface for arbitrary wave incidence, polarization, and frequency can be derived.

In MicPIC, an alternative but equivalent formulation is utilized, the so-called uniaxial perfectly matched layer (UPML) introduced by Stephen Gedney in 1996 [18]. The main idea behind the UPML concept is to achieve perfect matching at the interface via an uniaxial anisotropic absorption layer. Therefore, the isotropic material parameters λ_ε and λ_μ in (32) are replaced by uniaxial tensors. Considering an interface perpendicular to the x-axis, similar to the isotropic case discussed above, the tensor parameters exhibit a form that is rotationally symmetric about the x-axis

$$\bar{\bar{\lambda}}_\varepsilon = \begin{pmatrix} a & 0 & 0 \\ 0 & b & 0 \\ 0 & 0 & b \end{pmatrix}, \quad \bar{\bar{\lambda}}_\mu = \begin{pmatrix} c & 0 & 0 \\ 0 & d & 0 \\ 0 & 0 & d \end{pmatrix}. \tag{39}$$

Maxwell's curl equations in the frequency domain then read as

$$\nabla \times \check{\mathbf{H}} = -i\omega\varepsilon\bar{\bar{\lambda}}_\varepsilon \check{\mathbf{E}}, \tag{40a}$$

$$\nabla \times \check{\mathbf{E}} = i\omega\mu\bar{\bar{\lambda}}_\mu \check{\mathbf{H}}. \tag{40b}$$

The examination of an incident plane wave onto the region interface (analogous to the case discussed above) shows that reflectionless transmission of plane waves from the main simulation volume into the UPML region can be achieved for

$$\bar{\bar{\lambda}}_\varepsilon = \bar{\bar{\lambda}}_\mu = \bar{\bar{s}} = \begin{pmatrix} s_x^{-1} & 0 & 0 \\ 0 & s_x & 0 \\ 0 & 0 & s_x \end{pmatrix}, \tag{41}$$

irrespective of the angle of incidence, polarization, and frequency of the incident wave (for more details, see [17, 18, 45]). The reflectionless property of the interface holds for any s_x. Defining it similar to the isotropic absorbing layer case discussed above

$$s_x = \left(1 - \frac{\sigma_x}{i\omega\varepsilon}\right) \tag{42}$$

creates a reflectionless absorbing layer as intended. Considering a plane wave incident on an interface at $x = 0$ analogous to the isotropic case from (33) leads to the transmitted fields

$$\check{\mathbf{H}}^{\mathrm{UPML}} = \mathbf{e}_z H_0\, e^{i\beta_{1x}x + i\beta_{1y}y} e^{-\sigma_x\sqrt{\mu/\varepsilon}\cos\Theta\, x}, \tag{43a}$$

$$\check{\mathbf{E}}^{\mathrm{UPML}} = \left[\mathbf{e}_x s_x \frac{\beta_{1y}}{\omega\varepsilon} - \mathbf{e}_y \frac{\beta_{1x}}{\omega\varepsilon}\right] H_0 e^{i\beta_{1x}x + i\beta_{1y}y} e^{-\sigma_x\sqrt{\mu/\varepsilon}\cos\Theta\, x}. \tag{43b}$$

The transmitted waves propagate with the same phase velocity as the incident wave but undergo exponential decay along the axis normal to the region interface. The magnitude of the decay depends on the angle of incidence Θ and can be additionally adjusted by an appropriate choice of the UPML parameter σ_x.

So far, only the construction of a reflectionless planar interface between two half-spaces has been discussed. To truncate a finite three-dimensional simulation volume, absorption layers adjacent to all outer lattice boundaries are needed. To this end, a general material tensor can be defined

$$\begin{aligned} \bar{\bar{s}} &= \begin{pmatrix} s_x^{-1} & 0 & 0 \\ 0 & s_x & 0 \\ 0 & 0 & s_x \end{pmatrix} \begin{pmatrix} s_y & 0 & 0 \\ 0 & s_y^{-1} & 0 \\ 0 & 0 & s_y \end{pmatrix} \begin{pmatrix} s_z & 0 & 0 \\ 0 & s_z & 0 \\ 0 & 0 & s_z^{-1} \end{pmatrix} \\ &= \begin{pmatrix} s_x^{-1} s_y s_z & 0 & 0 \\ 0 & s_x s_y^{-1} s_z & 0 \\ 0 & 0 & s_x s_y s_z^{-1} \end{pmatrix}, \end{aligned} \tag{44}$$

where

$$s_x = 1 - \frac{\sigma_x}{i\omega\varepsilon}; \quad s_y = 1 - \frac{\sigma_y}{i\omega\varepsilon}; \quad s_z = 1 - \frac{\sigma_z}{i\omega\varepsilon}. \tag{45}$$

This tensor is sufficient to describe the anisotropic PML medium in the entire FDTD simulation volume. By properly choosing the spatial dependence of σ_x, σ_y, and σ_z,

$$\sigma_x(x) = \begin{cases} \sigma'_x(x) & x \le x_{\min}, \ge x_{\max} \\ 0 & x_{\min} < x < x_{\max} \end{cases}, \tag{46a}$$

$$\sigma_y(y) = \begin{cases} \sigma'_y(y) & y \le y_{\min}, \ge y_{\max} \\ 0 & y_{\min} < y < y_{\max} \end{cases}, \tag{46b}$$

$$\sigma_z(z) = \begin{cases} \sigma'_z(z) & z \le z_{\min}, \ge z_{\max} \\ 0 & z_{\min} < z < z_{\max} \end{cases}, \tag{46c}$$

the tensor becomes the unit dyad in the main simulation volume, while it is still properly expressed in the PML regions. The final equations to solve read as

$$\begin{bmatrix} \frac{\partial \check{H}_z}{\partial y} - \frac{\partial \check{H}_y}{\partial z} \\ \frac{\partial \check{H}_x}{\partial z} - \frac{\partial \check{H}_z}{\partial x} \\ \frac{\partial \check{H}_y}{\partial x} - \frac{\partial \check{H}_x}{\partial y} \end{bmatrix} = -i\omega\varepsilon \begin{bmatrix} \frac{s_y s_z}{s_x} & 0 & 0 \\ 0 & \frac{s_x s_z}{s_y} & 0 \\ 0 & 0 & \frac{s_x s_y}{s_z} \end{bmatrix} \begin{bmatrix} \check{E}_x \\ \check{E}_y \\ \check{E}_z \end{bmatrix}, \tag{47}$$

and

$$\begin{bmatrix} \frac{\partial \check{E}_z}{\partial y} - \frac{\partial \check{E}_y}{\partial z} \\ \frac{\partial \check{E}_x}{\partial z} - \frac{\partial \check{E}_z}{\partial x} \\ \frac{\partial \check{E}_y}{\partial x} - \frac{\partial \check{E}_x}{\partial y} \end{bmatrix} = i\omega\mu \begin{bmatrix} \frac{s_y s_z}{s_x} & 0 & 0 \\ 0 & \frac{s_x s_z}{s_y} & 0 \\ 0 & 0 & \frac{s_x s_y}{s_z} \end{bmatrix} \begin{bmatrix} \check{H}_x \\ \check{H}_y \\ \check{H}_z \end{bmatrix}. \tag{48}$$

However, the direct transformation of these equations back into the time domain would lead to a convolution of the tensor coefficients and the magnetic and electric fields,[6] respectively. This would be computationally very expensive but can be circumvented by the definition of the relationships [18]

$$\check{D}_x = \varepsilon \frac{s_z}{s_x} \check{E}_x \quad , \quad \check{D}_y = \varepsilon \frac{s_x}{s_y} \check{E}_y \quad , \quad \check{D}_z = \varepsilon \frac{s_y}{s_z} \check{E}_z. \tag{49}$$

Inserting (49) into (47) leads to a decoupling of the frequency-dependent terms [17, 18]. Subsequent backtransformation into the time domain yields

$$\begin{bmatrix} \frac{\partial H_z}{\partial y} - \frac{\partial H_y}{\partial z} \\ \frac{\partial H_x}{\partial z} - \frac{\partial H_z}{\partial x} \\ \frac{\partial H_y}{\partial x} - \frac{\partial H_x}{\partial y} \end{bmatrix} = \frac{\partial}{\partial t} \begin{bmatrix} D_x \\ D_y \\ D_z \end{bmatrix} + \frac{1}{\varepsilon} \begin{bmatrix} \sigma_y & 0 & 0 \\ 0 & \sigma_z & 0 \\ 0 & 0 & \sigma_x \end{bmatrix} \begin{bmatrix} D_x \\ D_y \\ D_z \end{bmatrix}. \tag{50}$$

6 The Fourier transform of a product of functions $\check{h}(\omega) = \check{f}(\omega)\check{g}(\omega)$ is given by the convolution of its constituents $h(t) = \int_{-\infty}^{\infty} f(t')g(t-t')dt'$.

Analogous definitions for the magnetic fields

$$\breve{B}_x = \mu \frac{s_z}{s_x} \breve{H}_x \quad , \quad \breve{B}_y = \mu \frac{s_x}{s_y} \breve{H}_y \quad , \quad \breve{B}_z = \mu \frac{s_y}{s_z} \breve{H}_z \tag{51}$$

yield

$$\begin{bmatrix} \frac{\partial E_z}{\partial y} - \frac{\partial E_y}{\partial z} \\ \frac{\partial E_x}{\partial z} - \frac{\partial E_z}{\partial x} \\ \frac{\partial E_y}{\partial x} - \frac{\partial E_x}{\partial y} \end{bmatrix} = -\frac{\partial}{\partial t} \begin{bmatrix} B_x \\ B_y \\ B_z \end{bmatrix} - \frac{1}{\varepsilon} \begin{bmatrix} \sigma_y & 0 & 0 \\ 0 & \sigma_z & 0 \\ 0 & 0 & \sigma_x \end{bmatrix} \begin{bmatrix} B_x \\ B_y \\ B_z \end{bmatrix} . \tag{52}$$

As a last step, definitions (49) and (51) need to be transformed to the time domain as well. In the following, the equation for $\breve{D}_x$ is considered exemplarily. Multiplying both sides with s_x and transforming them back leads to

$$s_x \breve{D}_x = \varepsilon s_z \breve{E}_x , \tag{53a}$$

$$\left(1 - \frac{\sigma_x}{i\omega\varepsilon}\right) \breve{D}_x = \varepsilon \left(1 - \frac{\sigma_z}{i\omega\varepsilon}\right) \breve{E}_x , \tag{53b}$$

$$\left(-i\omega + \frac{\sigma_x}{\varepsilon}\right) \breve{D}_x = \varepsilon \left(-i\omega + \frac{\sigma_z}{\varepsilon}\right) \breve{E}_x , \tag{53c}$$

$$\frac{\partial}{\partial t}(D_x) + \frac{\sigma_x}{\varepsilon} D_x = \varepsilon \left[\frac{\partial}{\partial t}(E_x) + \frac{\sigma_z}{\varepsilon} E_x\right] . \tag{53d}$$

Repeating this procedure for the other five relations finally yields the 12 equations that need to be discretized and solved on a numerical grid:

$$\frac{\partial D_x}{\partial t} = \left[\frac{\partial H_z}{\partial y} - \frac{\partial H_y}{\partial z} - \frac{\sigma_y}{\varepsilon} D_x\right] , \qquad \frac{\partial}{\partial t} D_x + \frac{\sigma_x}{\varepsilon} D_x = \varepsilon \left[\frac{\partial}{\partial t} E_x + \frac{\sigma_z}{\varepsilon} E_x\right] , \tag{54a}$$

$$\frac{\partial D_y}{\partial t} = \left[\frac{\partial H_x}{\partial z} - \frac{\partial H_z}{\partial x} - \frac{\sigma_z}{\varepsilon} D_y\right] , \qquad \frac{\partial}{\partial t} D_y + \frac{\sigma_y}{\varepsilon} D_y = \varepsilon \left[\frac{\partial}{\partial t} E_y + \frac{\sigma_x}{\varepsilon} E_y\right] , \tag{54b}$$

$$\frac{\partial D_z}{\partial t} = \left[\frac{\partial H_y}{\partial x} - \frac{\partial H_x}{\partial y} - \frac{\sigma_x}{\varepsilon} D_z\right] , \qquad \frac{\partial}{\partial t} D_z + \frac{\sigma_z}{\varepsilon} D_z = \varepsilon \left[\frac{\partial}{\partial t} E_z + \frac{\sigma_y}{\varepsilon} E_z\right] , \tag{54c}$$

$$\frac{\partial B_x}{\partial t} = \left[\frac{\partial E_y}{\partial z} - \frac{\partial E_z}{\partial y} - \frac{\sigma_y}{\varepsilon} B_x\right] , \qquad \frac{\partial}{\partial t} B_x + \frac{\sigma_x}{\varepsilon} B_x = \mu \left[\frac{\partial}{\partial t} H_x + \frac{\sigma_z}{\varepsilon} H_x\right] , \tag{54d}$$

$$\frac{\partial B_y}{\partial t} = \left[\frac{\partial E_z}{\partial x} - \frac{\partial E_x}{\partial z} - \frac{\sigma_z}{\varepsilon} B_y\right] , \qquad \frac{\partial}{\partial t} B_y + \frac{\sigma_y}{\varepsilon} B_y = \mu \left[\frac{\partial}{\partial t} H_y + \frac{\sigma_x}{\varepsilon} H_y\right] , \tag{54e}$$

$$\frac{\partial B_z}{\partial t} = \left[\frac{\partial E_x}{\partial y} - \frac{\partial E_y}{\partial x} - \frac{\sigma_x}{\varepsilon} B_z\right] , \qquad \frac{\partial}{\partial t} B_z + \frac{\sigma_z}{\varepsilon} B_z = \mu \left[\frac{\partial}{\partial t} H_z + \frac{\sigma_y}{\varepsilon} H_z\right] . \tag{54f}$$

Note that in regions with $\sigma_x = \sigma_y = \sigma_x = 0$, equations (54) reduce to the equations for the main simulation volume (22) without currents. Therefore, the scheme (54) is used throughout the whole simulation volume, while currents are only assigned within the main region.

The final discretized expressions

Utilizing the finite difference expressions, the set of final equations (54) can be discretized on the computational grid. For brevity, the resulting expressions are given here only for the propagation of E_x; they read as

$$D_x\big|^{n+\frac{1}{2}}_{i+\frac{1}{2},j,k} = C^{1,E_x}_{i+\frac{1}{2},j,k}\, D_x\big|^{n-\frac{1}{2}}_{i+\frac{1}{2},j,k} + C^{2,E_x}_{i+\frac{1}{2},j,k}\left[\frac{H_z\big|^{n}_{i+\frac{1}{2},j+\frac{1}{2},k} - H_z\big|^{n}_{i+\frac{1}{2},j-\frac{1}{2},k}}{\Delta y} - \frac{H_y\big|^{n}_{i+\frac{1}{2},j,k+\frac{1}{2}} - H_y\big|^{n}_{i+\frac{1}{2},j,k-\frac{1}{2}}}{\Delta z}\right]$$

and

$$E_x\big|^{n+\frac{1}{2}}_{i+\frac{1}{2},j,k} = C^{3,E_x}_{i+\frac{1}{2},j,k}\, E_x\big|^{n-\frac{1}{2}}_{i+\frac{1}{2},j,k} + C^{4,E_x}_{i+\frac{1}{2},j,k}\, C^{5,E_x}_{i+\frac{1}{2},j,k}\, D_x\big|^{n+\frac{1}{2}}_{i+\frac{1}{2},j,k} - C^{4,E_x}_{i+\frac{1}{2},j,k}\, C^{6,E_x}_{i+\frac{1}{2},j,k}\, D_x\big|^{n-\frac{1}{2}}_{i+\frac{1}{2},j,k}$$

with the position-dependent coefficients defined by

$$C^{1,E_x}_{i+\frac{1}{2},j,k} = \frac{2\varepsilon_0 - \Delta t \sigma_{y,i+\frac{1}{2},j,k}}{2\varepsilon_0 + \Delta t \sigma_{y,i+\frac{1}{2},j,k}}, \tag{55a}$$

$$C^{2,E_x}_{i+\frac{1}{2},j,k} = \frac{2\varepsilon_0 \Delta t}{2\varepsilon_0 + \Delta t \sigma_{y,i+\frac{1}{2},j,k}}, \tag{55b}$$

$$C^{3,E_x}_{i+\frac{1}{2},j,k} = \frac{2\varepsilon_0 - \Delta t \sigma_{z,i+\frac{1}{2},j,k}}{2\varepsilon_0 + \Delta t \sigma_{z,i+\frac{1}{2},j,k}}, \tag{55c}$$

$$C^{4,E_x}_{i+\frac{1}{2},j,k} = \frac{1}{2\varepsilon_0 \varepsilon + \Delta t \varepsilon \sigma_{z,i+\frac{1}{2},j,k}}, \tag{55d}$$

$$C^{5,E_x}_{i+\frac{1}{2},j,k} = 2\varepsilon_0 + \Delta t \sigma_{x,i+\frac{1}{2},j,k}, \tag{55e}$$

$$C^{6,E_x}_{i+\frac{1}{2},j,k} = 2\varepsilon_0 - \Delta t \sigma_{x,i+\frac{1}{2},j,k}. \tag{55f}$$

Note that the material parameters σ_x, σ_y, and σ_z have a spatial dependence. The perfectly reflectionless character of the interface between the main simulation volume and the UPML layer only applies to the analytic description. When the discretized expressions are evaluated, every discontinuity in the material parameters causes some reflection. In order to reduce these reflections to a minimum, these parameters are gradually increased toward the outer boundary of the UPML layer starting with a value of zero directly at the interface. In MicPIC, a polynomial ramping according [6, 45] is used. The UPML layer itself is terminated with PBCs. The fields that reach this outer boundary are already strongly damped, are then mapped onto the opposite side of the numerical grid, and further damped on their way through the second UPML layer.

2.1.4 Treatment of external fields: The total-field–scattered-field scheme

In MicPIC, the external laser field is not explicitly propagated on the main grid. Instead, the total-field–scattered-field scheme is utilized. Because of the linearity of Maxwell's equations, the total electric and magnetic fields can be decomposed into

$$\mathbf{E}_{\text{total}} = \mathbf{E}_{\text{inc}} + \mathbf{E}_{\text{scatt}}, \tag{56a}$$

$$\mathbf{B}_{\text{total}} = \mathbf{B}_{\text{inc}} + \mathbf{B}_{\text{scatt}}, \tag{56b}$$

where $\mathbf{E}_{\text{inc}}$ and $\mathbf{B}_{\text{inc}}$ are the incident electric and magnetic wave fields. These are the fields that would exist in vacuum and are therefore assumed to be known at all lattice points and all time steps. They satisfy

$$\nabla \times \mathbf{E}_{\text{inc}} = -\dot{\mathbf{B}}_{\text{inc}}, \tag{57a}$$

$$\nabla \times \mathbf{B}_{\text{inc}} = \mu_0 \varepsilon_0 \dot{\mathbf{E}}_{\text{inc}}. \tag{57b}$$

From (56) and (22) follow directly that the scattered fields $\mathbf{E}_{\text{scatt}}$ and $\mathbf{B}_{\text{scatt}}$ have to satisfy

$$\nabla \times \mathbf{E}_{\text{scatt}} = -\dot{\mathbf{B}}_{\text{scatt}}, \tag{58a}$$

$$\nabla \times \mathbf{B}_{\text{scatt}} = \mu_0 \left(\mathbf{j} + \varepsilon_0 \dot{\mathbf{E}}_{\text{scatt}}\right). \tag{58b}$$

These are the microscopic Maxwell equations with scattered instead of total fields. The incident waves do not have to be explicitly propagated on the numerical grid as long as they satisfy Maxwell's equations in vacuum. Instead, they can be calculated analytically or numerically on lower-dimensional auxiliary grids. Note that for the particle propagation, the total fields at the respective particle position have to be used.

2.2 Particle representation on the PIC level

Within the MicPIC framework, the field equations discussed above are coupled to the dynamics of the charged plasma particles via the electric current density $\mathbf{j}$. The currents are determined by the plasma particle velocities, which are in turn driven by the electromagnetic fields, closing the self-consistent description. In order to establish this connection numerically, the particle shape needs to be linked to the discrete FDTD mesh. Typically, this is done via relatively simple weighting schemes, like the cloud-in-cell (CIC) scheme where particles are represented by a top-hat charge distribution [7]. This allows their mapping to the grid with relatively small numerical effort. However, these low-order schemes suffer from strong force anisotropies, which precludes their use in MicPIC. The application of a gridless correction of the short-range forces (the "Mic" force) requires a well-defined, isotropic, and low-noise interparticle force on the PIC level.

In MicPIC, this is achieved by utilizing a Gaussian shape function, as originally proposed by Eastwood and Hockney [12] for an electrostatic description,

$$S(x) = \frac{1}{w_{\text{pic}}\sqrt{\pi}} \exp\left(-\frac{x^2}{w_{\text{pic}}^2}\right), \tag{59}$$

satisfying the normalization

$$\int_{-\infty}^{\infty} S(x)dx = 1. \tag{60}$$

The corresponding three-dimensional representation then reads as

$$S(x, y, z) = S(x)S(y)S(z). \tag{61}$$

Usually, the shape function is sampled onto the discrete numerical grid by calculating the amount of charge within or the amount of charge that travels into/out of the cells touched by the particle, respectively. For the low-order weighting schemes mentioned above, the number of cells that are touched by the particle is well defined by the finite size of the particle distribution. However, the Gaussian shape function in principle extends to infinity, which means that a crucial property of the model, the charge conservation, strongly depends on the number of sampling points as well as the width of the Gaussian distribution. The charge conservation properties for Gaussian shape functions are shown in Figure 2 in terms of the relative charge error as a function of the particle width with respect to the grid spacing Δx. The curves shown in the figure have been obtained for seven sampling points per space dimension,[7] as this turned out to be sufficient for our calculations.

The different colors correspond to two different ways of sampling the Gaussians: (i) The blue curves show results where the charge in every cell is calculated analytically and then summed up over all touched cells. Here, a decreasing particle size results in better charge conservation. Asymptotically, for infinitely small particle width, all charge is contained in the center cell and the error vanishes. (ii) The red curves correspond to a much simpler method, where the charge density is sampled locally at each cell center and summed up afterward. Surprisingly, this method results in even

7 Starting from the nearest cell center, the shape is sampled onto all grid points within ±3 cells in each direction, resulting in a total number $7^3 = 343$ involved cells.

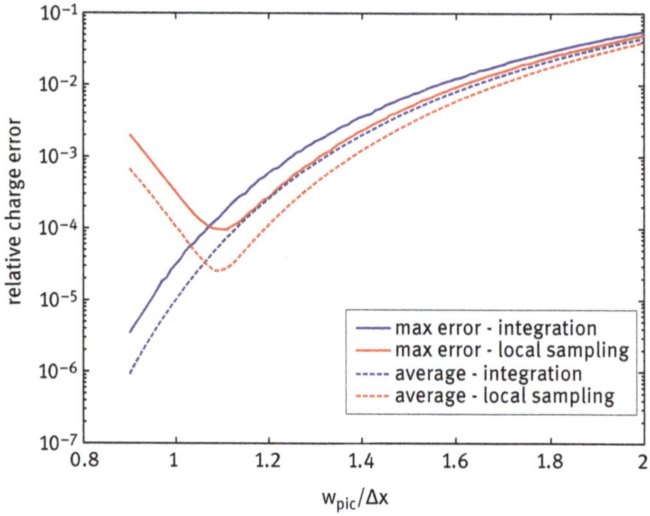

Fig. 2. Relative charge error for integral (blue) and local (red) sampling of the charge density onto the grid as a function of the relative Gaussian width $w_{\text{pic}}^{rel} = w_{\text{pic}}/\Delta x$. The maximal and average charge errors have been determined by evaluating a statistical ensemble of 10,000 random positions of the particle center within a grid cell. Using seven grid points per dimension for the local sampling yields an optimal Gauss width of $w_{\text{pic}}^{rel} = 1.12$ that on average conserves 99.9999% of the particle charge.

better charge conservation for particle sizes larger than roughly one cell and shows an optimal particle size of about 1.1 grid cells. For smaller particle sizes, the charge error increases again as most of the charge is confined to one cell, such that the local sampling is no longer describing a Gaussian shape.

Following this scheme, the charge and current density contributions from particle i at grid point $\mathbf{r}_g$ can be obtained from

$$\rho_i(\mathbf{r}_g) = q_i S(\mathbf{r}_i - \mathbf{r}_g) \tag{62}$$

and

$$\mathbf{j}_i(\mathbf{r}_g) = q_i \mathbf{v}_i S(\mathbf{r}_i - \mathbf{r}_g). \tag{63}$$

To ensure a self-consistent description of the particle dynamics on the PIC level, the electromagnetic PIC fields acting on particle i have to be obtained in the same way as the current densities. The resulting effective electric and magnetic PIC fields are then given by the weighted average over all touched grid points g, using the same shape function and the same sampling method,

$$\mathbf{E}_i = \sum_g \mathbf{E}(\mathbf{r}_g) S(\mathbf{r}_i - \mathbf{r}_g) \tag{64}$$

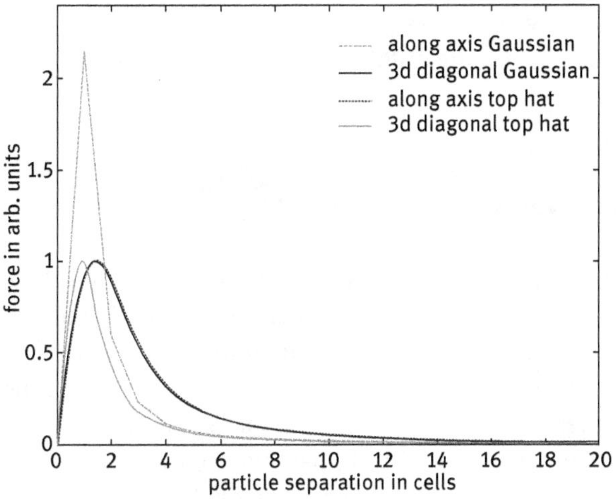

Fig. 3. Interparticle forces on the PIC level, produced by Gaussian (black) and top-hat (gray) distribution functions in the electrostatic limit (for particles at rest). The dotted lines correspond to particle trajectories along a natural grid axis and the solid lines to trajectories along the three-dimensional grid diagonal. While the top-hat shape function results in strongly anisotropic forces, the Gaussian profiles show only negligible anisotropy.

and

$$\mathbf{B}_i = \sum_g \mathbf{B}(\mathbf{r}_g) S(\mathbf{r}_i - \mathbf{r}_g). \tag{65}$$

Together with the field propagation algorithm described in Section 2.1, these expressions allow the calculation of the particle-particle forces on the PIC level. Figure 3 shows the resulting forces in the electrostatic limit, i.e., for particles at rest, as a function of the particle separation. The black dotted and solid lines show the interaction forces for particle trajectories along a natural grid axis and along the three-dimensional grid diagonal, respectively. For comparison, also the corresponding forces for top-hat distributions are shown in gray. The Gaussian shape functions result in negligible force anisotropy, which is essential for the application of the force decomposition scheme. In contrast to that, the top-hat distributions exhibit strongly anisotropic forces, which shows that they are not suited for the MicPIC scheme.

2.3 Local correction

Because of the short-range nature of the microscopic correction $\mathbf{f}_i^{\text{mic}}$, the corresponding binary forces need to be evaluated only for a small subset of particle pairs. In MicPIC, this local correction is done in a similar way as in MD codes with short-range binary interactions. The binary correction forces given by (19) are evaluated for each plasma particle within a sphere with finite cutoff radius r_{cut} around its particle center $\mathbf{r}_i$.

2.3.1 Local correction and cutoff radius

The cutoff radius r_{cut} is one of the key parameters in MicPIC as it determines both the accuracy and the numerical workload of the microscopic correction. In order to pick a reasonable value for r_{cut}, the corresponding force error introduced by the finite cutoff radius has to be estimated. Figure 4 shows the force composition (total, PIC, and microscopic) for the idealized example of two-point-like plasma particles ($w_0 \to 0$) at rest as a function of their separation in units of the PIC-particle size w_{pic}.

The black and red curves in Figure 4(a) correspond to the total and microscopic forces according to (20), normalized to the force at $f(r = w_{pic})$. The gray area denotes the contribution from the PIC term, which has only an electrostatic component, as the particles are at rest. The red and blue areas show the contributions from the microscopic force for distances below and above $r_{cut} = 3w_{pic}$. The fact that Mic and PIC

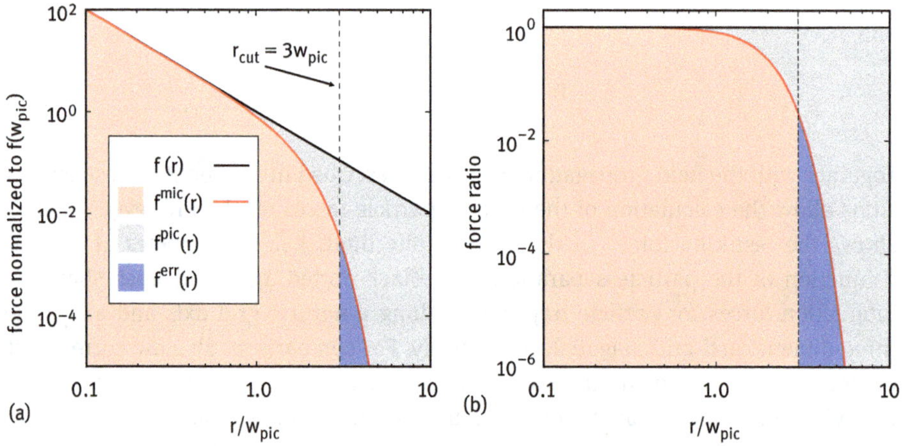

Fig. 4. MicPIC force decomposition for two interacting point charges vs interparticle distance. (a) Black curve: total force $f(r)$ normalized to the force at $f(r = w_{pic})$; red line: Mic force f_{mic}; gray area: PIC force contribution f_{PIC}; red and blue shaded areas: microscopic force contribution for radii below and above a cutoff radius $r_{cut} = 3w_{pic}$, respectively. (b) Forces in (a) are normalized to $f(r)$; hence, the total force becomes $f(r)/f(r) = 1$ (black curve); red curve: $f_{mic}(r)/f(r)$. As f_{mic} is neglected for $r > r_{cut}$, the red- and blue-shaded areas denote the parts of the relative microscopic force correction that are included and neglected, respectively. The gray area gives the relative PIC contribution. Reproduced from [50]. Copyright © 2012 American Physical Society (APS).

forces dominate for either $r < w_{pic}$ or $r > w_{pic}$ nicely illustrates the main idea behind the MicPIC force decomposition.

For easier evaluation of the respective contributions of Mic and PIC forces, they are normalized to the total force $f(r)$ in the right panel of Figure 4. Considering a cutoff radius of $r_{cut} = 3w_{pic}$, the red and blue areas show the part of the microscopic correction that is taken into account or neglected. The relative error drops rapidly with increasing r_{cut}, e.g., more than one order of magnitude when increasing it from $3w_{pic}$ to $4w_{pic}$. However, this would also lead to approximately twice the numerical workload due to the higher number of particles in the correction sphere. The experience of operating MicPIC for the last years has shown that a cutoff radius of $r_{cut} = 3w_{pic}$ is sufficient for most calculations.

2.3.2 Cell indexing

Theoretically, the numerical workload connected with the local correction scales linearly with the total number of plasma particles N, as the binary forces for every particle i have to be evaluated only for a small number of pairs, namely, with respect to particles within the correction sphere around particle i (on average M particles). However, to identify these pairs, the corresponding particle-particle distances have to be determined, which would, if directly evaluated for all N particles, result in an N^2 operation. To sustain the linear scaling, MicPIC makes use of the cell indexing scheme introduced by Allen et al. [2] in 1989. The key idea behind this scheme is to assign all particles to cells of an auxiliary grid that allows to backtrack the number and indices of all particles in a specific cell. This way, only particles in the neighboring cells have to be touched. A sketch of the cell indexing procedure is given in Figure 5.

In a first step, an auxiliary grid **PC** (particle count) is created, where the total number of particles in each cell is stored. This action requires a single loop through

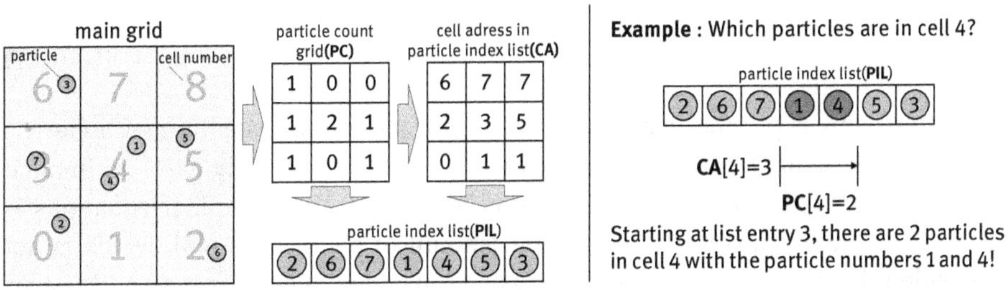

Fig. 5. Sketch of the cell-indexing procedure. All particles are assigned to cells in the main computational grid with the help of auxiliary grids. A detailed description of the complete procedure can be found in the text. The right part shows how the indices of particles located in a specific cell can be retrieved with this method.

all particle positions. Next, a second auxiliary grid **CA** (cell address) is created, where each cell contains the number of particles in all previous cells.[8] Finally, a particle index list (**PIL**) is created, where every particle index is stored in the order of their position in the main grid. This requires a second loop through the particle list.

How the indices of all particles contained in a specific cell can be retrieved from these auxiliary grids is shown for an example in the right part of Figure 5. First, one has to find the corresponding cell address in the **PIL** from the cell address grid **CA**. Next, the number of particles contained in this cell N_C is given by the corresponding entry in the particle count grid (**PC**). The wanted N_C particle indices can then be found in the particle index list **PIL**, starting with the position given by **CA**.

2.4 Particle propagation

In the final step of the MicPIC propagation scheme, the particle positions and velocities need to be advanced in time according to the electromagnetic fields acting on them. To propagate the particle positions, a simple finite difference expression according to

$$\mathbf{r}^{n} = \mathbf{r}^{n-1} + \dot{\mathbf{r}}^{n-\frac{1}{2}} \Delta t \tag{66}$$

can be applied. Advancing the particle velocities on the basis of the Lorentz force is more complicated as it involves a rotation. A very efficient and accurate method to achieve this has been introduced by J. P. Boris in 1970 [8] and is now briefly discussed.

2.4.1 Boris scheme

The basis of the Boris scheme is a centered finite difference expression of the Lorentz force, which is given by

$$\frac{\dot{\mathbf{r}}^{n+\frac{1}{2}} - \dot{\mathbf{r}}^{n-\frac{1}{2}}}{\Delta t} = \frac{q}{m}\left(\mathbf{E}^{n} + \frac{\dot{\mathbf{r}}^{n+\frac{1}{2}} + \dot{\mathbf{r}}^{n-\frac{1}{2}}}{2} \times \mathbf{B}^{n}\right). \tag{67}$$

From this expression, the main challenge in the particle propagation with electric and magnetic fields becomes already evident. Equation (67) only gives an implicit expression for the new particle velocity $\dot{\mathbf{r}}^{n+\frac{1}{2}}$. Solving these implicit equations is not impossible but involves a significant amount of calculation [46], which is not convenient if millions of particles have to be propagated in each time step. However, J. P. Boris found an elegant way to derive explicit expressions for the velocities by

8 This corresponds to the accumulative sum of **PC**.

separating the electric and magnetic forces. Substituting

$$\dot{\mathbf{r}}^{n-\frac{1}{2}} = \mathbf{v}_- - \alpha \mathbf{E}^n, \tag{68}$$

$$\dot{\mathbf{r}}^{n+\frac{1}{2}} = \mathbf{v}_+ + \alpha \mathbf{E}^n \tag{69}$$

with $\alpha = \frac{q}{m}\frac{\Delta t}{2}$ into (67) yields an expression where $\mathbf{E}^n$ cancels out,

$$\mathbf{v}_+ - \mathbf{v}_- = \alpha\,(\mathbf{v}_+ + \mathbf{v}_-) \times \mathbf{B}^n. \tag{70}$$

Further, substituting $\mathbf{t}^n = \alpha \mathbf{B}^n$ leaves a compact expression that describes a rotation

$$\mathbf{v}_+ = \mathbf{v}_- + (\mathbf{v}_+ + \mathbf{v}_-) \times \mathbf{t}^n. \tag{71}$$

Together, equations (68), (69), and (71) describe a three-step process: First, half of the electric impulse is added to the old velocity, then the intermediate velocity is rotated, and finally, the second half of the electric impulse is added.

The expression for the rotation is still implicit. To retrieve an explicit expression, an additional vector $\mathbf{v}'$ is introduced that corresponds to only a partial rotation from that given in (71)

$$\mathbf{v}' = \mathbf{v}_- + \mathbf{v}_- \times \mathbf{t}^n. \tag{72}$$

Using (71), one can also express this vector by

$$\mathbf{v}' = \mathbf{v}_+ - \mathbf{v}_+ \times \mathbf{t}^n. \tag{73}$$

Evaluation of the cross product with $\mathbf{t}$ yields

$$\left(\mathbf{v}' \times \mathbf{t}\right) = \left(\mathbf{v}_- \times \mathbf{t}^n\right) - \left|\mathbf{t}^n\right|^2 \mathbf{v}_- + \left(\mathbf{v}_- \cdot \mathbf{t}^n\right) \mathbf{t}^n \tag{74}$$

$$= \left(\mathbf{v}_+ \times \mathbf{t}^n\right) + \left|\mathbf{t}^n\right|^2 \mathbf{v}_+ - \left(\mathbf{v}_+ \cdot \mathbf{t}^n\right) \mathbf{t}^n. \tag{75}$$

Finally, subtraction yields the desired explicit expression for the full rotation,

$$\mathbf{v}_+ = \mathbf{v}_- + \left(\mathbf{v}' \times \mathbf{s}^n\right) \tag{76}$$

with

$$\mathbf{s}^n = \frac{2\mathbf{t}^n}{1 + (t^n)^2}. \tag{77}$$

The complete explicit propagation scheme for the particle velocities is now given by the set of equations

$$\mathbf{v}_- = \dot{\mathbf{r}}^{n-\frac{1}{2}} + \alpha \mathbf{E}^n, \tag{78}$$

$$\mathbf{v}_+ = \mathbf{v}_- + \left[\left(\mathbf{v}_- + \left(\mathbf{v}_- \times \mathbf{t}^n\right)\right) \times \mathbf{s}^n\right], \tag{79}$$

$$\dot{\mathbf{r}}^{n+\frac{1}{2}} = \mathbf{v}_+ + \alpha \mathbf{E}^n. \tag{80}$$

2.5 Implementation of ionization

The numerical details discussed so far complete the main part of MicPIC that is necessary to model the classical laser-driven dynamics of a plasma. In order to enable MicPIC to also describe the plasma formation process, atomic ionization mechanisms have to be implemented. This section briefly reviews the corresponding atomic models and shows how they can be modified to be applicable to many-particle systems.

2.5.1 Tunnel ionization

When exposed to strong electric fields, bound electrons have a probability to tunnel through the generated potential barrier. The corresponding tunnel rate for an atom in electromagnetic fields can be calculated quantum mechanically (see [38, 39] for reviews). In atomic units, the rates read as [3]

$$W_{\text{tunnel}}^{\text{au}} = I_p^{\text{au}} C_{n^*l^*}^2 A_{lm} \left(\frac{2\kappa^3}{\mathcal{E}^{\text{au}}}\right)^{2n^*-|m|-1} \exp\left(-\frac{2\kappa^3}{3\mathcal{E}^{\text{au}}}\right) \tag{81}$$

with

$$A_{lm} = \frac{(2l+1)(l+|m|)!}{2^{|m|}(|m|)!(l-|m|)!} \quad , \quad \kappa = \sqrt{2I_p^{\text{au}}} \tag{82}$$

and

$$C_{n^*l^*}^2 = \frac{2^{2n^*}}{n^*\Gamma(n^*+l^*+1)\Gamma(n^*-l^*)} \quad , \quad n^* = Z/\kappa \quad , \quad l^* = n^* - 1 \,, \tag{83}$$

where I_p^{au} is the ionization potential[9] and $\mathcal{E}^{\text{au}}$ is the electric field strength, Z the resulting charge state of the ion, and m and l the magnetic and angular momentum quantum number, respectively. The auxiliary parameters l^* and m^* are referred to as effective quantum numbers. The resulting tunneling rates exhibit an extremely nonlinear intensity dependence, e.g., rising by almost ten orders of magnitude when increasing the laser intensity from $I = 1 \times 10^{13}$ W/cm^2 to $I = 1 \times 10^{14}$ W/cm^2 for the ionization of neutral xenon atoms.

To take many-particle effects into account, these rates are evaluated for the total electric fields on the PIC level, i.e., the sum of the laser field and the fields created by all other charged particles. The microscopic fields associated with the local correction are neglected to avoid double counting of electrons with trajectories close to atoms or ions, which will be accounted for in the routine for electron impact ionization.

9 Values for I_p^{au} used in MicPIC have been calculated with the relativistic Dirac-LDA code from [4].

2.5.2 Electron impact ionization

The treatment of electron impact ionization, i.e., the liberation of secondary electrons as a result of inelastic electron-atom/ion collisions, is restricted to sequential ionization

$$X^{j+} + e \rightarrow X^{(j+1)+} + 2e, \tag{84}$$

while nonsequential ionization or ionization via excited intermediate states is so far neglected. In contrast to the typical treatment of impact ionization in PIC codes, these inelastic collisions are not evaluated via rates and Monte Carlo schemes. In MicPIC, the microscopic character of this ionization mechanism is effectively preserved, i.e., every electron-ion collision is tested for ionization using impact ionization cross sections. To that end, the well-known parameterized empiric formula introduced by W. Lotz [28] is utilized,

$$\sigma_j(E) = \sum_i a_i q_i \frac{\ln(E/P_i)}{EP_i} \left\{1 - b_i \exp\left[-c_i(E/P_i - 1)\right]\right\}, \qquad E \geq P_i. \tag{85}$$

Here, E is the kinetic energy of the impinging electron, P_i the ionization potential of the ith electronic shell, q_i the number of electrons in the ith shell, and a_i, b_i, and c_i are empirical parameters. With these parameters, the calculated cross sections can be fitted to experimental data over a wide range of elements, charge states, and projectile energies [1, 5, 19, 20, 22].

Many-particle effects can be accounted for by employing effective ionization potentials. For an atomic ion within a plasma environment, neighboring ions and electronic screening by quasi-free plasma electrons lead to an effective ionization threshold $E^{\star}_{nl} = E_{nl} - \Delta_{\text{env}}$. The pure atomic value E_{nl} is lowered by an environmental shift Δ_{env}. While E_{nl} corresponds to the energy that is needed to completely remove an electron with principal and angular quantum numbers n and l from the atom or ion, $E^{\star}_{nl}$ specifies the corresponding minimal energy that is required to lift the electron into the quasi-continuum within the plasma environment. The shift Δ_{env} is evaluated directly from the plasma fields in the simulation, following the scheme in [14].

However, using effective ionization potentials $P^{\star}_i = P_i - \Delta_{\text{env}}$ leads to a continuous spectrum of ionization potentials, which makes it impractical to adopt the parameters a_i, b_i, and c_i to specific charge states. Therefore, a simplified version of (85) is used in the current implementation

$$\sigma_j(E) = \sum_i a q_i \frac{\ln(E/P^{\star}_i)}{EP^{\star}_i}, \qquad E \geq P^{\star}_i, \tag{86}$$

with the fixed parameter $a = 450 \times 10^{-16}\ \text{cm}^2(\text{eV})^2$, which satisfactorily reproduces experimental data [29].

2.6 MicPIC parameters and scaling

A number of simulation parameters have been introduced in the discussion of the numerical implementation of the MicPIC approach. A list of the main simulation parameters and their meaning is given in Table 1 as a reminder. This section discusses how to choose the values of these parameters in an optimal way under existing physical and technical constraints. Additionally, also their influence on the performance of the code is discussed in terms of a general scaling analysis.

Tab. 1. List of the main simulation parameters.

Parameter	Description
Δx	Cell width on FDTD grid
Δt	Time step
w_{pic}	Particle width on PIC-level
r_{cut}	Cutoff radius for local correction
w_0	Actual particle width

First, the cell width on the PIC level Δx determines the resolution of wave propagation phenomena on the numerical grid. It has to be chosen small enough to resolve all relevant scales of the radiated fields, i.e., the skin depth and the wavelength of the laser and also possibly generated higher harmonics. Typically, a grid spacing of $\Delta x \leq \lambda/20$ is sufficient for that task, which leads to values of a few to a few tens nanometers in the optical excitation regime. To ensure convergence of the corresponding FDTD solution, the time step has to fulfill the Courant stability criterion, which imposes an upper limit for the time step according to $\Delta t \leq \Delta x/(\sqrt{3}c)$ [45], with c being the vacuum speed of light.

Further, the particle width on the PIC level has to fulfill $w_{\text{pic}} \simeq 1.1\Delta x$ for optimal charge conservation properties (see Section 2.2). Next, to ensure an accurate evaluation of the microscopic correction (Section 2.3), the cutoff radius r_{cut} has to be in the range of $r_{\text{cut}} \simeq 3w_{\text{pic}}$. Finally, the actual particle width has to be chosen such that classical electron recombination below quantum-mechanical energy levels is precluded, i.e., the classical binding energy resulting from (18) has to be smaller or equal to the quantum mechanical energy levels used for ionization.

In summary, choosing a grid resolution compatible with the upper limit given by the relevant scales of the radiated fields determines every other major parameter, except for the actual particle width. As smaller values of Δx leave the physics unchanged, this freedom can be utilized to balance the numerical work load between the microscopic and PIC parts of the code.

To evaluate how this influences the performance of the code, a scaling analysis is performed under the assumption that the excitation of a system with the total

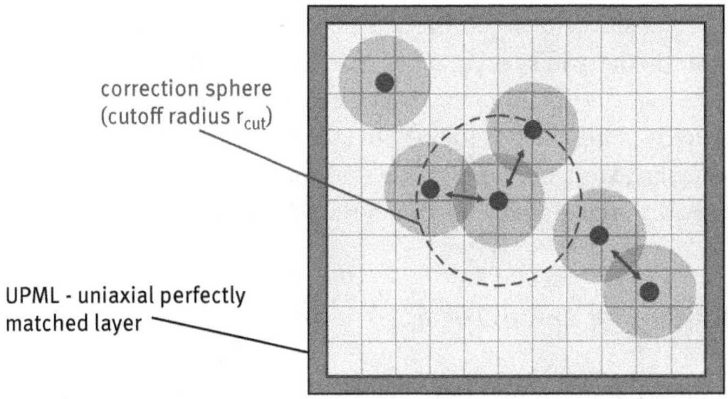

Fig. 6. Illustration of the local correction scheme. Only interactions for particles within a finite cutoff radius have to be corrected.

particle number N by an external laser of wavelength λ has to be modeled in a simulation volume V. The workload associated with the microscopic correction is then determined by the total number of particles N times the number of particles within the correction sphere $M = (4\pi/3)r_{\text{cut}}^3 N/V$ (see Figure 6). As a result, advancing the microscopic part one time step scales as $O_{\text{step}}^{\text{mic}} = \alpha N(N/V)r_{\text{cut}}^3$. Advancing the PIC part one time step requires the calculation of the currents and forces for all particles, which scales with the particle number N, and the update of the electromagnetic fields, which scales with the number of grid points, $V/(\Delta x)^3$. Together, this results in a scaling of the PIC part given by $O_{\text{step}}^{\text{pic}} = \beta N + \gamma V/(\Delta x)^3$. Putting both together yields the total workload scaling for one time step

$$O_{\text{step}}^{\text{MicPIC}} = \alpha N(N/V)r_{\text{cut}}^3 + \beta N + \gamma V/(\Delta x)^3, \tag{87}$$

where the parameters α, β, and γ are prefactors corresponding to the microscopic correction, current/force calculation, and the field update, respectively. Exploiting the fact that $\Delta x \simeq w_{\text{pic}} \propto r_{\text{cut}}$ and assuming a constant particle density (N/V =const.) allows to rewrite expression (87) in terms of the cutoff radius

$$O_{\text{step}}^{\text{MicPIC}} = \alpha' N r_{\text{cut}}^3 + \beta' N + \gamma' N/r_{\text{cut}}^3, \tag{88}$$

where the parameters α', β', and γ' are modified prefactors. Eventually, MicPIC will be used to model the plasma dynamics for certain time intervals, i.e., the quantity of interest is the workload per unit time. The corresponding translation of the above result can be done by dividing it by $\Delta t \propto r_{\text{cut}}$ and yields

$$O^{\text{MicPIC}}(N, r_{\text{cut}}) \propto N(\alpha' r_{\text{cut}}^2 + \beta'/r_{\text{cut}} + \gamma'/r_{\text{cut}}^4), \tag{89}$$

where the first term comes from the Mic part and the second and third from PIC. This shows that MicPIC scales linearly with the total particle number N, as desired. The workload distribution between Mic and PIC part can be balanced by choosing the cutoff radius (or equivalently the grid spacing Δx) within its physical constraints. For very small/large values of r_{cut}, the dominant load is produced on the PIC or Mic parts. In the limiting case where r_{cut} approaches the box length $V^{1/3}$, all particles need to be corrected, the time step is no longer bound to the grid spacing, and (89) yields the well-known $O(N) = N^2$ scaling associated with MD codes.

2.7 MicPIC system energy calculation

An important observable for the evaluation of MicPIC calculations is the energy absorption by the system contained in the numerical box. It is given by the total energy difference before and after laser excitation. The total box energy reads as

$$\begin{aligned} E_{\text{tot}} = & \sum_i \frac{m_i}{2}\dot{\mathbf{r}}_i^2 + \frac{1}{2}\int \left[\varepsilon_0(\mathbf{E}^{\text{pic}})^2 + \frac{1}{\mu_0}(\mathbf{B}^{\text{pic}})^2\right] d^3r \\ & + \sum_{i<j} V_{ij}^{\text{mic}}(r_{ij}) - \frac{1}{2}\sum_i V_{ii}(0, w_{\text{pic}}), \end{aligned} \tag{90}$$

where the individual terms describe the kinetic energy, the electromagnetic energy on the PIC level, the energy resulting from the microscopic correction, and the energy renormalization to remove the spurious self-energy of the particles on the PIC grid, respectively.

3 Applications

After introducing MicPIC's basic concepts and the most important numerical aspects, we are now ready to apply MicPIC to real physical problems. The main goals of this section are (i) to provide a simple example that can be calculated with regular desktop hardware to enable the interested reader with some reference data for comparison and (ii) to present a study that demonstrates the unique capabilities of MicPIC for the description of strong-field physics.

3.1 Laser excitation of a solid-density foil: A simple MicPIC example

First, we consider the irradiation of a thin foil (d = 400 nm) at solid density with a short (τ = 16 fs) and moderately intense infrared laser pulse ($I_0 = 1 \times 10^{14}\,\text{W/cm}^2$, $\lambda = 800$ nm) with flat beam profile (no focusing) and propagation along the z-axis. The basic simulation setup is schematically drawn in Figure 7. The foil extends to infinity in

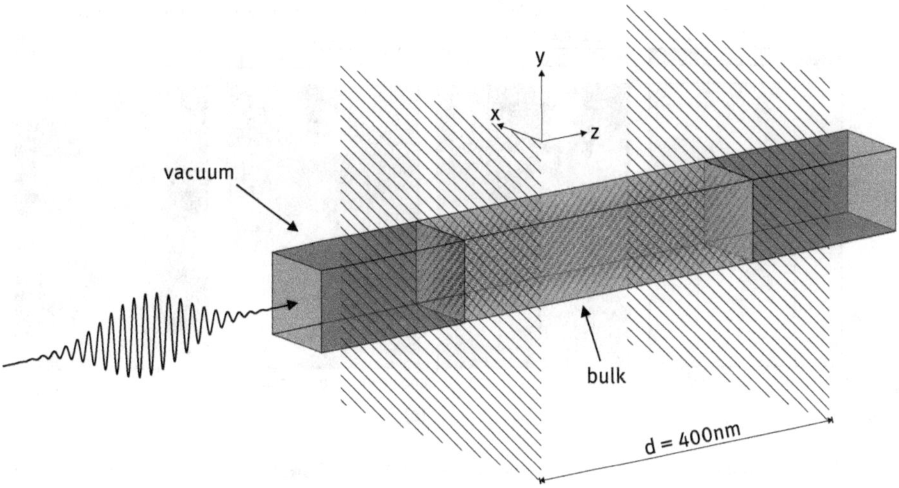

Fig. 7. Schematics of the simulation setup for the laser excitation of a foil at normal incidence. We emulate an infinitely wide foil by terminating the numerical box with ABCs along the laser propagation direction (z-direction) and with PBCs perpendicular to it.

the other two directions, which results in an effectively one-dimensional problem. To represent this geometry properly in the numerical simulation, ABCs are only applied along the laser propagation direction (z-direction), while in x and y direction, the numerical box is terminated with PBCs. The numerical grid has a size of $15 \times 15 \times 2999$ grid points with a grid spacing of $\Delta x = 3$. The foil itself is modeled with atoms of atomic mass $m_a = 60\,\text{amu}$ distributed in fcc crystal structure with an average number density of $n_a = 0.022^{-3}$. Every model atom holds one (initially bound) electron with an ionization potential of $I_p = 9\,\text{eV}$, which results in a corresponding actual particle width of $w_0 = 1.8$. The above parameters describe a medium similar to SiO_2 at solid density.

The corresponding simulation results are shown in Figure 8 in terms of the space- and time-resolved evolution of (a) the total electric field in polarization direction, (b) the density of liberated electrons, and (c) and (d) the ionization rates for tunneling and electron impact, respectively. The time evolution of the interaction can be divided into three major stages.

In the early stage of the interaction (up to $t = -10\,\text{fs}$), the laser electric field is not strong enough to drive significant tunnel ionization (see Figure 8(c)). Therefore, the foil stays neutral and transparent such that the laser field can propagate through it without attenuation or reflection, as can be seen from Figure 8(a). The fact that the lines of maximum field strength are tilted with respect to the horizontal axis indicates the finite propagation velocity of the light.

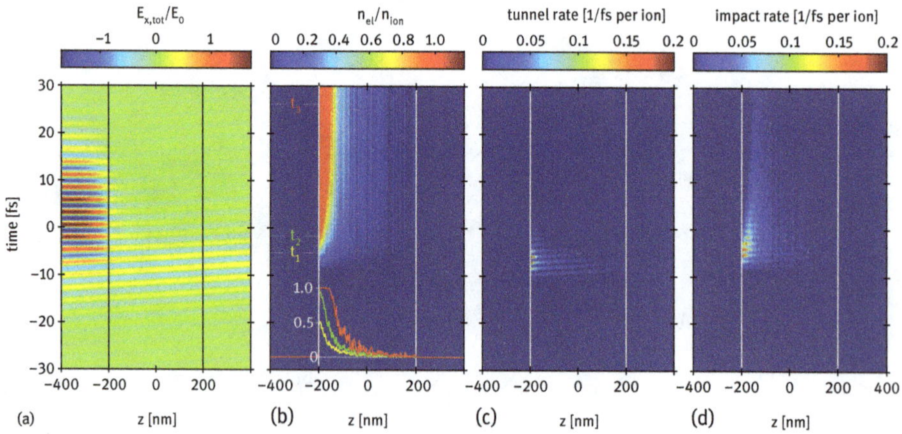

Fig. 8. Interaction dynamics of a 400-nm thick foil with a 16-fs infrared (λ = 800 nm) laser pulse at $I_0 = 1 \times 10^{14}\,\mathrm{W/cm^2}$. The corresponding panels show the time- and space-resolved (a) total electric field in polarization direction normalized to the peak laser electric field, (b) electron density normalized to the ion background density, (c) tunnel ionization rate, and (d) electron impact ionization rate. Note that all space dependencies are averaged along the x and y direction (perpendicular to the laser propagation direction).

In the second stage, starting at around $t = -10\,\mathrm{fs}$, this behavior changes dramatically. The highly nonlinear dependence of the tunneling probability on the actual field strength leads to an abrupt onset of tunnel ionization once the field strength of a laser subcycle is sufficiently strong (see Figure 8(c)). The resulting first plasma electrons are then accelerated by the laser electric field and reach kinetic energies that are sufficient to drive additional electron impact ionization (see Figure 8(d)). Together, tunnel and impact ionization rapidly generate a fully inner ionized[10] surface layer. The line profiles of the electron density plotted at the bottom of Figure 8(b) for different time steps facilitate this observation and show that this process takes only a few laser cycles.

For the considered scenario, full inner ionization corresponds to a plasma electron density, that is more than 10 times overcritical, where the critical density is given by

$$n_c = \frac{\varepsilon_0 m_e}{e^2}\omega^2. \tag{91}$$

As electromagnetic waves cannot propagate in overdense plasmas, this layer acts as a plasma mirror, preventing further tunnel ionization deeper in the medium. A direct signature of this laser-induced reflectivity is the observation of standing waves in front

10 "Inner ionization" means the electrons are liberated from their parent atom but not from the target as a whole.

of the surface layer. After plasma generation (see $t > -5$ fs in Figure 8(a)), the lines of maximum field strength are perfectly horizontal. Standing waves do not propagate in space, instead we only observe an oscillation in time.

After the second stage where the surface plasma layer has been generated, tunnel ionization essentially stops (cf. Figure 8(c)). However, Figure 8(b) clearly shows that this does not immediately stop the spreading of the plasma layer. Instead, the plasma generation continues in the vicinity of the already existent plasma because of electron impact ionization (see Figure 8(d)). Close to the surface, electrons are still subject to the incident laser field. Though the resulting acceleration is perpendicular to the foil plane, momentum redistribution due to electron-electron collisions allows sufficiently energetic electrons to enter deeper parts of the foil and drive further electron impact ionization. As a result, the thickness of the plasma layer slowly increases till the end of the pulse.

3.2 Time-resolved x-ray imaging

The scenario discussed above already indicated MicPIC's potential to describe many-particle strong-field physics. However, for a numerical description of this particular scenario, other methods may be applicable as well. In this section, we want to discuss a setting where the unique capabilities of MicPIC are essential to appropriately describe the underlying physics. The problem we want to discuss here is closely related to the rapid advances in laser technology, namely, x-ray free electron lasers (XFELs). During the past decade, a number of such facilities started operation, others are currently under construction. The available pulses of coherent x-ray radiation with extremely high peak powers opened up new possibilities in a wide range of research areas [9]. One particularly interesting and promising application is single-shot diffractive imaging. In principle, it allows the determination of the shape and structure of finite-size systems with atomic resolution without immobilization, which, for example, is of great interest for studies of biologically relevant systems that are difficult to crystallize, such as large biomolecules or viruses. The interpretation of the recorded scattering images, however, remains a challenging task for various reasons. First, in the experiment, only the intensities of the scattered fields can be recorded while the phase information is lost. Here, theory plays an important role as it allows to establish a connection between signatures in the scattering image and the corresponding particle properties, e.g., its shape, composition, or electronic structure, which are generally unknown in the experiment. Second, because of the extremely high intensities, the target is subject to structural changes that occur on the time scale of the imaging pulse length. This problem has been identified and studied with the help of particle-based numerical simulations long before the first x-ray FEL started operation [34].

While these laser-induced structural changes are problematic for identifying ground-state structures, they also define a new interesting research field. Time-resolved experiments, where structural changes are induced by a pump pulse and imaged by a time delayed probe pulse, offer a route to making molecular movies of the femtosecond dynamics. MicPIC is perfectly suited for corresponding theoretical studies as it allows simultaneous access to the particle dynamics and the resulting scattering image. In the following, we will discuss a full numerical time-resolved x-ray imaging experiment, where a solid-density cluster is first turned into a plasma via an intense near-infrared (NIR) laser pulse, and its time evolution is subsequently imaged with a second x-ray pulse.

The basic numerical setup for this scenario is depicted in Figure 9. As a model system, we consider a $R = 25$nm hydrogen cluster at solid density with atoms/ions initialized in fcc structure with Wigner-Seitz radius of $r_s = 1.79$. The fact that hydrogen has only one electron contributing to the x-ray scattering significantly simplifies the

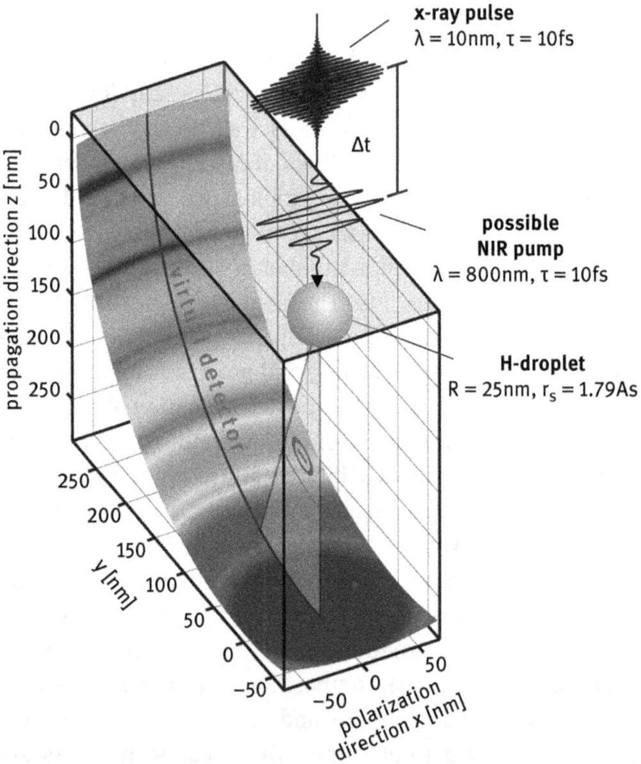

Fig. 9. Sketch of the x-ray simulation setup with the $R = 25-$nm hydrogen cluster, the virtual detector (defined by a sphere with radius $R_d = 290$ nm around the cluster center), and the incident laser fields. Reproduced from [37]. Copyright © 2014 American Physical Society (APS).

analysis, making it the ideal model system. For laser excitation, we consider an intense 10 – fs NIR pump pulse ($\lambda = 800\,\text{nm}$) and a soft x-ray ($\lambda = 10\text{nm}$) probe pulse. The corresponding time-dependent scattered fields are then recorded on a virtual detector located on a virtual sphere within the simulation box.

Ideally, the radius of the detection sphere is large enough to enter the far field region, where the fields include only propagating radiation, and nonradiating near-field contributions have already died out. The corresponding required detector distance is not exactly defined, but as a rule of thumb, the far-field region for an antenna with a dimension of the order of the emitted wavelength starts at the Fraunhofer distance $d_\text{f} = \frac{2D^2}{\lambda}$ where D is the spatial dimension of the emitter. For the scenario discussed here, this would result in a detector distance of about $R_\text{d} \simeq 500\,\text{nm}$. A corresponding MicPIC simulation run, however, would require a very large computational box, which is inconvenient from a technical point of view. Instead, a slightly smaller value of $R_\text{d} = 290\,\text{nm}$ is utilized that allows the use of a smaller numerical box. It can be shown that the finite detector distance has only minor impact on the shape of the scattering signal, essentially only affecting the sharpness of the features in the scattering pattern.

The scattering signal is evaluated here in terms of the scattered fraction

$$S(\Theta) = \frac{\varepsilon_0 c R_\text{d}^2}{\pi R^2 F_0} \int \left[\mathbf{E}_\perp(\Theta, t)\right]^2 \text{d}t, \tag{92}$$

with ε_0 the vacuum permittivity, c the vacuum speed of light, $F_0 \simeq I_0\tau$ the fluence of the incident x-ray field, and $\mathbf{E}_\perp$ the transverse electric field.[11] In the limit $R_\text{d} \to \infty$, $S(\Theta)$ specifies the number of x-ray photons scattered into an element of solid angle per incident photon impinging on the initial geometric cluster cross section.

First, we want to concentrate on the dynamics induced by the NIR pump pulse. The time evolution of selected observables during and after this initial excitation is shown in the left part of Figure 10. It reveals the following picture: Plasma formation is triggered by tunnel ionization in the rising edge of the pulse. Subsequently, laser heating of the first liberated electrons drives additional electron impact ionization. The combined action of both contributions creates a fully inner-ionized cluster near the pulse peak, see Figure 10(c). Because of the low proton mass, the cluster starts to expand almost immediately (cf. Figure 10(c)). The expansion is driven by a mixture of hydrodynamic expansion and Coulomb explosion. Signatures of both can be found in the time evolution of the energy contributions in Figure 10(a). The observed conversion of electron kinetic energy into ion kinetic energy clearly indicates hydrodynamic expansion, while the fact that the ion kinetic energy gain is stronger than the electron energy loss shows that Coulomb explosion is also contributing.

11 Only transverse electric fields, i.e., with electric field vector perpendicular to the propagation direction, can contribute to the final scattering pattern.

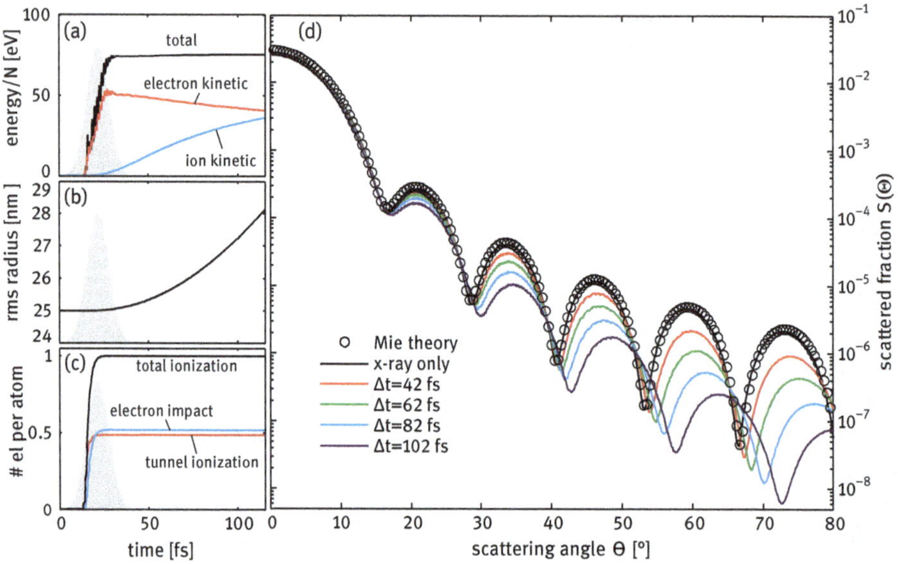

Fig. 10. (a)–(c) Time evolution of selected observables during and after the NIR pump excitation. (d) Resulting elastic scattered fraction (detected in the *yz*-plane) for different x-ray probe delays and the Mie-theory result for an unexpanded cluster for comparison. Adapted from [37].

In the next step, the NIR-driven cluster dynamics is imaged by a soft x-ray probe pulse for various pulse delays. Note that the NIR-induced dynamics shown in Figure 10(a–c) is not affected by the x-ray pulse because of the negligible inverse bremsstrahlung in the x-ray regime. The resulting delay-dependent elastic scattering signals shown in Figure 10(d) exhibit two major pump-probe effects. First, the slope of the scattering signal increases with pulse delay, such that the signal drops by more than one order of magnitude for the largest scattering angles. And second, the separation of the fringes increases continuously with the pulse delay.

For a first interpretation of these observations, it is helpful to recall some basic scattering relations. The far-field scattering intensity is essentially given by the Fourier transform of the projected density of the scatterer. In this picture, the separation of the typically found Mie rings is a measure of the objects size, where smaller ring separations in **k**-space correspond to larger scales in real space. If we now assume that the cluster keeps the shape of a homogeneous sphere during expansion, the trend observed in Figure 10(d) would correspond to a reduction of cluster size. This behavior, however, seems to be in stark contradiction with the observed increasing cluster radius shown in Figure 10(b). To shed light on this apparent contradiction, a more detailed analysis of the expansion dynamics itself is needed.

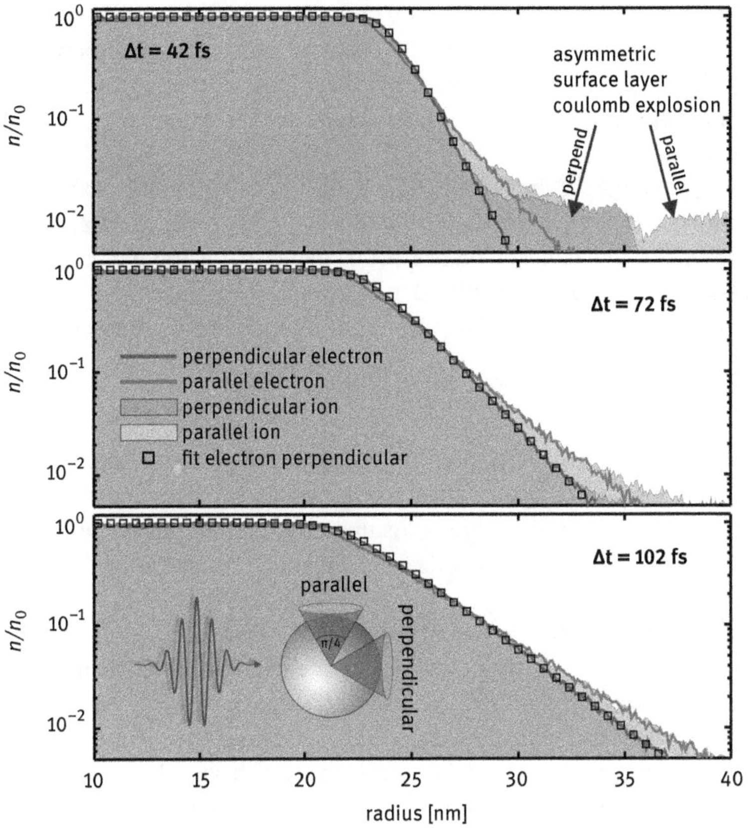

Fig. 11. Radial electron and ion density profiles for different pulse delays (as indicated). Solid lines (electrons) and shaded areas (ions) show density profiles extracted directly from the MicPIC simulation for directions parallel to (light gray) and perpendicular to (gray) the polarization direction. Open squares correspond to electron density fits perpendicular to the NIR polarization using the profile given in (93). Reproduced from [37]. Copyright © 2014 American Physical Society (APS).

Figure 11 shows direction-resolved radial density profiles[12] of electrons and ions for directions parallel (green) and perpendicular (blue) to the laser polarization for different times. In early stages, surface ions are unscreened because of partial outer ionization and consequently undergo rapid Coulomb explosion (lines and colored areas in 11(a)). The fact that the ion spectra are anisotropic can be attributed to the collective electron motion during the NIR excitation. During the oscillation of

12 The density profiles have been averaged over cones with $\pi/4$ full opening angle and cone apex at the cluster center.

the electron cloud, ions at the cluster poles are periodically exposed, which results in an effectively reduced charge screening and stronger ion acceleration along the polarization axis [10, 15, 26]. Once the surface ions have been ejected, the remaining ion and electron densities overlap and screen each other. The cluster continues to expand hydrodynamically.

During the hydrodynamic expansion phase, the density profiles exhibit a self-similar shape with a core region of constant density and an exponential decay of the surface layer. A similar behavior has been predicted for the expansion of semi-infinite plasmas [21, 31]. The radial density profiles observed here can be accurately described in all expansion stages by a sharpened Fermi distribution

$$n_e(r) = \frac{n_c}{[\exp(\frac{r-r_c}{ds}) + 1]^s}, \tag{93}$$

where n_c is the core density, r_c the core radius, and d the decay length as a measure of the surface width. The sharpness factor s ensures the correct transition behavior between the two asymptotic limits. Examples of corresponding fits of the electron density perpendicular to the polarization direction are depicted as squares in Figure 11.

Application of the fit procedure to all available density snapshots yields the time evolution of the density profile parameters shown in Figure 13(a–b) as solid lines. The blue and green curves correspond to directions perpendicular and parallel to the polarization direction, respectively. Though the actual values for both directions differ due to the anisotropic expansion, they show the same general behavior. The core radius r_c is linearly decreasing with time while the surface width d is linearly increasing. The sharpness factor s (not shown here) is converging rapidly and is therefore only of minor importance for the expansion dynamics. To substantiate this statement, fits for $s \to 0$ have been performed (dashed lines in Figure 13(a and b)), which corresponds to the limit of a sharp edge between core and surface region. The fact that only minor changes in core radius and surface width are observed shows that the complex plasma expansion dynamics can be sufficiently described by these two parameters.

To connect the evolution of the density profile parameters with the features observed in the delay-dependent scattering images shown in Figure 10, it is helpful to study the effect of the profile parameters on the scattering pattern separately. Unfortunately, Mie theory is no longer applicable because of the anisotropic and inhomogeneous density profile. An alternative is to calculate the scattered fields in first-order Born approximation, where higher-order scattering by neighboring scatterers is not taken into account. This is justified in the considered scenario as light absorption can be neglected at the given wavelength for hydrogen. The scattered field in first-order Born approximation including polarization effects is determined by

$$\mathbf{E}(\mathbf{r}_d) = \int_V \frac{\tilde{\mathbf{r}} \times (\mathbf{E}_0 \times \tilde{\mathbf{r}})\, r_e e^{i(\mathbf{kr}+k\tilde{r})}}{\tilde{r}^3} n_e(\mathbf{r})\, d^3r, \tag{94}$$

with $\mathbf{k}$ the incident wavevector, r_e the classical electron radius, $n_e(\mathbf{r})$ an arbitrarily shaped electron density, and $\mathbf{E}_0$ and $\mathbf{E}$ the complex field amplitudes of the incident field and the scattered field at detector position $\mathbf{r}_d$. The detector position in the frame of each scattering subvolume is denoted by $\tilde{\mathbf{r}} = \mathbf{r}_d - \mathbf{r}$. The corresponding scattered fraction in Born approximation is then given by

$$S_B(\Theta) = \frac{R_d^2 E^2}{\pi R^2 E_0^2}. \tag{95}$$

Inserting the parametric density profile from (93) into the Born expression above allows a selective analysis of the influence of the core radius and surface width on the scattering pattern by fixing the respective other parameter. The results shown in Figure 12 reveal that the core radius r_c affects only the fringe spacing without changing the envelope of the scattering signal. The decay length d on the other hand mainly modifies the slope of the envelope and hardly changes the fringe positions. As a result, the growing fringe separation and increasing envelope slope observed in the delay-dependent scattering pattern in Figure 10 can be attributed to the shrinking core radius and the cluster surface expansion.

The above parameter analysis already allows one to qualitatively explain the observed scattering features. However, the ultimate goal is to quantitatively recon-

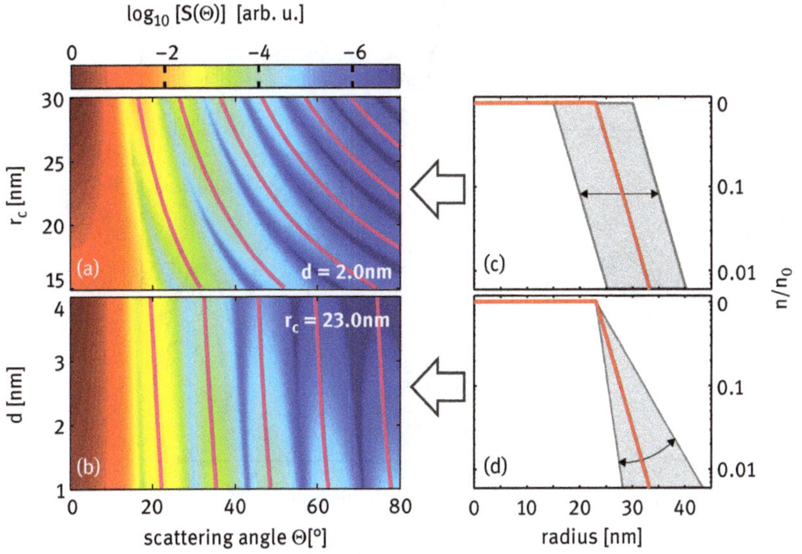

Fig. 12. Selective density profile parameter analysis with the Born method in the sharp edge limit (for $s \to 0$). Panels (a) and (b) show scattering spectra when only varying the core radius r_c or the surface width d, respectively. The corresponding one-dimensional density profiles are shown in panels (c) and (d). Gray areas illustrate the range of variation of the profiles. Adapted from [37].

struct the anisotropic plasma expansion dynamics from experimentally measured angular-resolved scattering images. In the following, it will be discussed how this can be achieved with the above-introduced tools. First, the anisotropic character of the expansion needs to be taken into account. This can be done by using anisotropic values for the density profile parameters $r_c(\theta)$ and $d(\theta)$ in (93). The sharpness factor can be neglected, as it is irrelevant for the dynamics. To model an ellipsoidal density, an angular dependence according to

$$\alpha(\theta) = \alpha_{\mathrm{perp}} + (\alpha_{\mathrm{par}} - \alpha_{\mathrm{perp}}) \cos^2 \theta \tag{96}$$

has been chosen for both parameters, where θ is the angle with respect to the polarization axis. Using this parametric form yields an angle-dependent density profile $n(r, \theta)$ with a total of four free parameters. To mimic the full two-dimensional scattering images available in typical x-ray scattering experiments, additional MicPIC

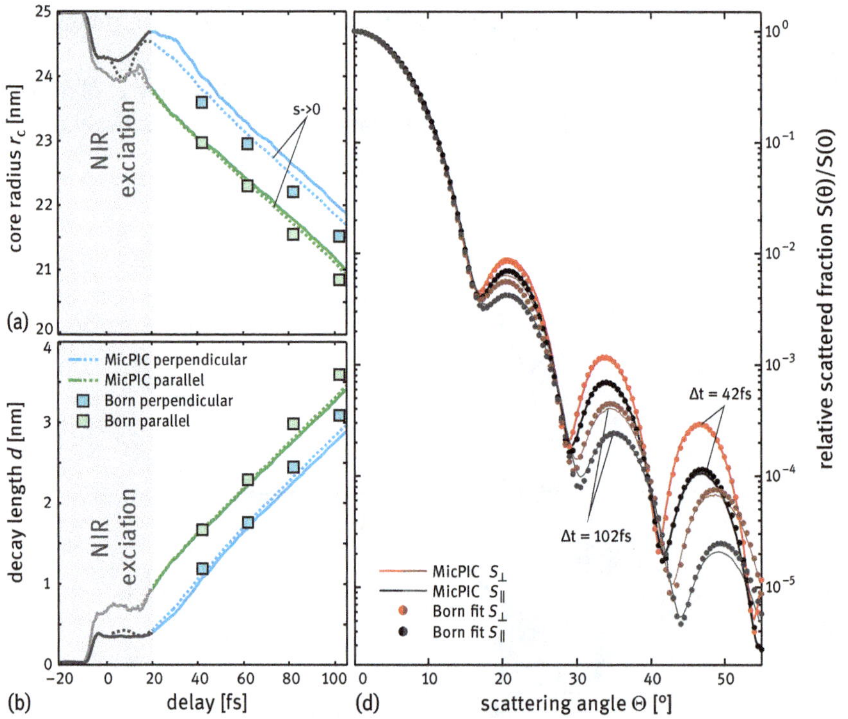

Fig. 13. (a), (b) Time evolution of density profile parameters; lines correspond to direct fits of the MicPIC density for directions parallel to (green) and perpendicular to (blue) the polarization direction using all parameters (solid) or the sharp-edge limit $s \to 0$ (dashed); squares denote parameters reconstructed from the scattering pattern in the sharp-edge limit (see text). (d) Full NIR polarization–dependent MicPIC scattering pattern (solid lines) and corresponding Born fits (circles) for two delays (as indicated). Adapted from [37].

runs have been performed with a rotated virtual detector (perpendicular to the polarization direction $y = 0$).

The corresponding Born scattering patterns calculated from the angular dependent density profile $n(r, \theta)$ can now be compared to the actual MicPIC scattering pattern. Optimal values of the four free parameters are then determined by simultaneously fitting the Born images in the $x = 0$ ($S_\perp$) and $y = 0$ ($S_\parallel$) planes to the corresponding MicPIC results via simplex optimization. Figure 13(c) shows that the resulting Born fits accurately describe the actual direction–resolved MicPIC scattering data.

The resulting optimal parameter values retrieved from the delay-dependent MicPIC scattering images are compared to the parameters directly extracted from the MicPIC electron density profiles in Figure 13(a and b) as squares and solid lines, respectively. The evolution of both anisotropic core radius and surface width can be reconstructed quantitatively with only small deviations that are attributed to the simplified four-parameter geometry model. During the hydrodynamic expansion, electron and ion density profiles evolve together so that the reconstructed profiles also describe the evolution of the ion density.

4 Summary

In this chapter, we reviewed the MicPIC method, a model that connects electrostatic MD and PIC in a two-level approach. It allows a self-consistent classical description of laser-matter interaction in the regime of strong coupling with full account of field propagation effects like field attenuation, reflection, etc. After a brief discussion of the basic physical concept behind MicPIC, we examined the most important technical aspects of its implementation. We first presented a relatively simple application example, i.e., the laser excitation of a solid density foil, which indicates the capability of this method and might serve as a reference scenario for self-employed implementations. To demonstrate the full potential of MicPIC, we further discussed a scenario where MicPIC's unique capabilities are essential to describe the underlying physics, namely, time-resolved single-shot coherent diffractive imaging. This research field is still in its infancy, and theoretical modeling is highly desired for the interpretation of experiments. The results presented here clearly demonstrate that MicPIC is a powerful tool to drive substantial advances in this field.

The methodical details described in this chapter represent the backbone implementation of the method, which means that there is plenty of room for improvement. One particularly interesting direction for future development is the effective treatment of quantum effects, which are so far only accounted for in terms of incoherent effective ionization rates. For example, a proper description of the effective band structure in dielectrics would enable MicPIC to directly model the generation of high harmonic radiation in solids, which is of great interest for future applications like lightwave electronics [30, 49].

Bibliography

[1] C. Achenbach, A. Müller, E. Salzborn, and R. Becker. Single ionization of multiply charged xenon ions by electron-impact. *Journal of Physics B: Atomic and Molecular Physics*, 17(7): 1405–1425, 1984.

[2] M. P. Allen and D. J. Tildesle *Computer Simulation of Liquids*. Oxford University Press, New York, 1989.

[3] M. V. Ammosov, N. B. Delone, and V. P. Krainov. Tunnel ionization of complex atoms and of atomic ions in an alternating electromagnetic field. *Soviet Physics JETP*, 64:1191–4, 1986.

[4] A. L. Ankudinov, S. I. Zabinsky, and J. J. Rehr. Single configuration Dirac-Fock atom code. *Computer Physics Communications*, 98(3):359–364, 1996.

[5] M. E. Bannister, D. W. Mueller, L. J. Wang, M. S. Pindzola, D. C. Griffin, and D. C. Gregory. Cross sections for electron-impact single ionization of Kr^{8+} and Xe^{8+}. *Physical Review A*, 38:38–43, 1988.

[6] J. Berenger. A perfectly matched layer for the absorption of electromagnetic waves. *Journal of Computational Physics*, 114:185, 1994.

[7] C. K. Birdsall and A. B. Langdon. *Plasma Physics via Computer Simulation*. McGraw-Hill, New York, 1985.

[8] J. P. Boris and R. A. Shanny, eds. *Relativistic plasma simulations – Optimization of a hybrid code*, Proceedings of the conference on the numerical simulation of plasmas (4th), Naval Research Laboratory, Washington D.C., 1970.

[9] C. Bostedt, S. Boutet, D. M. Fritz, Z. Huang, H. J. Lee, H. T. Lemke, A. Robert, W. F. Schlotter, J. J. Turner, and G. J. Williams. Linac coherent light source: The first five years. *Reviews of Modern Physics*, 88:015007, 2016.

[10] B. N. Breizman, A. V. Arefiev, and M. V. Fomyts'kyi. Nonlinear physics of laser-irradiated microclusters. *Physics of Plasmas*, 12(5):056706, 2005.

[11] J. M. Dawson. Particle simulation of plasmas. *Reviews of Modern Physics*, 55(2):403–447, 1983.

[12] J. W. Eastwood and R. W. Hockney. Shaping force law in 2-dimensional particle-mesh models. *Journal of Computational Physics*, 16(4):342–359, 1974.

[13] L. Englert, M. Wollenhaupt, L. Haag, C. Sarpe-Tudoran, B. Rethfeld, and T. Baumert. Material processing of dielectrics with temporally asymmetric shaped femtosecond laser pulses on the nanometer scale. *Applied Physics A*, 92(4):749–753, 2008.

[14] T. Fennel, L. Ramunno, and T. Brabec. Highly charged ions from laser-cluster interactions: Local-field-enhanced impact ionization and frustrated electron-ion recombination. *Physical Review Letters*, 99:233401, 2007.

[15] T. Fennel, G. F. Bertsch, and K.-H. Meiwes-Broer. Ionization dynamics of simple metal clusters in intense fields by the Thomas-Fermi-Vlasov method. *European Physical Journal D*, 29(3):367, 2004.

[16] T. Fennel, K.-H. Meiwes-Broer, J. Tiggesbaeumker, P.-G. Reinhard, P. M. Dinh, and E. Suraud. Laser-driven nonlinear cluster dynamics. *Reviews of Modern Physics*, 82(2):1793–1842, 2010.

[17] S. D. Gedney. An anisotropic PML absorbing media for the FDTD simulation of fields in lossy and dispersive media. *Electromagnetics*, 16(4):399–415, 1996.

[18] S. D. Gedney An anisotropic perfectly matched layer absorbing media for the truncation of FDTD latices. *IEEE Transactions on Antennas and Propagation*, 44:1630, 1996.

[19] D. C. Gregory and D. H. Crandall. Measurement of the cross section for electron-impact ionization of Xe^{6+} ions. *Physical Review A*, 27:2338–2341, 1983.

[20] D. C. Griffin, C. Bottcher, M. S. Pindzola, S. M. Younger, D. C. Gregory, and D. H. Crandall. Electron-impact ionization in the xenon isonuclear sequence. *Physical Review A*, 29:1729–1741, 1984.
[21] A. V. Gurevich, L. V. Pariiska, and L. P. Pitaevskii. Self-similar motion of rarefied plasma. *Soviet Physics JETP*, 22(2):449, 1966.
[22] A. Heidenreich, I. Last, and J. Jortner. Electron impact ionization of atomic clusters in ultraintense laser fields. *European Physical Journal D*, 35:567–577, 2005.
[23] R. Holland and J. W. Williams. Total-field versus scattered-field finite-difference codes - A comparative-assessment. *IEEE Transactions on Nuclear Science*, 30(6):4583–4588, 1983.
[24] J. D. Jackson. *Classical Electrodynamics*. John Wiley & Sons, New York, 1962.
[25] S. Kim, J. Jin, Y.-J. Kim, I.-Y. Park, Y. Kim, and S.-W. Kim. High-harmonic generation by resonant plasmon field enhancement. *Nature*, 453:757–760, 2008.
[26] V. Kumarappan, M. Krishnamurthy, and D. Mathur. Two-dimensional effects in the hydrodynamic expansion of xenon clusters under intense laser irradiation. *Physical Review A*, 66:033203, 2002.
[27] T. V. Liseykina, S. Pirner, and D. Bauer. Relativistic attosecond electron bunches from laser-illuminated droplets. *Physical Review Letters*, 104:095002, 2010.
[28] W. Lotz. An empirical formula for electron-impact ionization cross-section. *Zeitschrift für Physik*, 206(2):205–211, 1967.
[29] W. Lotz. Electron-impact ionization cross-sections and ionization rate coefficients for atoms and ions from hydrogen to calcium. *Zeitschrift für Physik*, 216(3):241–247, 1968.
[30] T. T. Luu, M. Garg, S. Yu Kruchinin, A. Moulet, M. Th. Hassan, and E. Goulielmakis. Extreme ultraviolet high-harmonic spectroscopy of solids. *Nature*, 521(7553):498–502, 2015.
[31] P. Mora. Plasma expansion into a vacuum. *Physical Review Letters*, 90:185002, 2003.
[32] T. Nakamura, Y. Fukuda, and Y. Kishimoto. Ionization dynamics of cluster targets irradiated by x-ray free-electron-laser light. *Physical Review A*, 80:053202, 2009.
[33] M. Nantel, G. Ma, S. Gu, C. Y. Côté, J. Itatani, and D. Umstadter. Pressure ionization and line merging in strongly coupled plasmas produced by 100-fs laser pulses. *Physical Review Letters*, 80:4442–4445, 1998.
[34] R. Neutze, R. Wouts, D. van der Spoel, E. Weckert, and J. Hajdu. Potential for biomolecular imaging with femtosecond x-ray pulses. *Nature*, 406:752–757, 2000.
[35] I.-Y. Park, S. Kim, J. Choi, D.-H. Lee, Y.-J. Kim, M. F. Kling, M. I. Stockman, and S.-W. Kim. Plasmonic generation of ultrashort extreme-ultraviolet light pulses. *Nature Photonics*, 5: 677–681, 2011.
[36] C. Peltz. *Fully microscopic analysis of laser-driven finite plasmas*. Ph.D. thesis, University of Rostock, 2015.
[37] C. Peltz, C. Varin, T. Brabec, and T. Fennel. Time-resolved x-ray imaging of anisotropic nanoplasma expansion. *Physical Review Letters*, 113:133401, 2014.
[38] V. S. Popov. Tunnel and multiphoton ionization of atoms and ions in a strong laser field (Keldysh theory). *Physics-Uspekhi*, 47:855, 2004.
[39] S. V. Popruzhenko. Keldysh theory of strong field ionization: history, applications, difficulties and perspectives. *Journal of Physics B: Atomic, Molecular and Optical Physics*, 47:204001, 2014.
[40] A. Pukhov. Strong field interaction of laser radiation. *Reports on Progress in Physics*, 66: 47–101, 2003.
[41] A. Pukhov and J. Meyer-ter Vehn. Relativistic magnetic self-channeling of light in near-critical plasma: Three-dimensional particle-in-cell simulation. *Physical Review Letters*, 76:3975–3978, 1996.

[42] L. Ramunno, C. Jungreuthmayer, H. Reinholz, and T. Brabec. Probing attosecond kinetic physics in strongly coupled plasmas. *Journal of Physics B: Atomic Molecular and Optical Physics*, 39: 4923–4931, 2006.
[43] U. Saalmann, C. Siedschlag, and J. M. Rost. Mechanisms of cluster ionization in strong laser pulses. *Journal of Physics B: Atomic Molecular and Optical Physics*, 39:R39, 2006.
[44] M. I. Stockman, M. F. Kling, U. Kleineberg, and F. Krausz. Attosecond nanoplasmonic-field microscope. *Nature Photonics*, 1:539–544, 2007.
[45] A. Taflove and S. C. Hagness. *Computational Electrodynamics: The Finite-Difference Time-Domain Method*. Artech House, INC./Norwood, MA, 2005.
[46] T. Tajima. *Computational Plasma Physics*. Westview Press, Westview Press/Boulder, CO, 2004.
[47] T. Takizuka and H. Abe. Binary collision model for plasma simulation with a particle code. *Journal of Computational Physics*, 25:205–219, 1977.
[48] R. R. Thomson, T. A. Birks, S. G. Leon-Saval, A. K. Kar, and J. Bland-Hawthorn. Ultrafast laser inscription of an integrated photonic lantern. *Optics Express*, 19:5698–5705, 2011.
[49] G. Vampa, T. J. Hammond, N. Thiré, B. E. Schmidt, F. Légaré, C. R. McDonald, T. Brabec, and P. B. Corkum. Linking high harmonics from gases and solids. *Nature*, 522:462–464, 2015.
[50] C. Varin, C. Peltz, T. Brabec, and T. Fennel. Attosecond plasma wave dynamics in laser-driven cluster nanoplasmas. *Physical Review Letters*, 108:175007, 2012.
[51] J. P. Verboncoeur. Particle simulation of plasmas: Review and advances. *Plasma Physics and Controlled Fusion*, 47:A231–A260, 2005.
[52] B.-B. Xu, Y.-L. Zhang, H. Xia, W.-F. Dong, H. Ding, and H.-B. Sun. Fabrication and multifunction integration of microfluidic chips by femtosecond laser direct writing. *Lab on a Chip*, 13: 1677–1690, 2013.
[53] K. S. Yee. Numerical solution of initial boundary value problems involving Maxwell's equations in isotropic media. *IEEE Transactions on Antennas and Propagation*, 14:302, 1966.
[54] S. Zherebtsov, T. Fennel, J. Plenge, E. Antonsson, I. Znakovskaya, A. Wirth, O. Herrwerth, F. Suessmann, C. Peltz, I. Ahmad, S. A. Trushin, V. Pervak, S. Karsch, M. J. J. Vrakking, B. Langer, C. Graf, M. I. Stockman, F. Krausz, E. Ruehl, and M. F. Kling. Controlled near-field enhanced electron acceleration from dielectric nanospheres with intense few-cycle laser fields. *Nature Physics*, 7:656–662, 2011.

Index

www.ingramcontent.com/pod-product-compliance
Lightning Source LLC
Chambersburg PA
CBHW061345210326
41598CB00035B/5888
* 9 7 8 3 1 1 0 4 1 7 2 5 8 *